Bosch Professional Automotiv

Bosch Professional Automotive Information is a definitive reference for automotive engineers. The series is compiled by one of the world´s largest automotive equipment suppliers. All topics are covered in a concise but descriptive way backed up by diagrams, graphs, photographs and tables enabling the reader to better comprehend the subject.
There is now greater detail on electronics and their application in the motor vehicle, including electrical energy management (EEM) and discusses the topic of intersystem networking within vehicle. The series will benefit automotive engineers and design engineers, automotive technicians in training and mechanics and technicians in garages.

Konrad Reif
Editor

Brakes, Brake Control and Driver Assistance Systems

Function, Regulation and Components

Editor
Prof. Dr.-Ing. Konrad Reif
Duale Hochschule Baden-Württemberg
Friedrichshafen, Germany
reif@dhbw-ravensburg.de

ISBN 978-3-658-03977-6 ISBN 978-3-658-03978-3 (eBook)
DOI 10.1007/978-3-658-03978-3

Library of Congress Control Number: 2014945109

Springer Vieweg

Printed on acid-free paper

Springer is part of Springer Science+Business Media
www.springer.com

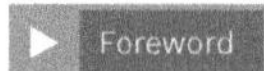

Braking systems have been continuously developed and improved throughout the last years. Major milestones were the introduction of antilock braking system (ABS) and electronic stability program. This reference book provides a detailed description of braking components and how they interact in electronic braking systems.

Complex technology of modern motor vehicles and increasing functions need a reliable source of information to understand the components or systems. The rapid and secure access to these informations in the field of Automotive Electrics and Electronics provides the book in the series "Bosch Professional Automotive Information" which contains necessary fundamentals, data and explanations clearly, systematically, currently and application-oriented. The series is intended for automotive professionals in practice and study which need to understand issues in their area of work. It provides simultaneously the theoretical tools for understanding as well as the applications.

Motor-vehicle safety
Dipl.-Ing. Wulf Post.

Basic principles of vehicle dynamics
Dipl.-Ing. Friedrich Kost.

Car braking systems
Dipl.-Ing. Wulf Post.

Car braking-system components
Dipl.-Ing. Wulf Post.

Wheel brakes
Dipl.-Ing. Wulf Post.

Antilock braking system (ABS)
Dipl.-Ing. Heinz-Jürgen Koch-Dücker,
Dipl.-Ing. (FH) Ulrich Papert.

Traction control system (TCS)
Dr.-Ing. Frank Niewels,
Dipl.-Ing. Jürgen Schuh.

Electronic stability program (ESP)
Dipl.-Ing. Thomas Ehret.

Automatic brake functions
Dipl.-Ing. (FH) Jochen Wagner.

Hydraulic modulator
Dr.-Ing. Frank Heinen,
Peter Eberspächer.

Sensors for brake control
Dr.-Ing. Erich Zabler.

Sensotronic brake control (SBC)
Dipl.-Ing. Bernhard Kant.

Active steering
Dipl.-Ing. (FH) Wolfgang Rieger,
ZF Lenksysteme, Schwäbisch Gmünd, Germany.

Occupant protection systems
Dipl.-Ing. Bernhard Mattes.

Driving assistance systems
Prof. Dr.-Ing. Peter Knoll.

Adaptive cruise control (ACC)
Prof. Dr. rer. nat. Hermann Winner,
Dr.-Ing. Klaus Winter,
Dipl.-Ing. (FH) Bernhard Lucas,
Dipl.-Ing. (FH) Hermann Mayer,
Dr.-Ing. Albrecht Irion,
Dipl.-Phys. Hans-Peter Schneider,
Dr.-Ing. Jens Lüder.

Parking systems
Prof. Dr.-Ing. Peter Knoll.

Instrumentation
Dr.-Ing. Bernhard Herzog.

Orientation methods
Dipl.-Ing. Gerald Spreitz,
S. Rehlich,
M. Neumann,
Dipl.-Ing. Marcus Risse,
Dipl.-Ing. Wolfgang Baierl.

Navigation systems
Dipl.-Ing. Ernst-Peter Neukirchner,
Dipl.-Kaufm. Ralf Kriesinger,
Dr.-Ing. Jürgen Wazeck.

Workshop technology
Dipl.-Wirtsch.-Ing. Stephan Sohnle,
Dipl.-Ing. Rainer Rehage,
Rainer Heinzmann.

and the editorial team in cooperation with the responsible in-house specialist departments.

Unless otherwise stated, the authors are all employees of Robert Bosch GmbH.

Basics

Motor-vehicle safety

In addition to the components of the drivetrain (engine, transmission), which provide the vehicle with its means of forward motion, the vehicle systems that limit movement and retard the vehicle also have an important role to play. Without them, safe use of the vehicle in road traffic would not be possible. Furthermore, systems that protect vehicle occupants in the event of an accident are also becoming increasingly important.

Safety systems

There are a many factors that affect vehicle safety in road traffic situations:

- the condition of the vehicle (e.g. level of equipment, condition of tires, component wear),
- the weather, road surface and traffic conditions (e.g. side winds, type of road surface and density of traffic), and
- the capabilities of the driver, i.e. his/her driving skills and physical and mental condition.

In the past, it was essentially only the braking system (apart, of course, from the vehicle lights) consisting of brake pedal, brake lines and wheel brakes that contributed to vehicle safety. Over the course of time though, more and more systems that actively intervene in braking-system operation have been added. Because of their active intervention, these safety systems are also referred to as *active safety systems.*

The motor-vehicle safety systems that are found on the most up-to-date vehicles substantially improve their safety.

The brakes are an essential component of a motor vehicle. They are indispensable for safe use of the vehicle in road traffic. At the slow speeds and with the small amount of traffic that were encountered in the early years of motoring, the demands placed on the braking system were far less exacting than they are today. Over the course of time, braking systems have become more and more highly developed. In the final analysis, the high speeds that cars can be driven at today are only possible because there are reliable braking systems which are capable of slowing down the vehicle and bringing it safely to a halt even in hazardous situations. Consequently, the braking system is a key part of a vehicle's safety systems.

As in all other areas of automotive engineering, electronics have also become established in the safety systems. The demands now placed on safety systems can only be met with the aid of electronic equipment.

Table 1

1 Safety when driving on roads (concepts and influencing variables)

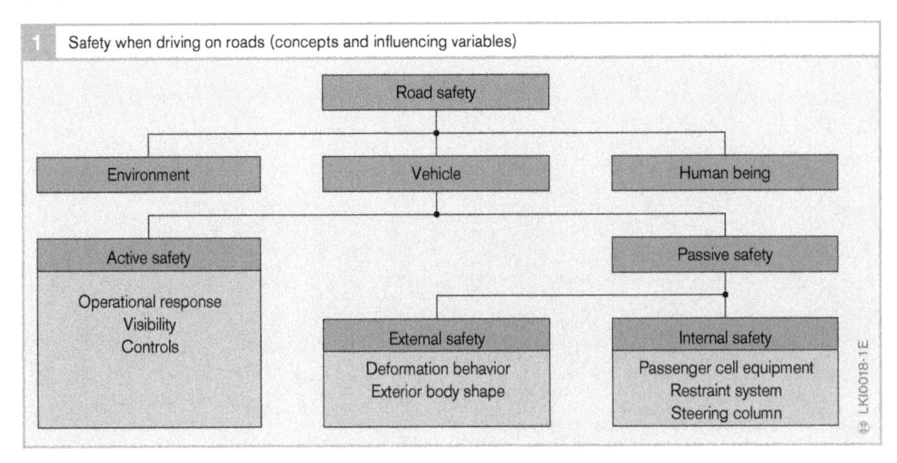

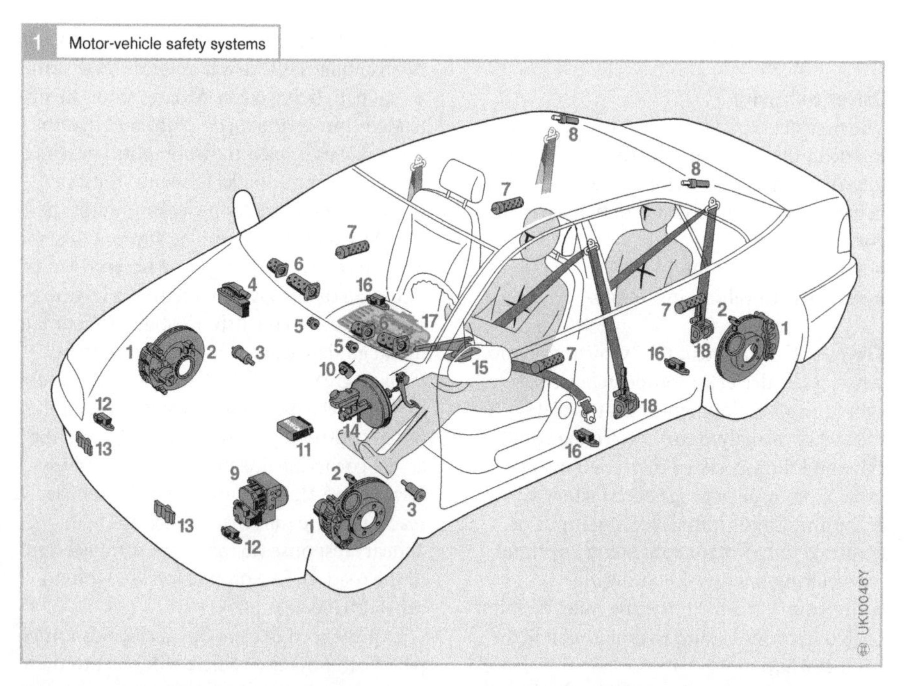

Fig. 1
1 Wheel brake with brake disk
2 Wheel-speed sensor
3 Gas inflator for foot airbag
4 ESP control unit (with ABS and TCS function)
5 Gas inflator for knee airbag
6 Gas inflators for driver and passenger airbags (2-stage)
7 Gas inflator for side airbag
8 Gas inflator for head airbag
9 ESP hydraulic modulator
10 Steering-angle sensor
11 Airbag control unit
12 Upfront sensor
13 Precrash sensor
14 Brake booster with master cylinder and brake pedal
15 Parking brake lever
16 Acceleration sensor
17 Sensor mat for seat-occupant detection
18 Seat belt with seat-belt tightener

Active safety systems

These systems help to prevent accidents and thus make a preventative contribution to road safety. Examples of active vehicle safety systems include

- ABS (Antilock Braking System),
- TCS (Traction Control System), and
- ESP (Electronic Stability Program).

These safety systems stabilize the vehicle's handling response in critical situations and thus maintain its steerability.

Apart from their contribution to vehicle safety, systems such as Adaptive Cruise Control (ACC) essentially offer added convenience by maintaining the distance from the vehicle in front by automatically throttling back the engine or applying the brakes.

Passive safety systems

These systems are designed to protect vehicle occupants from serious injury in the event of an accident. They reduce the risk of injury and thus the severity of the consequences of an accident.

Examples of passive safety systems are the seat-belts required by law, and airbags – which can now be fitted in various positions inside the vehicle such as in front of or at the side of the occupants.

Fig. 1 illustrates the safety systems and components that are found on modern-day vehicles equipped with the most advanced technology.

Basics of vehicle operation

Driver behavior

The first step in adapting vehicle response to reflect the driver and his/her capabilities is to analyze driver behavior as a whole. Driver behavior is broken down into two basic categories:

- vehicle guidance, and
- response to vehicle instability.

The essential feature of the "vehicle guidance" aspect is the driver's aptitude in anticipating subsequent developments; this translates into the ability to analyze current driving conditions and the associated interrelationships in order to accurately gauge such factors as:

- the amount of initial steering input required to maintain consistently optimal cornering lines when cornering,
- the points at which braking must be initiated in order to stop within available distances, and
- when acceleration should be started in order to overtake slower vehicles without risk.

Steering angle, braking and throttle application are vital elements within the guidance process. The precision with which these functions are discharged depends upon the driver's level of experience.

While stabilizing the vehicle (response to vehicle instability), the driver determines that the actual path being taken deviates from the intended course (the road's path) and that the originally estimated control inputs (steering angle, accelerator pedal pressure) must be revised to avoid traction loss or prevent the vehicle leaving the road. The amount of stabilization (correction) response necessary after initiation of any given maneuver is inversely proportional to the driver's ability to estimate initial guidance inputs; more driver ability leads to greater vehicle stability. Progressively higher levels of correspondence between the initial control input (steering angle) and the actual cornering line produce progressively lower correction requirements; the vehicle reacts to these minimal corrections with "linear" response (driver input is transferred to the road surface proportionally, with no substantial deviations).

Experienced drivers can accurately anticipate both how the vehicle will react to their control inputs and how this reactive motion will combine with predictable external factors and forces (when approaching curves and road works etc.). Novices not only need more time to complete this adaptive process, their results will also harbor a greater potential for error. The conclusion is that inexperi-

1 Overall system of "Driver – Vehicle – Environment"

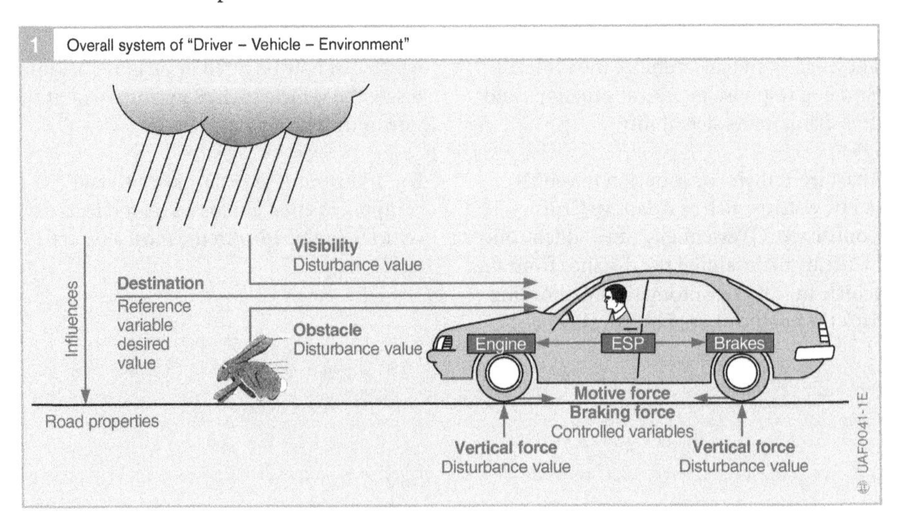

enced drivers concentrate most of their attention on the stabilization aspect of driving.

When an unforeseen development arises for driver and vehicle (such as an unexpectedly sharp curve in combination with restricted vision, etc.), the former may react incorrectly, and the latter can respond by going into a skid. Under these circumstances, the vehicle responds non-linearly and transgresses beyond its physical stability limits, so that the driver can no longer anticipate the line it will ultimately take. In such cases, it is impossible for either the novice or the experienced driver to retain control over his/her vehicle.

Accident causes and prevention

Human error is behind the vast majority of all road accidents resulting in injury. Accident statistics reveal that driving at an inappropriate speed is the primary cause for most accidents. Other accident sources are

- incorrect use of the road,
- failure to maintain the safety margin to the preceding vehicle,
- errors concerning right-of-way and traffic priority,
- errors occurring when making turns, and
- driving under the influence of alcohol.

Technical deficiencies (lighting, tires, brakes, etc.) and defects related to the vehicle in general are cited with relative rarity as accident sources. Accident causes beyond the control of the driver more frequently stem from other factors (such as weather).

These facts demonstrate the urgency of continuing efforts to enhance and extend the scope of automotive safety technology (with special emphasis on the associated electronic systems). Improvements are needed to

- provide the driver with optimal support in critical situations,
- prevent accidents in the first place, and
- reduce the severity of accidents when they do occur.

The designer's response to critical driving conditions must thus be to foster "predictable" vehicle behavior during operation at physical limits and in extreme situations. A range of parameters (wheel speed, lateral acceleration, yaw velocity, etc.) can be monitored for processing in one or several electronic control units (ECUs). This capability forms the basis of a concept for virtually immediate implementation of suitable response strategies to enhance driver control of critical processes.

The following situations and hazards provide examples of potential "limit conditions":

- changes in prevailing road and/or weather conditions,
- "conflicts of interest" with other road users,
- animals and/or obstructions on the road, and
- a sudden defect (tire blow-out, etc.) on the vehicle.

Critical traffic situations

The one salient factor that distinguishes critical traffic situations is abrupt change, such as the sudden appearance of an unexpected obstacle or a rapid change in road-surface conditions. The problem is frequently compounded by operator error. Owing to lack of experience, a driver who is travelling too fast or is not concentrating on the road will not be able to react with the judicious and rational response that the circumstances demand.

Because drivers only rarely experience this kind of critical situation, they usually fail to recognize how close evasive action or a braking maneuver has brought them to the vehicle's physical limits. They do not grasp how much of the potential adhesion between tires and road surface has already been "used up" and fail to perceive that the vehicle may be at its maneuverability limit or about to skid off the road. The driver is not prepared for this and reacts either incorrectly or too precipitously. The ultimate results are accidents and scenaria that pose threats to other road users.

These factors are joined by still other potential accident sources including outdated technology and deficiencies in infrastructure (badly designed roads, outdated traffic-guidance concepts).

Terms such as "improvements in vehicle response" and "support for the driver in critical situations" are only meaningful if they refer to mechanisms that produce genuine long-term reductions in both the number and severity of accidents. Lowering or removing the risk from these critical situations entails executing difficult driving maneuvers including

- rapid steering inputs including countersteering,
- lane changes during emergency braking,
- maintaining precise tracking while negotiating curves at high speeds and in the face of changes in the road surface.

These kinds of maneuvers almost always provoke a critical response from the vehicle, i.e., lack of tire traction prevents the vehicle reacting in the way that the driver would normally expect; it deviates from the desired course.

Due to lack of experience in these borderline situations, the driver is frequently unable to regain active control of the vehicle, and often panics or overreacts. Evasive action serves as an example. After applying excessive steering input in the moment of initial panic, this driver then countersteers with even greater zeal in an attempt to compensate for his initial error. Extended sequences of steering and countersteering with progressively greater input angles then lead to a loss of control over the vehicle, which responds by breaking into a skid.

Driving behavior

A vehicle's on-the-road handling and braking response are defined by a variety of influences. These can be roughly divided into three general categories:

- vehicle characteristics,
- the driver's behavior patterns, ability and reflexes, and
- peripheral circumstances/or influences from the surroundings or from outside.

A vehicle's handling, braking and overall dynamic response are influenced by its structure and design.

Handling and braking responses define the vehicle's reactions to driver inputs (at steering wheel, accelerator pedal, brakes, etc.) as do external interference factors (road-surface condition, wind, etc.).

Good handling is characterized by the ability to precisely follow a given course and thus comply in full with driver demand.
The driver's responsibilities include:

- adapting driving style to reflect traffic and road conditions,
- compliance with applicable traffic laws and regulations,
- following the optimal course as defined by the road's geometry as closely as possible, and
- guiding the vehicle with foresight and circumspection.

The driver pursues these objectives by continuously adapting the vehicle's position and motion to converge with a subjective conception of an ideal status. The driver relies upon personal experience to anticipate developments and adapt to instantaneous traffic conditions.

2 Overall system of "driver – vehicle – environment" as a closed control loop

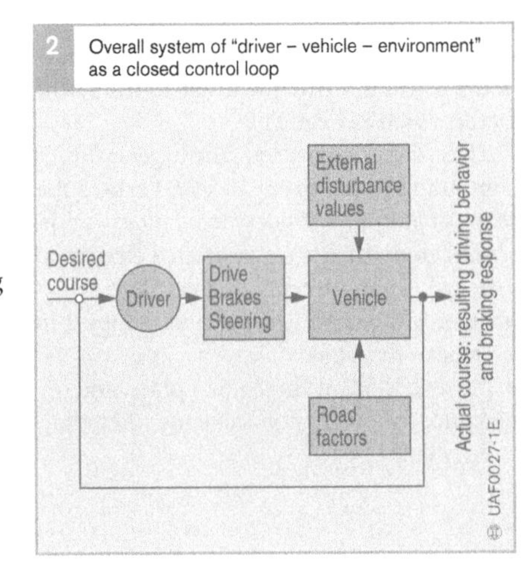

Evaluating driver behavior

Subjective assessments made by experienced drivers remain the prime element in evaluations of vehicle response. Because assessments based on subjective perceptions are only relative and not absolute, they cannot serve as the basis for defining objective "truths". As a result, subjective experience with one vehicle can be applied to other vehicles only on a comparative, relative basis.

Test drivers assess vehicle response using selected maneuvers conceived to reflect "normal" traffic situations. The overall system (including the driver) is judged as a closed loop. While the element "driver" cannot be precisely defined, this process provides a replacement by inputting objective, specifically defined interference factors into the system. The resulting vehicular reaction is then analyzed and evaluated. The following maneuvers are either defined in existing ISO standards or currently going through the standardization process. These dry-surface exercises serve as recognized procedures for assessing vehicular stability:

- steady-state skid-pad circulation,
- transition response,
- braking while cornering,
- sensitivity to crosswinds,
- Straight-running properties (tracking stability), and
- load change on the skid pad.

In this process, prime factors such as road geometry and assignments taken over by the driver assume vital significance. Each test driver attempts to gather impressions and experience in the course of various prescribed vehicle maneuvers; the subsequent analysis process may well include comparisons of the impressions registered by different drivers. These often hazardous driving maneuvers (e.g. the standard VDA evasive-action test, also known as the "elk test") are executed by a series of drivers to generate data describing the dynamic response and general handling characteristics of the test vehicle. The criteria include:

- stability,
- steering response and brake performance, and
- handling at the limit. The tests are intended to describe these factors as a basis for implementing subsequent improvements.

The advantages of this procedure are:

- it allows assessment of the overall, synergistic system ("driver–vehicle– environment") and
- supports realistic simulation of numerous situations encountered under everyday traffic conditions.

The disadvantages of this procedure are:

- the results extend through a broad scatter range, as drivers, wind, road conditions and initial status vary from one maneuver to the next,
- subjective impressions and experience are colored by the latitude for individual interpretation, and
- the success or failure of an entire test series can ultimately be contingent upon the abilities of a single driver.

Table 1 (next page) lists the essential vehicle maneuvers for evaluating vehicle response within a closed control loop.

Owing to the subjective nature of human behavior, there are still no definitions of dynamic response in a closed control loop that are both comprehensive and objectively grounded (closed-loop operation, meaning with driver, Fig. 2).

Despite this, the objective driving tests are complimented by various test procedures capable of informing experienced drivers about a vehicle's handling stability (example: slalom course).

1 Evaluating driver behavior					
Vehicle response	**Driving maneuver** (Driver demand and current conditions)	**Driver makes continuous corrections**	**Steering wheel firmly positioned**	**Steering wheel released**	**Steering angle input**
Linear response	Straight-running stability – stay in lane	•	•	•	
	Steering response/turning	•			
	Sudden steering – releasing the steering			•	
	Load-change reaction	•	•	•	
	Aquaplaning	•	•	•	
	Straight-line braking	•	•	•	
	Crosswind sensitivity	•	•	•	
	High-speed aerodynamic lift		•		
	Tire defect	•	•	•	
Transition input/ transmission response	Sudden steering-angle change				•
	Single steering and countersteering inputs				•
	Multiple steering and countersteering inputs				•
	Single steering impulse				•
	"Random" steering-angle input	•			•
	Driving into a corner	•			
	Driving out of a corner	•			
	Self-centering			•	
	Single lane change	•			
	Double lane change	•			
Cornering	Steady-state skid-pad circulation		•		
	Dynamic cornering	•	•		
	Load-change reaction when cornering	•	•		
	Steering release			•	
	Braking during cornering	•	•		
	Aquaplaning in curve	•	•		
Alternating directional response	Slalom course around marker cones	•			
	Handling test (test course with sharp corners)	•			
	Steering input/acceleration			•	
Overall characteristics	Tilt resistance	•			•
	Reaction and evasive action tests	•			

Table 1

Driving maneuvers

Steady-state skid-pad circulation

Steady-state cornering around the skid pad is employed to determine maximum lateral acceleration. This procedure also provides information on the transitions that dynamic handling undergoes as cornering forces climb to their maximum. This information can be used to define the vehicle's intrinsic handling (self-steering) properties (oversteer, understeer, neutral cornering response).

3 Evasive maneuver ("Elk Test")

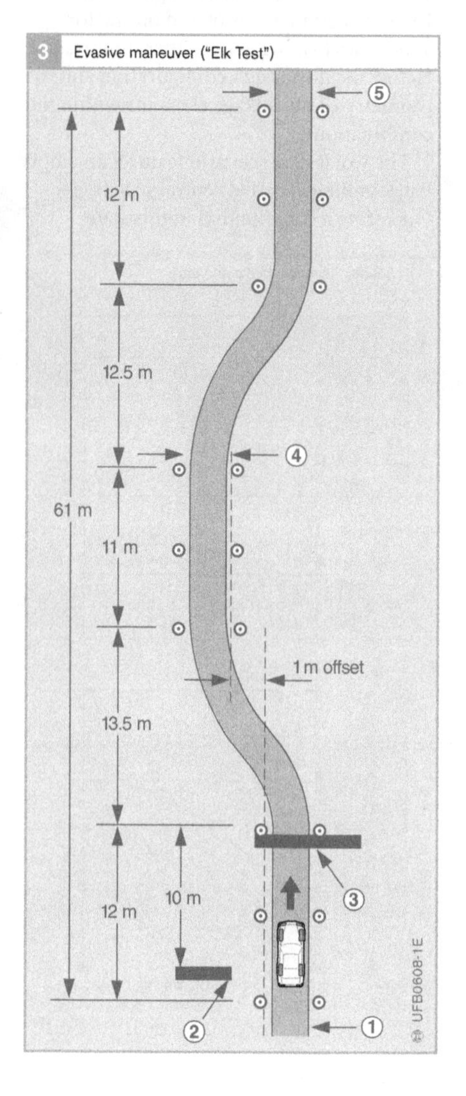

Fig. 3
Test start:
Phase 1:
Top gear (manual transmission)
Position D at 2,000 rpm (automatic transmission)

Phase 2:
Accelerator released

Phase 3:
Speed measurement with photoelectric light barrier

Phase 4:
Steering to the right

Phase 5:
End of test

Transition response

Transition response joins steady-state self-steering properties (during skid-pad circulation) as a primary assessment parameter. This category embraces such maneuvers as suddenly taking rapid evasive action when driving straight ahead.

The "elk test" simulates an extreme scenario featuring sudden evasive action to avoid an obstacle. A vehicle traveling over a 50 meter stretch of road must safely drive around an obstacle 10 meters in length projecting outward onto the track by a distance of 4 meters (Fig. 3).

Braking during cornering – load-change reactions

One of the most critical situations encountered in every-day driving – and thus one of the most vital considerations for vehicle design – is braking during cornering.

From the standpoint of the physical forces involved, whether the driver simply releases the accelerator or actually depresses the brake pedal is irrelevant; the physical effects will not differ dramatically. The resulting load shift from rear to front increasing the rear slip angle while reducing that at the front, and since neither the given cornering radius nor the vehicle speed modifies the lateral force requirement, the vehicle tends to adopt an oversteering attitude.

With rear-wheel drive, tire slip exerts less influence on the vehicle's intrinsic handling response than with front-wheel drive; this means that RWD vehicles are more stable under these conditions.

Vehicle reaction during this maneuver must represent the optimal compromise between steering response, stability and braking efficiency.

Parameters

The primary parameters applied in the assessment of dynamic handling response are:
- steering-wheel angle,
- lateral acceleration,
- longitudinal acceleration or longitudinal deceleration,
- yaw velocity,
- side-slip angle and roll angle.

Additional data allow more precise definition of specific handling patterns as a basis for evaluating other test results:
- longitudinal and lateral velocity,
- steering angles of front/rear wheels,
- slip angle at all wheels,
- steering-wheel force.

Reaction time

Within the overall system "driver-vehicle-environment", the driver's physical condition and state of mind, and thus his/her reaction times, join the parameters described above as decisive factors. This lag period is the time that elapses between perception of an obstacle and initial application of pressure to the brake pedal. The decision to act and the foot movement count as intermediate stages in this process. This period is not consistent; depending upon personal factors and external circumstances it is at least 0.3 seconds.

Special examinations are required to quantify individual reaction patterns (as conducted by medical/psychological institutes).

Motion

Vehicle motion may be consistent in nature (constant speed) or it may be inconsistent (during acceleration from a standing or rolling start, or deceleration and braking with the accompanying change in velocity).

The engine generates the kinetic energy required to propel the vehicle. Forces stemming either from external sources or acting through the engine and drivetrain must always be applied to the vehicle as a basic condition for changes in the magnitude and direction of its motion.

Handling and braking response in commercial vehicles

Objective evaluation of handling and braking response in heavy commercial vehicles is based on various driving maneuvers including steady-state skid-pad cornering, abrupt steering-angle change (vehicle reaction to "tugging" the steering wheel through a specified angle) and braking during cornering.

The dynamic lateral response of tractor and trailer combinations generally differs substantially from that of single vehicles. Particular emphasis is placed on tractor and trailer loading, while other important factors include design configuration and the geometry of the linkage elements within the combination.

The worst-case scenario features an empty truck pulling a loaded central-axle trailer. Operating a combination in this state

4 Actions: reaction, brake and stop

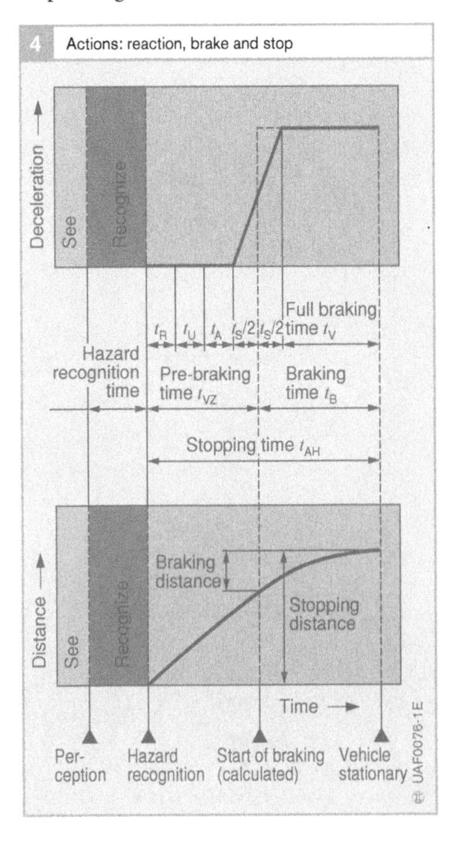

Fig. 4
t_R Reaction time
t_U Conversion time
t_A Response time
t_S Pressure buildup

requires a high degree of skill and circumspection on the part of the driver.

Jack-knifing is also a danger when tractor-trailer combinations are braked in extreme situations. This process is characterized by a loss of lateral traction at the tractor's rear axle and is triggered when "overbraking" on slippery road surfaces, or by extreme yaw rates on μ-split surfaces (with different friction coefficients at the center and on the shoulder of the lane). Jack-knifing can be avoided with the aid of antilock braking systems (ABS).

2 Personal reaction-time factors

→	Psychophysical reaction		← ¦ →	Muscular reaction →		
Perceived object	Perception	Comprehension	Decision	Mobilization	Motion	Object of action
(e.g. traffic sign)	Visual acuity	Perception and registration	Processing	Movement apparatus	Personal implementation speed	(e.g. brake pedal)

Table 2

3 Reaction time as a function of personal and external factors

Short reaction time ←	→ Long reaction time
Personal factors, driver	
Trained reflex action	Ratiocinative reaction
Good condition, optimal performance potential	Poor condition, e.g. fatigue
High level of driving skill	Low level of driving skill
Youth	Advanced age
Anticipatory attitude	Inattention, distraction
Good physical and mental health	Physical or mental impediment
	Panic, alcohol
External Factors	
Simple, unambiguous, predicable and familiar traffic configuration	Complex, unclear, incalculable and unfamiliar traffic conditions
Conspicuous obstacle	Inconspicuous obstacle
Obstacle in line of sight	Obstacle on visual periphery
Logical and effective arrangement of the controls in the vehicle	Illogical and ineffective control arrangement in vehicle

Table 3

Basic principles of vehicle dynamics

A body can only be made to move or change course by the action of forces. Many forces act upon a vehicle when it is being driven. An important role is played by the tires as any change of speed or direction involves forces acting on the tires.

Tires

Task

The tire is the connecting link between the vehicle and the road. It is at that point that the safe handling of a vehicle is ultimately decided. The tire transmits motive, braking and lateral forces within a physical environment whose parameters define the limits of the dynamic loads to which the vehicle is subjected. The decisive criteria for the assessment of tire quality are:

- Straight-running ability
- Stable cornering properties
- Ability to grip on a variety of road surfaces
- Ability to grip in a variety of weather conditions
- Steering characteristics
- Ride comfort (vibration absorption and damping, quietness)
- Durability and
- Economy

Design

There are a number of different tire designs that are distinguished according to the nature and sophistication of the technology employed. The design of a conventional tire is determined by the characteristics required of it in normal conditions and emergency situations.

Legal requirements and regulations specify which tires must be used in which conditions, the maximum speeds at which different types of tire may be used, and the criteria by which tires are classified.

Radial tires

In a radial tire, the type which has now become the standard for cars, the cords of the tire-casing plies run radially, following the shortest route from bead to bead (Fig. 1).
A reinforcing belt runs around the perimeter of the relatively thin, flexible casing.

1 Structure of a radial car tire

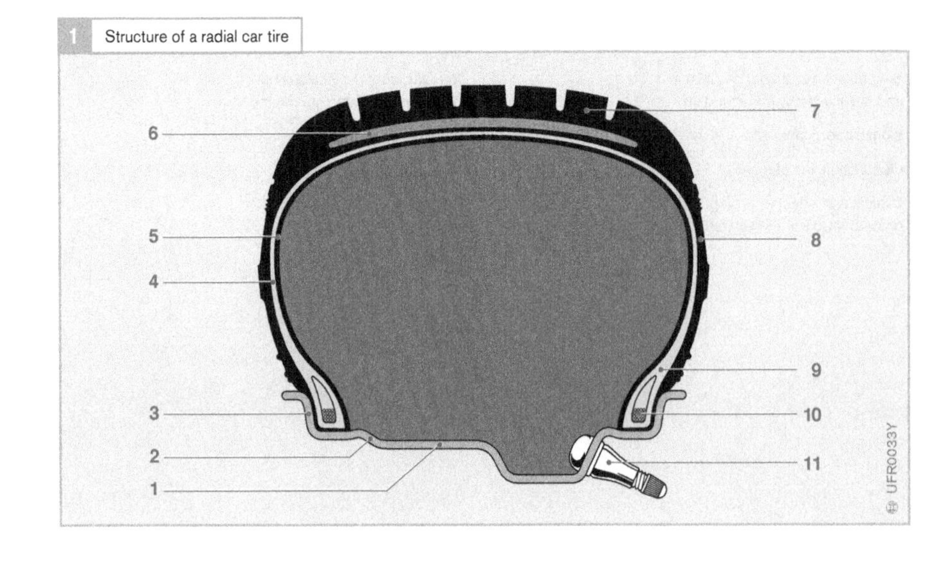

Fig. 1
1 Rim bead seat
2 Hump
3 Rim flange
4 Casing
5 Air-tight rubber layer
6 Belt
7 Tread
8 Sidewall
9 Bead
10 Bead core
11 Valve

Cross-ply tires

The cross-ply tire takes its name from the fact that the cords of alternate plies of the tire casing run at right angles to one another so that they cross each other. This type of tire is now only of significance for motorcycles, bicycles, and industrial and agricultural vehicles. On commercial vehicles it is increasingly being supplanted by the radial tire.

Regulations

In Europe, the Council Directives, and in the USA the *FMVSS (Federal Motor Vehicle Safety Standard)* require that motor vehicles and trailers are fitted with pneumatic tires with a tread pattern consisting of grooves with a depth of at least 1.6 mm around the entire circumference of the tire and across the full width of the tread.

Cars and motor vehicles with a permissible laden weight of less than 2.8 tonnes and designed for a maximum speed of more than 40 km/h, and trailers towed by them, must be fitted either with cross-ply tires all round or with radial tires all round; in the case of vehicle-and-trailer combinations the requirement applies individually to each unit of the combination. It does not apply to trailers towed by vehicles at speeds of up to 25 km/h.

Application

To ensure correct use of tires, it is important the correct tire is selected according to the recommendations of the vehicle or tire manufacturer. Fitting the same type of tire to all wheels of a vehicle ensures the best handling results. The specific instructions of the tire manufacturer or a tire specialist regarding tire care, maintenance, storage and fitting should be followed in order to obtain maximum durability and safety.

2 Increase in braking distance on wet road surface as a function of tread depth at 100 km/h

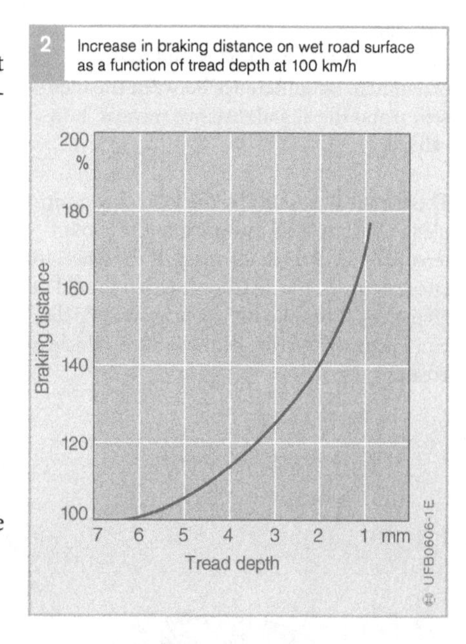

When the tires are in use, i.e. when they are fitted to the wheel, care should be taken to ensure that

- the wheels are balanced so as to guarantee optimum evenness of running,
- all wheels are fitted with the same type of tire and the tires are the correct size for the vehicle,
- the vehicle is not driven at speeds in excess of the maximum allowed for the tires fitted, and
- the tires have sufficient depth of tread.

The less tread there is on a tire, the thinner is the layer of material protecting the belt and the casing underneath it. And particularly on cars and fast commercial vehicles, insufficient tread depth on wet road surfaces has a decisive effect on safe handling characteristics due to the reduction in grip. Braking distance increases disproportionately as tread depth reduces (Fig. 2). An especially critical handling scenario is aquaplaning in which all adhesion between tires and road surface is lost and the vehicle is no longer steerable.

Tire slip

Tire slip, or simply "slip", is said to occur when there is a difference between the theoretical and the actual distance traveled by a vehicle.

This can be illustrated by the following example in which we will assume that the circumference of a car tire is 2 meters. If the wheel rotates ten times, the distance traveled should be 20 meters. If tire slip occurs, however, the distance actually traveled by the braked vehicle is greater.

3 Effect of braking on a rolling wheel

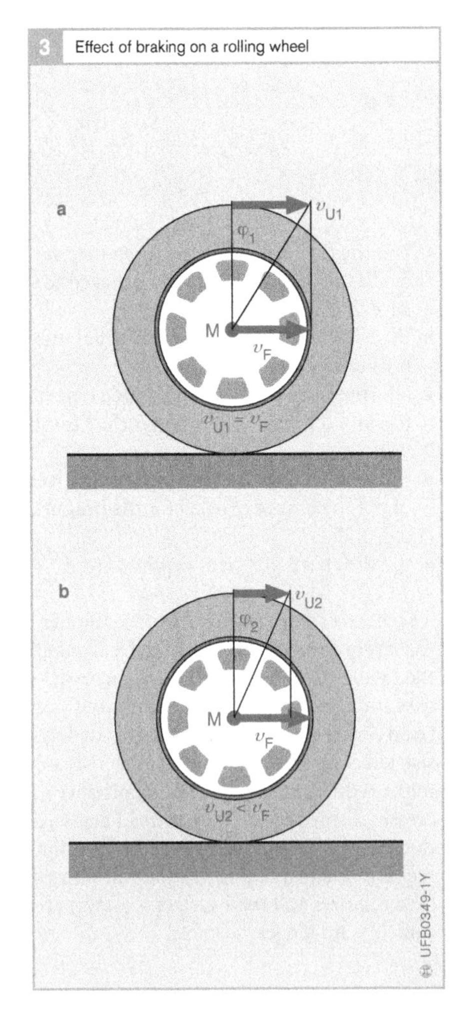

Fig. 3
a Rolling wheel (unbraked)
b Braked wheel
v_F Vehicle speed at wheel center, M
v_U Circumferential speed

On a braked wheel, the angle of rotation, φ, per unit of time is smaller (slip)

Causes of tire slip

When a wheel rotates under the effect of power transmission or braking, complex physical processes take place in the contact area between tire and road which place the rubber parts under stress and cause them to partially slide, even if the wheel does not fully lock. In other words, the elasticity of the tire causes it to deform and "flex" to a greater or lesser extent depending on the weather conditions and the nature of the road surface. As the tire is made largely of rubber, only a proportion of the "deformation energy" is recovered as the tread moves out of the contact area. The tire heats up in the process and energy loss occurs.

Illustration of slip

The slip component of wheel rotation is referred to by λ, where

$$\lambda = (v_F - v_U)/v_F$$

The quantity v_F is the vehicle road speed, v_U is the circumferential velocity of the wheel (Fig. 3). The formula states that brake slip occurs as soon as the wheel is rotating more slowly than the vehicle road speed would normally demand. Only under that condition can braking forces or acceleration forces be transmitted.

Since the tire slip is generated as a result of the vehicle's longitudinal movement, it is also referred to as "longitudinal slip". The slip generated during braking is usually termed "brake slip".

If a tire is subjected to other factors in addition to slip (e.g. greater weight acting on the wheels, extreme wheel positions), its force transmission and handling characteristics will be adversely affected.

Forces acting on a vehicle

Theory of inertia

Inertia is the property possessed by all bodies, by virtue of which they will naturally maintain the status in which they find themselves, i.e. either at rest or in motion. In order to bring about a change to that status, a force has to be applied to the body. For example, if a car's brakes are applied when it is cornering on black ice, the car will carry on in a straight line without altering course and without noticeably slowing down. That is because on black ice, only very small tire forces can be applied to the wheels.

Turning forces

Rotating bodies are influenced by turning forces. The rotation of the wheels, for example, is slowed down due to the braking torque and accelerated due to the drive torque.

Turning forces act on the entire vehicle. If the wheels on one side of the vehicle are on a slippery surface (e.g. black ice) while the wheels on the other side are on a road surface with normal grip (e.g. asphalt), the vehicle will slew around its vertical axis when the brakes are applied (μ-split braking). This rotation is caused by the yaw moment, which arises due to the different forces applied to the sides of the vehicle.

Distribution of forces

In addition to the vehicle's weight (resulting from gravitational force), various different types of force act upon it regardless of its state of motion (Fig. 1). Some of these are

- forces which act along the longitudinal axis of the vehicle (e.g. motive force, aerodynamic drag or rolling friction); others are
- forces which act laterally on the vehicle (e.g. steering force, centrifugal force when cornering or crosswinds). The tire forces which act laterally on the vehicle are also referred to as lateral forces.

The longitudinal and the lateral forces are transmitted either "downwards" or "sideways" to the tires and ultimately to the road. The forces are transferred through

- the chassis (e.g. wind),
- the steering (steering force),
- the engine and transmission (motive force), or
- the braking system (braking force).

Opposing forces act "upwards" from the road onto the tires and thence to the vehicle because every force produces an opposing force.

1 Forces acting on a vehicle

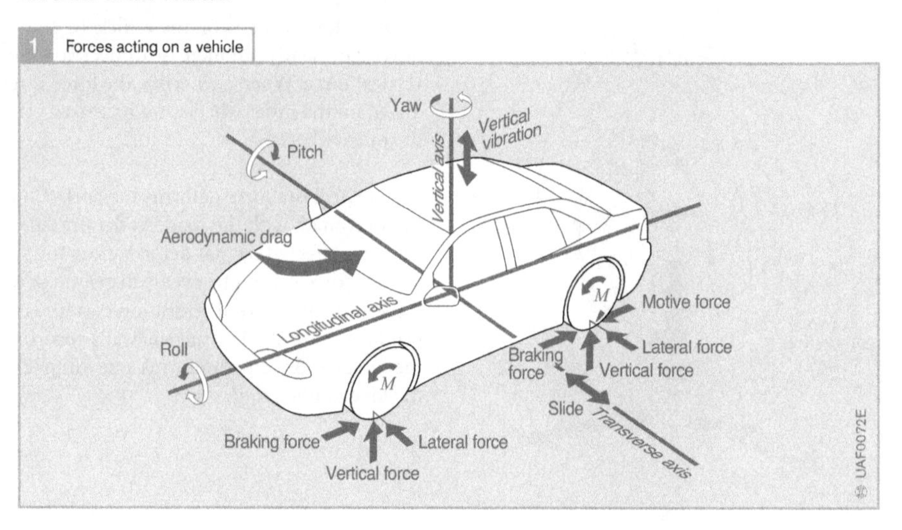

Basically, in order for the vehicle to move, the motive force of the engine (engine torque) must overcome all forces that resist motion (all longitudinal and lateral forces) such as are generated by road gradient or camber.

In order to assess the dynamic handling characteristics or handling stability of a vehicle, the forces acting between the tires and the road, i.e. the forces transmitted in the contact areas between tire and road surface (also referred to as "tire contact area" or "footprint"), must be known.

With more practice and experience, a driver generally learns to react more effectively to those forces. They are evident to the driver when accelerating or slowing down as well as in cross winds or on slippery road surfaces. If the forces are particularly strong, i.e. if they produce exaggerated changes in the motion of the vehicle, they can also be dangerous (skidding) or at least are detectable by squealing tires (e.g. when accelerating aggressively) and increased component wear.

2 Components of tire force and pressure distribution over the footprint of a radial tire

Fig. 2
F_N Vertical tire force, or normal force
F_U Circumferential force (positive: motive force; negative: braking force)
F_S Lateral force

Tire forces

A motor vehicle can only be made to move or change its direction in a specific way by forces acting through the tires. Those forces are made up of the following components (Fig. 2):

Circumferential force

The circumferential force F_U is produced by power transmission or braking. It acts on the road surface as a linear force in line with the longitudinal axis of the vehicle and enables the driver to increase the speed of the vehicle using the accelerator or slow it down with the brakes.

Vertical tire force (normal force)

The vertical force acting downwards between the tire and road surface is called the vertical tire force or normal force F_N. It acts on the tires at all times regardless of the state of motion of the vehicle, including, therefore, when the vehicle is stationary.

The vertical force is determined by the proportion of the combined weight of vehicle and payload that is acting on the individual wheel concerned. It also depends on the degree of upward or downward gradient of the road that the vehicle is standing on. The highest levels of vertical force occur on a level road.

Other forces acting on the vehicle (e.g. heavier payload) can increase or decrease the vertical force. When cornering, the force is reduced on the inner wheels and increased on the outer wheels.

The vertical tire force deforms the part of the tire in contact with the road. As the tire sidewalls are affected by that deformation, the vertical force cannot be evenly distributed. A trapezoidal pressure-distribution pattern is produced (Fig. 2). The tire sidewalls absorb the forces and the tire deforms according to the load applied to it.

Lateral force

Lateral forces act upon the wheels when steering or when there is a crosswind, for example. They cause the vehicle to change direction.

Braking torque

When the brakes are applied, the brake shoes press against the brake drums (in the case of drum brakes) or the brake pads press against the disks (in the case of disk brakes). This generates frictional forces, the level of which can be controlled by the driver by the pressure applied to the brake pedal.

The product of the frictional forces and the distance at which they act from the axis of rotation of the wheel is the braking torque M_B.

That torque is effective at the circumference of the tire under braking (Fig. 1).

Yaw moment

The yaw moment around the vehicle's vertical axis is caused by different longitudinal forces acting on the left and right-hand sides of the vehicle or different lateral forces acting at the front and rear axles. Yaw moments are required to turn the vehicle when cornering. Undesired yaw moments, such as can occur when braking on μ-split (see above) or if the vehicle pulls to one side when braking, can be reduced using suitable design measures. The kingpin offset is the distance between the point of contact between the tire and the road and the point at which the wheel's steering axis intersects the road surface (Fig. 3). It is negative if the point at which the steering axis intersects the road surface is on the outside of the point of contact between tire and road. Braking forces combine with positive and negative kingpin offset to create a lever effect that produces a turning force at the steering which can lead to a certain steering angle at the wheel. If the kingpin offset is negative, this steering angle counters the undesired yaw moment.

3 Kingpin offset

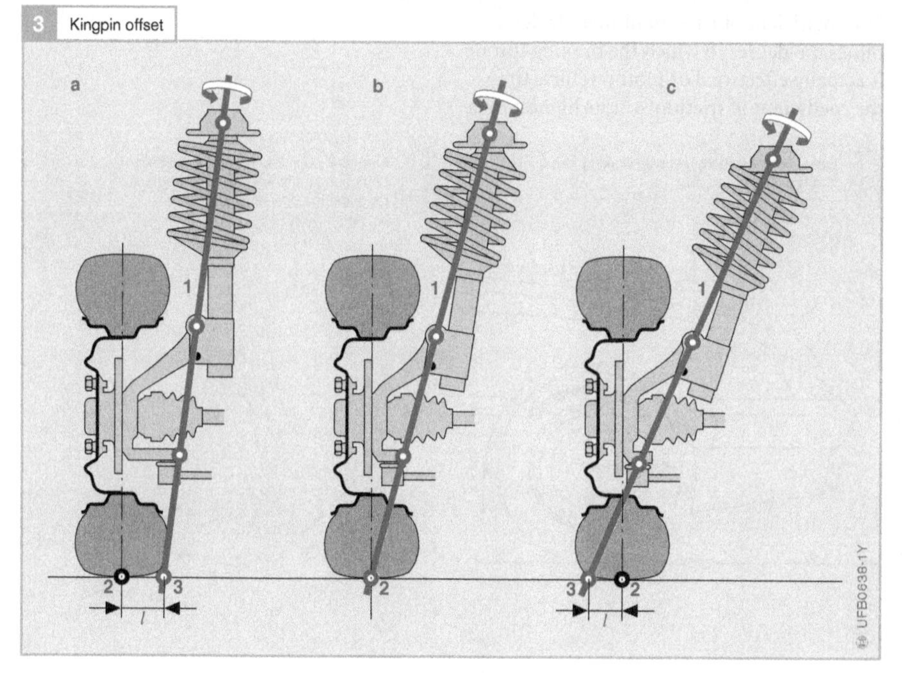

Fig. 3
a Positive kingpin offset: $M_{Ges} = M_T + M_B$
b Zero kingpin offset: no yaw moment
c Negative kingpin offset: $M_{Ges} = M_T - M_B$

1 Steering axis
2 Wheel contact point
3 Intersection point
l Kingpin offset
M_{Ges} Total turning force (yaw moment)
M_T Moment of inertia
M_B Braking torque

Friction force

Coefficient of friction

When braking torque is applied to a wheel, a braking force F_B is generated between the tire and the road surface that is proportional to the braking torque under stationary conditions (no wheel acceleration). The braking force transmitted to the road (frictional force F_R) is proportional to the vertical tire force F_N:

$$F_R = \mu_{HF} \cdot F_N$$

The factor μ_{HF} is the coefficient of friction. It defines the frictional properties of the various possible material pairings between tire and road surface and the environmental conditions to which they are exposed.

The coefficient of friction is thus a measure of the braking force that can be transmitted. It is dependent on

- the nature of the road surface,
- the condition of the tires,
- the vehicle's road speed, and
- the weather conditions.

The coefficient of friction ultimately determines the degree to which the braking torque is actually effective. For motor-vehicle tires, the coefficient of friction is at its highest on a clean and dry road surface; it is at its lowest on ice. Fluids (e.g. water) or dirt between the tire and the road surface reduce the coefficient of friction. The figures quoted in Table 1 apply to concrete and tarmacadam road surfaces in good condition.

On wet road surfaces in particular, the coefficient of friction is heavily dependent on vehicle road speed. At high speeds on less than ideal road surfaces, the wheels may lock up under braking because the coefficient of friction is not high enough to provide sufficient adhesion for the tires to grip the road surface. Once a wheel locks up, it can no longer transmit side forces and the vehicle is thus no longer steerable. Fig. 5 illustrates the frequency distribution of the coefficient of friction at a locked wheel at various road speeds on wet roads.

The friction or adhesion between the tire and the road surface determines the wheel's ability to transmit force. The ABS (Antilock Braking System) and TCS (Traction Control System) safety systems utilize the available adhesion to its maximum potential.

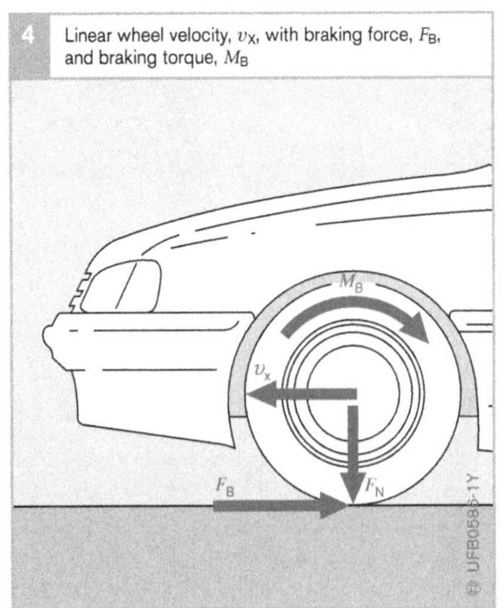

4 Linear wheel velocity, v_x, with braking force, F_B, and braking torque, M_B

Fig. 4
v_x Linear velocity of wheel
F_N Vertical tire force (normal force)
F_B Braking force
M_B Braking torque

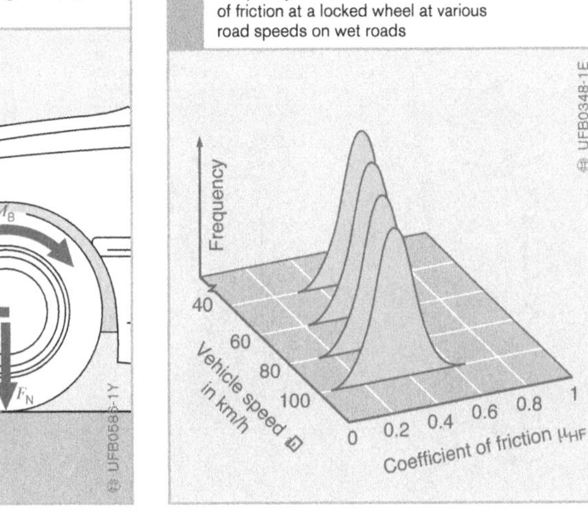

5 Frequency distribution of the coefficient of friction at a locked wheel at various road speeds on wet roads

Fig. 5
Source: Forschungsinstitut für Kraftfahrwesen und Fahrzeugmotoren, Stuttgart, Germany (research institute for automotive engineering and automotive engines)

Aquaplaning

The amount of friction approaches zero if rainwater forms a film on the road surface on which the vehicle then "floats". Contact between the tires and the road surface is then lost and the effect known as aquaplaning occurs. Aquaplaning is caused by a "wedge" of water being forced under the entire contact area of the tire with the road surface, thereby lifting it off the ground. Aquaplaning is dependent on:

- the depth of water on the road,
- the speed of the vehicle,
- the tire tread pattern, tire width and level of wear, and
- the force pressing the tire against the road surface.

Wide tires are particularly susceptible to aquaplaning. When a vehicle is aquaplaning, it cannot be steered or braked. Neither steering movements nor braking forces can be transmitted to the road.

Kinetic friction

When describing processes involving friction, a distinction is made between static friction and kinetic friction. With solid bodies, the static friction is greater than kinetic friction. Accordingly, for a rolling rubber tire there are circumstances in which the coefficient of friction is greater than when the wheel locks. Nevertheless, the tire can also slide while it is rolling, and on motor vehicles this is referred to as slip.

Effect of brake slip on coefficient of friction

When a vehicle is pulling away or accelerating – just as when braking or decelerating – the transmission of forces from tire to road depends on the degree of adhesion between the two. The friction of a tire basically has a constant relationship to the level of adhesion under braking or acceleration.

Fig. 6 shows the progression of the coefficient of friction μ_{HF} under braking. Starting from a zero degree of brake slip, is rises steeply to its maximum at between 10% and 40% brake slip, depending on the nature of the road surface and the tires, and then drops away again. The rising slope of the

6 Coefficient of friction, μ_{HF}, and lateral-force coefficient, μ_S, relative to brake slip

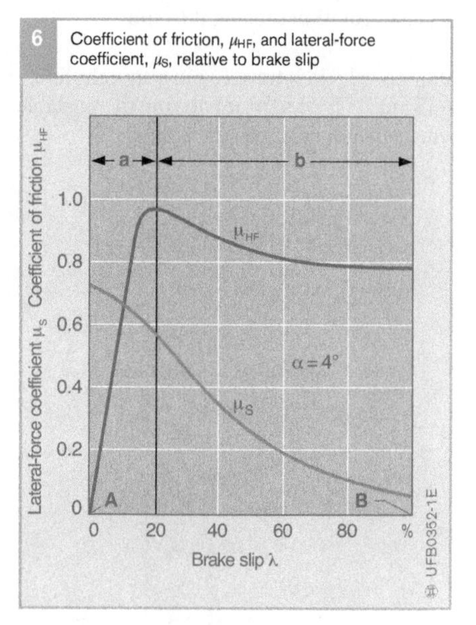

Fig. 6
a Stable zone
b Unstable zone
α Slip angle
A Rolling wheel
B Locked wheel

1 Coefficients of friction, μ_{HF}, for tires in various conditions of wear, on various road conditions and at various speeds

Vehicle road speed	Tire condition	Dry road	Wet road (depth of water 0.2 mm)	Heavy rain (depth of water 1 mm)	Puddles (depth of water 2 mm)	Icy (black ice)
km/h		μ_{HF}	μ_{HF}	μ_{HF}	μ_{HF}	μ_{HF}
50	new	0.85	0.65	0.55	0.5	0.1 and below
	worn out	1	0.5	0.4	0.25	
90	new	0.8	0.6	0.3	0.05	
	worn out	0.95	0.2	0.1	0.0	
130	new	0.75	0.55	0.2	0	
	worn out	0.9	0.2	0.1	0	

Table 1

curve represents the "stable zone" (partial-braking zone), while the falling slope is the "unstable zone".

Most braking operations involve minimal levels of slip and take place within the stable zone so that an increase in the degree of slip simultaneously produces an increase in the usable adhesion. In the unstable zone, an increase in the amount of slip generally produces a reduction in the level of adhesion. When braking in such situations, the wheel can lock up within a fraction of a second, and under acceleration the excess power-transmission torque rapidly increases the wheel's speed of rotation causing it to spin.

When a vehicle is traveling in a straight line, ABS and TCS prevent it entering the unstable zone when braking or accelerating.

Sideways forces

If a lateral force acts on a rolling wheel, the center of the wheel moves sideways. The ratio between the lateral velocity and the velocity along the longitudinal axis is referred to as "lateral slip". The angle between the resulting velocity, v_α, and the forward velocity, v_x, is called the "lateral slip angle α" (Fig. 7). The side-slip angle, γ, is the angle between the vehicle's direction of travel and its longitudinal axis. The side-slip angle encountered at high rates of lateral acceleration is regarded as an index of controllability, in other words the vehicle's response to driver input.

Under steady-state conditions (when the wheel is not being accelerated), the lateral force F_S acting on the center of the wheel is in equilibrium with the lateral force applied to the wheel by the road surface. The relationship between the lateral force acting through the center of the wheel and the wheel contact force F_N is called the "lateral-force coefficient μ_S".

7 Lateral slip angle, α, and the effect of lateral force, F_S, (overhead view)

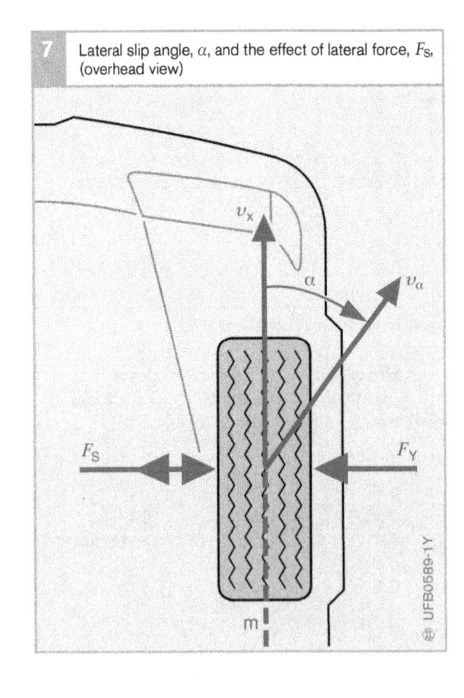

Fig. 7
v_α Velocity in lateral slip direction
v_x Velocity along longitudinal axis
F_S, F_y Lateral force
α Slip angle

8 Position of tire contact area relative to wheel in a right-hand bend showing lateral force, F_S, (front view)

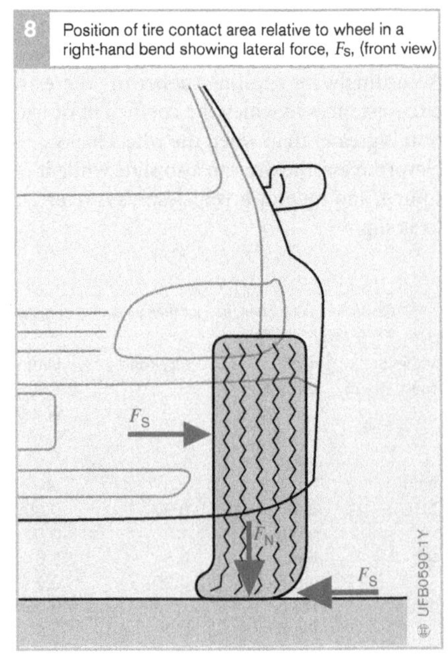

Fig. 8
F_N Vertical tire force (normal force)
F_S Lateral force

There is a nonlinear relationship between the slip angle α and the lateral-force coefficient μ_S that can be described by a lateral slip curve. In contrast with the coefficient of friction μ_{HF} that occurs under acceleration and braking, the lateral-force coefficient μ_S is heavily dependent on the wheel contact force F_N. This characteristic is of particular interest to vehicle manufacturers when designing suspension systems so that handling characteristics can be enhanced by stabilizers.

With a strong lateral force, F_S, the tire contact area (footprint) shifts significantly relative to the wheel (Fig. 8). This retards the buildup of the lateral force. This phenomenon greatly affects the transitional response (behavior during transition from one dynamic state to another) of vehicles under steering.

Effect of brake slip on lateral forces

When a vehicle is cornering, the centrifugal force acting outwards at the center of gravity must be held in equilibrium by lateral forces on all the wheels in order for the vehicle to be able to follow the curve of the road.

However, lateral forces can only be generated if the tires deform flexibly sideways so that the direction of movement of the wheel's center of gravity at the velocity, v_a, diverges from the wheel center plane "m" by the lateral slip angle, α (Fig. 7).

Fig. 6 shows the lateral-force coefficient, μ_S, as a function of brake slip at a lateral slip angle of 4°. The lateral-force coefficient is at its highest when the brake slip is zero. As brake slip increases, the lateral-force coefficient declines gradually at first and then increasingly rapidly until it reaches its lowest point when the wheel locks up. That minimum figure occurs as a result of the lateral slip angle position of the locked wheel, which at that point provides no lateral force whatsoever.

Friction – tire slip – vertical tire force

The friction of a tire depends largely on the degree of slip. The vertical tire force plays a subordinate role, there being a roughly linear relationship between braking force and vertical tire force at a constant level of slip.

The friction, however, is also dependent on the tire's lateral slip angle. Thus the braking and motive force reduces as the lateral slide angle is increased at a constant level of tire slip. Conversely, if the braking and motive force remains constant while the lateral slip angle is increased, the degree of tire slip increases.

Dynamics of linear motion

If the rim of a wheel is subjected both to a lateral force and braking torque, the road surface reacts to this by exerting a lateral force and a braking force on the tire. Accordingly, up to a specific limit determined by physical parameters, all forces acting on the rotating wheel are counterbalanced by equal and opposite forces from the road surface.

Beyond that limit, however, the forces are no longer in equilibrium and the vehicle's handling becomes unstable.

Total resistance to motion

The total resistance to vehicle motion, F_G, is the sum of the rolling resistance, aerodynamic drag and climbing resistance (Fig. 1). In order to overcome that total resistance, a sufficient amount of motive force has to be applied to the driven wheels. The greater the engine torque, the higher the transmission ratio between the engine and the driven wheels and the smaller the power loss through the drivetrain (efficiency η is approx. 0.88...0.92 with engines mounted in line, and approx. 0.91...0.95 with transversely mounted engines), the greater is the motive force available at the driven wheels.

A proportion of the motive force is required to overcome the total resistance to motion. It is adapted to suit the substantial increase in motion resistance on uphill gradients by the use of a choice of lower gearing ratios (multi-speed transmission). If there is a "surplus" of power because the motive force is greater than the resistance to motion, the vehicle will accelerate. If the overall resistance to motion is greater, the vehicle will decelerate.

Rolling resistance when traveling in a straight line

Rolling resistance is produced by deformation processes which occur where the tire is in contact with the road. It is the product of weight and rolling resistance coefficient and increases with a smaller wheel diameter and the greater the degree of deformation of the tire, e.g. if the tire is under-inflated. However, it also increases as the weight on the wheel and the velocity increases. Furthermore, it varies according to type of road surface – on asphalt, for example, it is only around 25% of what it is on a dirt track.

Fig. 1 Total resistance to motion, F_G

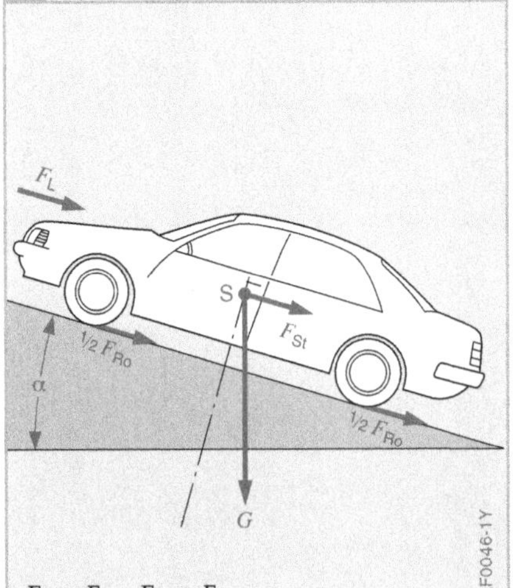

$$F_G = F_L + F_{St} + F_{Ro}$$

UAF0046-1Y

F_L Aerodynamic drag
F_{Ro} Rolling resistance
F_{St} Climbing resistance
F_G Total resistance to motion
G Weight
α Incline angle/gradient angle
S Center of gravity

Table 1 Examples of drag coefficient, c_W, for cars

Vehicle body shape	c_W
Convertible with top down	0.5...0.7
Box-type	0.5...0.6
Conventional saloon [1]	0.4...0.55
Wedge shape	0.3...0.4
Aerodynamic fairings	0.2...0.25
Tear-drop	0.15...0.2

[1] "Three-box" design

Table 2 Examples of drag coefficient, c_W, for commercial vehicles

Vehicle body shape	c_W
Standard tractor unit	
– without fairings	≥ 0.64
– with some fairings	0.54...0.63
– with all fairings	≤ 0.53

Rolling resistance when cornering

When cornering, the rolling resistance is increased by an extra component, cornering resistance, the coefficient of which is dependent on vehicle speed, the radius of the bend being negotiated, suspension characteristics, type of tires, tire pressure and lateral-slip characteristics.

Aerodynamic drag

The aerodynamic drag F_L is calculated from the air density ϱ, the drag coefficient c_W (dependent on the vehicle body shape, Tables 1 and 2), vehicle's frontal cross-sectional area A and the driving speed v (taking account of the headwind speed).

$$F_L = c_W \cdot A \cdot v^2 \cdot \varrho/2$$

Climbing resistance

Climbing resistance, F_{St} (if positive), or gravitational pull (if negative) is the product of the weight of the vehicle, G, and the angle of uphill or downhill gradient, α.

$$F_{St} = G \cdot \sin \alpha$$

Acceleration and deceleration

Steady acceleration or deceleration in a straight line occurs when the rate of acceleration (or deceleration) is constant. The distance required for deceleration is of greater significance than that required for acceleration because braking distance has direct implications in terms of vehicle and road safety.

The braking distance is dependent on a number of factors including

- Vehicle speed: at a constant rate of deceleration, braking distance increases quadratically relative to speed.
- Vehicle load: extra weight makes braking distances longer.
- Road conditions: wet roads offer less adhesion between road surface and tires and therefore result in longer braking distances.
- Tire condition: insufficient tread depth increases braking distances, particularly on wet road surfaces.
- Condition of brakes: oil on the brake pads/shoes, for example, reduces the friction between the pads/shoes and the disk/drum. The lower braking force thus available results in longer braking distances.
- Fading: The braking power also diminishes due to the brake components overheating.

The greatest rates of acceleration or deceleration are reached at the point when the motive or braking force is at the highest level possible without the tires starting to lose grip (maximum traction).

The rates actually achievable under real conditions, however, are always slightly lower because the vehicle's wheels are not all at the point of maximum adhesion at precisely the same moment. Electronic traction, braking and vehicle-handling control systems (TCS, ABS and ESP) are active around the point of maximum force transmission.

Dynamics of lateral motion

Response to crosswinds

Strong crosswinds can move a vehicle off course, especially if it is traveling at a high speed and its shape and dimensions present a large surface area for the wind to catch (Fig. 1). Sudden crosswind gusts such as may be encountered when exiting a road cutting can cause substantial sideways movement (yaw) of high-sided vehicles. This happens too quickly for the driver to react and may provoke incorrect driver response.

When a vehicle is driving through a crosswind, the wind force, F_W, produces a lateral component in addition to the longitudinal aerodynamic drag, F_L. Although its effect is distributed across the entire body surface, it may be thought of as a single force, the lateral wind force, F_{SW}, acting at a single point of action "D". The actual location of the point of action is determined by the vehicle's body shape and angle of incidence α of the wind.

The point of action is generally in the front half of the vehicle. On conventionally shaped saloon cars ("three-box" design) it is largely static and is closer to the center of the vehicle than on vehicles with a more streamlined body shape (sloping back), where it can move according to the angle of incidence of the wind.

The position of the center of gravity, S, on the other hand depends on the size and distribution of the vehicle load. In view of these variable factors, therefore, in order to arrive at a general representation of the effect of a crosswind (that is not affected by the relative position of the wheels and suspension to the body), a reference point 0 on the center line of the vehicle at the front is adopted.

When specifying lateral wind force at a reference point other than the true point of action, the turning force of the crosswind around the point of action, that is the yaw moment, M_Z, must also be considered. The crosswind force is resisted by the lateral cornering forces at the wheels. The degree of lateral cornering force which a pneumatic tire can provide depends on various factors in addition to lateral slip angle and wheel load, such as tire design and size, tire pressure and the amount of grip afforded by the road surface.

A vehicle will have good directional stability characteristics in a crosswind if the point of action is close to the vehicle's center of gravity. Vehicles that tend to oversteer will deviate less from their course in a crosswind if the point of action is forward of the center of gravity. The best position for the point of action on vehicles with a tendency to understeer is slightly behind the center of gravity.

1 Vehicle in crosswind

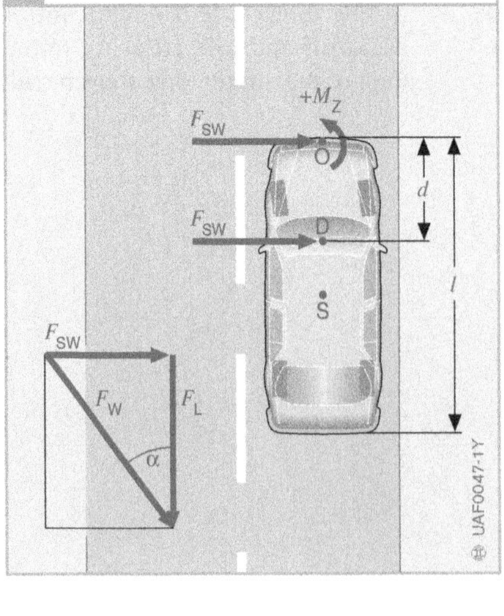

Fig. 1
D Point of action
O Reference point
S Center of gravity
F_W Wind force
F_L Aerodynamic drag
F_{SW} Lateral wind force
M_Z Yaw moment
α Angle of incidence
l Vehicle length
d Distance of point of action, D, from reference point, O
F_S and M_Z acting at O corresponds to F_S acting at D (in aerodynamics it is normal to refer to dimensionless coefficients instead of forces)

Understeer and oversteer

Cornering forces between a rubber-tired wheel and the road can only be generated when the wheel is rotating at an angle to its plane. A lateral slip angle must therefore be present. A vehicle is said to understeer when, as lateral acceleration increases, the lateral slip angle at the front axle increases more than it does at the rear axle. The opposite is true of a vehicle which oversteers (Fig. 2).

For safety reasons, vehicles are designed to slightly understeer. As a result of drive slip, however, a front-wheel drive vehicle can quickly change to sharply understeer or a rear-wheel drive vehicle to oversteer.

Centrifugal force while cornering

Centrifugal force, F_{cf}, acts at the center of gravity, S, (Fig. 3). Its effect depends on a number of factors such as

- the radius of the bend,
- the speed of the vehicle,
- the height of the vehicle's center of gravity,
- the mass of the vehicle,
- the track of the vehicle,
- the frictional characteristics of the tire and road surface (tire condition, type of surface, weather conditions), and
- the load distribution in the vehicle.

Potentially hazardous situations will occur when cornering if the centrifugal force reaches a point where it threatens to overcome the lateral forces at the wheels and the vehicle cannot be held on its intended course. This effect can be partially counteracted by positive camber or banked corners.

If the vehicle slips at the front wheel, it understeers; if it slips at the wheel axle, it oversteers. In both cases the Electronic Stability Program (ESP) detects an undesirable rotation about the vertical axle. By active intervention in the form of selective braking of individual wheels, it is then able to correct the imbalance.

2 Vehicle oversteer and understeer

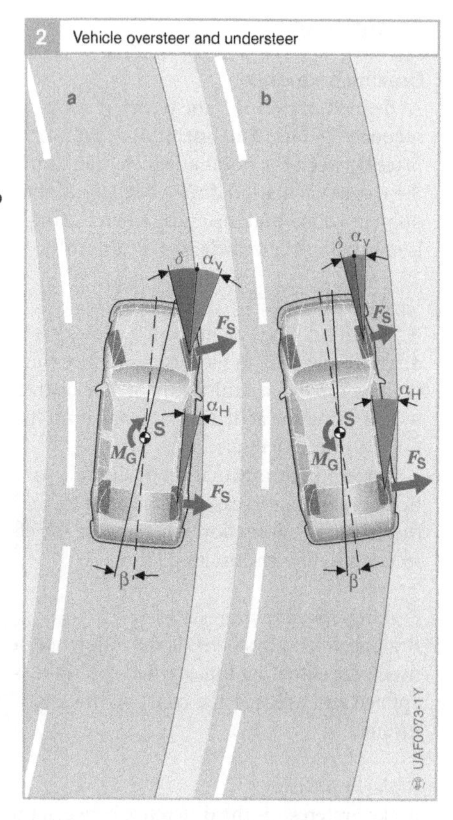

Fig. 2

- a Understeer
- b Oversteer
- α_V Front lateral slip angle
- α_H Rear lateral slip angle
- δ Steering angle
- β Side-slip angle
- F_S Lateral force
- M_G Yaw moment

3 Centrifugal force while cornering

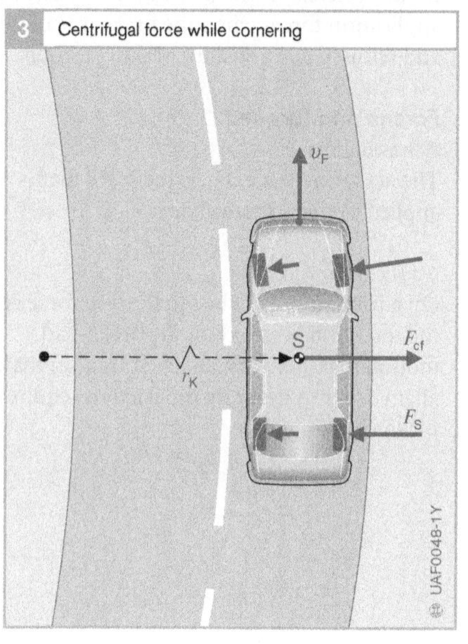

Fig. 3

- F_{cf} Centrifugal force
- v_F Vehicle speed
- F_S Lateral force at individual wheels
- r_K Radius of bend
- S Center of gravity

Definitions

Braking sequence

As defined in ISO 611, the term "braking sequence" refers to all operations that take place between the point at which operation of the (brake) actuation device begins and the point at which braking ends (when the brake is released or the vehicle is at a standstill).

Variable braking

A type of braking system which allows the driver at any time to increase or reduce the braking force to a sufficiently precise degree by operating the actuation device within its normal effective range.

If operating the actuation device in a particular manner increases the braking force, then the opposite action must reverse the effect and reduce the braking force.

Braking-system hysteresis

Braking system hysteresis is the difference between the actuating forces when the brake is applied and released at a constant braking torque.

Brake hysteresis

Brake hysteresis is the difference between the application forces when the brake is actuated and released at a constant braking torque.

Forces and torques

Actuating force

The actuating force, F_C, is the force that is applied to the actuation device.

Application force

On a friction brake, the application force is the total force exerted on the brake-pad mount, together with attached friction material, in order to generate the friction required for the braking force.

Total braking force

The total braking force, F_f, is the sum total of braking forces at each of the wheels that are produced by the effect of the braking system and which oppose the vehicle's motion or its tendency to move.

Braking torque

The braking torque is the product of the frictional forces generated in the brake by the application forces and the distance of the point of action of those forces from the axis of rotation of the wheel.

Braking-force distribution

The braking-force distribution indicates in terms of percentage share how the total braking force, F_f, is distributed between the front and rear wheels, e.g. front wheels 60%, rear wheels 40%.

External brake coefficient, C

The external brake coefficient, C, is the ratio of the output torque to the input torque or the output force to the input force of a brake.

Internal brake coefficient, C*

The internal brake coefficient, C*, is the ratio of the total tangential force acting at the effective radius of a brake to the application force, F_S.

Typical values: for drum brakes, values of up to $C^* = 10$ may be obtained, for disc brakes $C^* \approx 1$.

Time periods

The braking sequence is characterised by a number of time periods which are defined with reference to the ideal curves shown in Figure 1.

Period of movement of actuation device

The period of movement of the actuation device is the time from the point at which force is first applied to the actuation device (t_0), to the point at which it reaches its final position (t_3) as determined by the actuating force or the actuation travel. The same applies by analogy to the release of the brakes.

Response time

The response time, t_a, is the time that elapses from the point at which force is first applied to the actuation device to the point at which braking force is first produced (pressure generated in the brake lines) $(t_1 - t_0)$.

Pressure build-up time

The pressure build-up time, t_s, is the time from the point at which braking force is first produced to the point at which the pressure in the brake lines reaches its highest level $(t_5 - t_1)$.

Total braking time

The braking time, t_b, is the time that elapses from the point at which force is first applied to the actuation device to the point at which braking force ceases $(t_7 - t_0)$. If the vehicle comes to a halt, then the moment at which the vehicle is first stationary is the moment at which the braking time ends.

Effective braking time

The effective braking time, t_w, is the time that elapses from the moment at which braking force is first produced to the moment at which braking force ceases $(t_7 - t_2)$. If the vehicle comes to a halt, then the moment at which the vehicle is first stationary is the moment at which the effective braking time ends.

Distances

Braking distance

The braking distance, s_1, is the distance travelled by a vehicle during the period of the effective braking time $(t_7 - t_2)$.

Total braking distance

The total braking distance s_0 is the distance travelled by a vehicle during the period of the total braking time $(t_7 - t_0)$. That is the distance travelled from the point at which the driver first applies force to the actuation device to the point at which the vehicle is at a standstill.

Braking deceleration

Momentary deceleration

The momentary deceleration, a, is the quotient of the reduction in speed and the elapsed time.

$a = dv/dt$

Average deceleration over the total braking distance

From the vehicle speed v_0 at the time t_0, the average deceleration, a_{ms}, over the stopping distance, s_0, is calculated using the formula

$a_{ms} = v_0^2/2s_0$

Mean fully developed deceleration

The figure for mean fully developed deceleration, a_{mft}, represents the average deceleration during the period in which deceleration is at its fully developed level $(t_7 - t_6)$.

Braking factor

The braking factor, Z, is the ratio between total braking force, F_f, and total static weight, G_S, (vehicle weight) acting on the axle or axles of the vehicle. That is equivalent to the ratio of braking deceleration, a, to gravitational acceleration, g ($g = 9.81$ m/s^2).

1 Vehicle braking sequence to the point of standstill (ideal case)

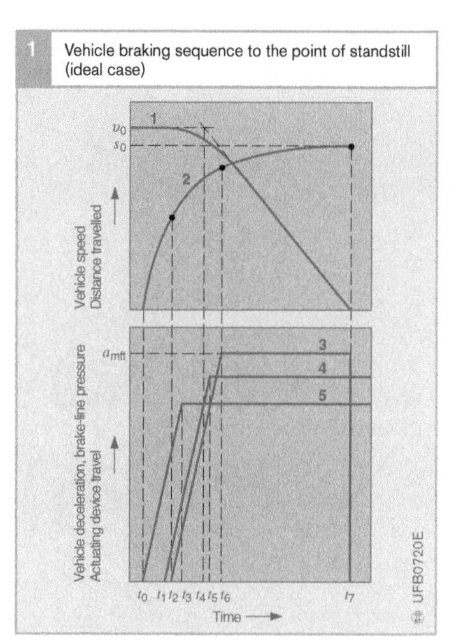

Fig. 1
1 Vehicle speed
2 Distance travelled while braking
3 Vehicle deceleration
4 Brake-line pressure (brake pressure)
5 Actuation device travel
t_0 Time at which the driver first applies force to actuation device
t_1 Brake-line pressure (brake pressure) starts to rise
t_2 Vehicle deceleration begins
t_3 Actuation device has reached intended position
t_4 Intersection of extended speed curve sections
t_5 Brake-line pressure has reached stabilised level
t_6 Vehicle deceleration has reached stabilised level
t_7 Vehicle comes to a halt

Car braking systems

Braking systems are indispensable for the roadworthiness and safe operation of a motor vehicle in road traffic conditions. They are therefore subject to strict legal requirements. The increasing effectiveness and sophistication demanded of braking systems over the course of time has meant that the mechanical systems have been continually improved. With the advent of microelectronics, the braking system has become a complex electronic system.

Overview

Car braking systems must perform the following fundamental tasks:

- Reduce the speed of the vehicle
- Bring the vehicle to a halt
- Prevent unwanted acceleration during downhill driving
- Keep the vehicle stationary when it is stopped

The first three of those tasks are performed by the service brakes. The driver controls the service brakes by operating the brake pedal. The parking brake ("hand brake") keeps the vehicle stationary once it is at a standstill.

Conventional braking systems

On conventional braking systems, the braking sequence is initiated exclusively by means of force applied to the brake pedal. In the braking system's master cylinder, that force is converted into hydraulic pressure. Brake fluid acts as the transmission medium between the master cylinder and the brakes (Figure 1).

On power-assisted braking systems such as are most frequently used on cars and light commercial vehicles, the actuation pressure is amplified by a brake servo unit (brake booster).

1 Example of a power-assisted car braking system

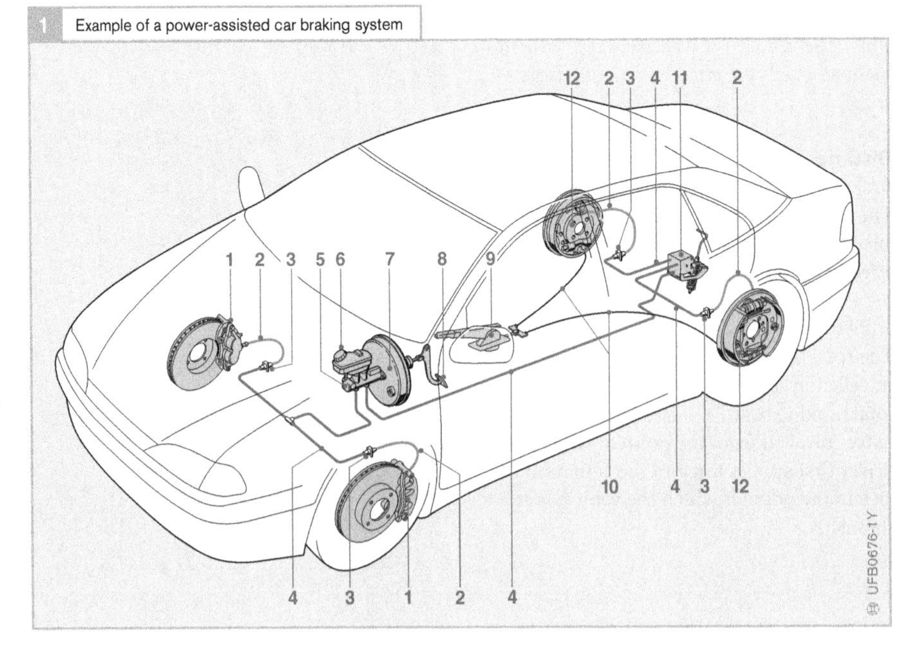

Fig. 1
1 Front brake (disc brake)
2 Brake hose
3 Connecting union between brake pipe and brake hose
4 Brake pipe
5 Master cylinder
6 Brake-fluid reservoir
7 Brake servo unit
8 Brake pedal
9 Handbrake lever
10 Brake cable (parking brake)
11 Braking-force reducer
12 Rear brake (drum brake in this case)

Electronic braking systems

Antilock braking system (ABS)

An electronic braking system was first used on a volume-production vehicle in 1978. ABS (Antilock Braking System) prevents the wheels locking up and the vehicle becoming uncontrollable under heavy braking.

As with conventional systems, an ABS system has a mechanical link between the brake pedal and the brakes. But it also incorporates an additional component, the hydraulic modulator. Solenoid valves in the hydraulic modulator are controlled in such a way that if the degree of wheel slip exceeds a certain amount, the brake pressure in the individual wheel cylinders is selectively limited to prevent the wheels locking.

ABS has been continually improved and developed to the extent that it is now standard equipment on virtually all new vehicles sold in western Europe.

Electrohydraulic brakes (SBC)

SBC (Sensotronic Brake Control) represents a new generation of braking systems. Under normal operating conditions, it has no mechanical link between the brake pedal and the wheel cylinders. The SBC electrohydraulic system detects the brake pedal travel electronically using duplicated sensor systems and analyses the sensor signals in an ECU. This method of operation is sometimes referred to as "brake by wire". The hydraulic modulator controls the pressure in the individual brakes by means of solenoid valves. Operation of the brakes is still effected hydraulically using brake fluid as the transmission medium.

Electromechanical brakes (EMB)

In the future there will be another electronic braking system, EMB (Electromechanical Brakes), which will operate electromechanically rather than employing hydraulics. Electric motors will force the brake pads against the discs in order to provide the braking action. The link between the brake pedal and the brakes will be purely electronic.

Electronic vehicle-dynamics systems

Continuing development of the ABS system led to the creation of TCS (Traction Control System). This system, which was first seen on volume-production cars in 1987, prevents wheel spin under acceleration and thus improves vehicle handling. Consequently, it is not a braking system in the strict sense of the word. Nevertheless, it makes use of and actively operates the braking system to prevent a wheel from spinning.

Another vehicle-dynamics system is the ESP (Electronic Stability Program), which prevents the vehicle entering a skid within physically determined parameters. It too makes use of and actively controls the braking system in order to stabilise the vehicle.

Ancillary functions of electronic systems

Electronic processing of data also makes possible a number of ancillary functions that can be integrated in the overall electronic braking and vehicle-dynamics systems.

- Brake Assistant (BA) increases brake pressure if the driver is hesitant in applying the full force of the brakes in an emergency.
- Electronic Braking-force distribution controls the braking force at the rear wheels so that the best possible balance between front and rear wheel braking is achieved.
- Hill Descent Control (HDC) automatically brakes the vehicle on steep descents.

History of the brake

Origin and development

The first use of the wheel is dated to 5,000 B.C. Usually, cattle were used as draft animals; later, horses and donkeys were also used. The invention of the wheel made it necessary to invent the brake. After all, a horse-drawn carriage traveling downhill had to be slowed down, not only to keep its speed within controllable limits, but also to prevent it running into the back of the horses. This was likely done using wooden rods braced against the ground or the wheel disks. Beginning around 700 B.C., wheels acquired iron tires to prevent premature wear of the wheel rim.

Beginning in 1690, coach drivers used a "chock" to brake their carriages. While driving downhill, they used its handle push it under a wheel, which then was immobilized and slid onto the chock.

In 1817, at the dawn of the industrial age, Baron Karl Drais rode from Karlsruhe in southern Germany to Kehl, proving to a stunned public that it is possible to ride on two wheels without falling over. As he surely had difficulty stopping when driving downhill, his last, 1820 model featured a friction brake on the rear wheel (Fig. 1).

Finally, in 1850, the iron axle was introduced in carriage construction, along with the shoe brake. In this type of brake, brake shoes were pressed against the metallic running surface of the iron-coated wooden wheels. The shoe brake could be operated from the driver's seat with the aid of a crank handle and a gear linkage (Fig. 2).

The low speed and sluggish drive train of the first automobiles did not place any great demands on the effectiveness of the brakes. In the early days, the shoe, band and wedge brakes, which were manually or foot-operated using levers, hinges and cables on the fixed rear axles, were sufficient for this purpose.

At first, the rear wheels were braked; occasionally, an intermediate shaft or only the cardan shaft was braked. Only about 35 years after the automobile was invented were the front wheels equipped with (cable-operated) brakes. Even more years passed before automobiles began to be equipped with hydraulically operated brakes, which, at the time, were only drum brakes. Use of the old method of cable activation continued in a few models, such as the VW Beetle, until after World War II. Other important milestones were the use of disk brakes and, in the present era, the introduction and incremental development of various driving stability systems.

1 Baron Karl Drais' wheel with friction brake, 1820

Fig. 1
1 Friction brake on wheel

2 Shoe brake with crank and linkage on a carriage (diagram)

Shoe and external shoe brakes on the wheel running surfaces

The first motor vehicles drove on wooden wheels with steel or rubber tires, or rubber-tired, spoked steel wheels. For braking, levers (as for the horse-drawn carriages) pushed brake shoes or external shoe brakes with friction linings against the running surfaces of the rear wheels. An initial example is the "riding carriage" developed by Gottlieb Daimler as an experimental vehicle in 1885 (the first motorcycle, with an engine performance of 0.5 horsepower and a top speed of 12 km/h). A cable led from the brake actuating lever, located at the front, close to the steering arm, to the *external shoe brake* on the rear wheel (Figures 3a, b).

In 1886, the first passenger cars with internal combustion engines were introduced: the Daimler motor carriage (1.1 hp, 16 km/h), which was derived from the horse carriage, and the Benz motorcar, which was newly designed as an automobile. Both of them had shoe brakes, as did the world's first truck, built in 1896. The shoe brake was installed in front of the rear wheels of each vehicle (Figures 3c, d, e, f.).

3 Historic motor vehicles and their wheel brakes (examples)

Fig. 3
a, b Daimler riding carriage 1885
1 Brake actuating lever
2 Cable to brake lever
3 Brake lever
4 External shoe brake on rear wheel

c Daimler motor carriage, 1886
1 Shoe brake, which also braked in "automatic" state when the flanged step was stepped on

d Daimler fire truck, 1890
1 Shoe brake

e Benz Viktoria, 1893
1 Shoe brake

f Benz Velo, 1894
1 Shoe brake

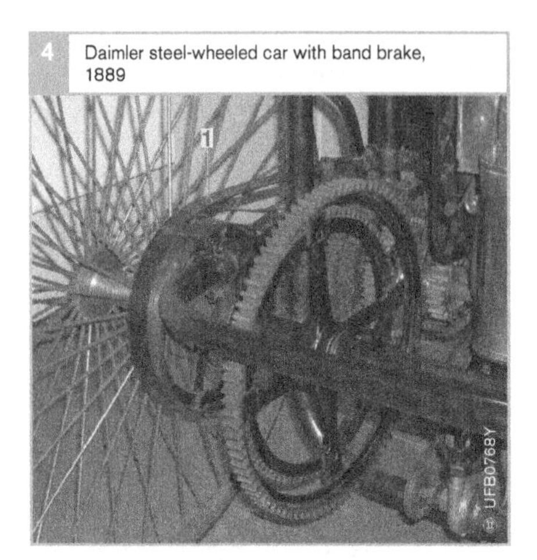

4 Daimler steel-wheeled car with band brake, 1889

Fig. 4
1 Band brake on the rear axle

5 Daimler Phoenix, 1889, drive shaft (front view)

Fig. 5
1 External shoe brake, front section
2 Brake lever and brake linkage

6 Daimler Phoenix, 1889, drive shaft (rear view)

Fig. 6
1 Brake rod
2 Brake lever
3 External shoe brake, rear section

Band and external shoe brakes

As solid rubber tires quickly became established for motor vehicles (beginning with the triangular Benz motorcar in 1886 and the Daimler steel-wheeled car of 1889) and were soon replaced by air-filled rubber tires for a more comfortable ride, the era of the shoe brake in automobiles had already come to an end.

From then on, *band brakes* (flexible steel brake bands that braked either directly or via several brake shoes riveted to the inside) or *external shoe brakes* (rigid cast iron or steel brake shoes with brake linings) began to be used. These pedal-operated brakes worked with external brake drums that were normally installed at the front on the intermediate drive shaft or on the drive axle in the rear wheel area.

For example, the Fahrzeugfabrik Eisenach produced the first Wartburg motorcar in 1898. Model 1 featured an exposed transmission and drive pinions. *Band brakes* braked both the axle drive and the two rear wheels.

In 1899, the Daimler steel-wheeled car had solid rubber tires and *steel band brakes* on the rear wheels (Fig. 4). The Daimler "Phoenix", also dating from 1899, still had solid rubber tires, but these were soon replaced by air-filled tires. A footbrake acted as an *external shoe brake* on the front drive shaft (Figures 5 and 6), and the handbrake acted on the rear wheels. In addition, this car featured – as did, for example, the Benz racing car of 1899 (Fig. 7) – a "sprag brake", a strong rod mounted on the rear that had the purpose of being driven into the (usually relatively soft) road.

An excerpt from the original text of the user manual for the "Phaeton" by Benz & Co. Rheinische Gasmotoren-Fabrik A.G. Mannheim from 1902 reads as follows: "In addition to a handbrake attached to its left side, the car is braked primarily by depressing the left foot pedal, which acts as a band brake on the brake disks fastened to the two rear wheels. Simultaneously, as mentioned above, the belt is automatically moved out. To stop the car immediately, both the left and the right pedals are depressed at the same time, which causes the band brake connected to the right pedal to act on the brake disk and thus brake the reduction gear."

Internal shoe drum brakes with mechanical cable activation

Over time, vehicles became faster and heavier. Therefore, they required a more effective brake system. Thus band and external shoe brakes soon gave way to the *internal shoe drum brake,* for which Louis Renault applied for a patent in 1902. A spreading mechanism pushed two crescent-shaped brake shoes against the inner surface of the cast iron or steel brake drums, which were connected to the wheel. Due to its self-reinforcing effect, the drum brake features low operating forces compared to the braking forces, long maintenance periods and long-lasting linings.

At first, the braking force was transmitted to the two drum brakes of the rear wheels using brake cables.

For example, the Mercedes Simplex already featured additional, *cable-operated rear wheel drum brakes* (Fig. 8) in addition to the cardan shaft band brake. Due to higher engine performance (40 horsepower), a second footbrake (Fig. 9) was added, which also acted as a *band brake* on the intermediate shaft of the chain drive. By the way, all four brakes were cooled by a water spray which, during braking, dripped onto the friction surfaces from a reservoir.

Beginning in about 1920, vehicles were fitted with *drum brakes on all four wheels.* The braking force was still transmitted using mechanical means, i.e. *levers, joints and brake cables.*

7 Benz racecar, 1899, with external band brake and suspended "sprag brake"

Fig. 7
1 "Sprag brake"
2 External band brake with brake shoes riveted to the inside

8 Daimler-Simplex, 1902, with cable-operated drum brake on the rear wheel

Fig. 8
1 Drum brake
2 Bowden cable

9 Daimler-Simplex, 1902, pedal and lever mechanism on the driver's seat

These cable-operated drum brakes remained in use for a long time. One example was the standard VW model of the 1950s (Fig. 10):

The primary element of this brake system was a brake pressure rail (Item 1). The four brake cables (2) attached to this element ran backwards through cable sleeves to the wheel brakes (drum brakes) of the four wheels (3). The rear part of the rail was supported by a short lever that sat on the brake pedal shaft. When the brake pedal of the footbrake (4) was depressed, the brake pressure rail was pushed forwards along with the four cables. The cables transmitted the force to the wheel brakes.

The lever for the handbrake (5) was further back in the car. However, via a decoupled rod, the handbrake ultimately acted on the same mechanism as the footbrake, and thus likewise acted on all four wheels.

10 Standard model VW, cable brake

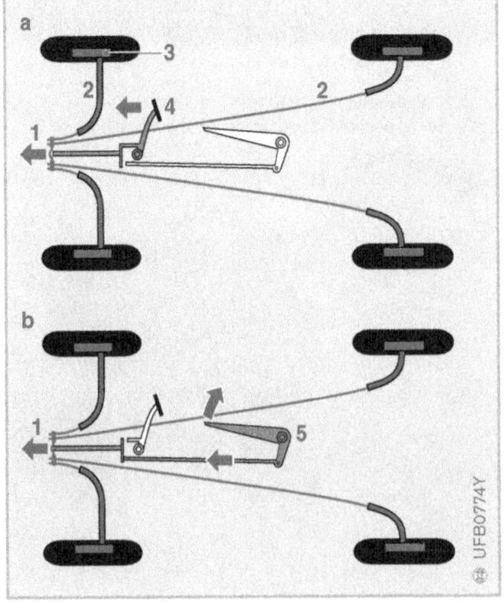

Fig. 10
a Activation of the footbrake
b Activation of the handbrake

1 Brake pressure rail
2 Brake cables
3 Wheel brakes
4 Brake pedal of the footbrake
5 Lever of the handbrake

Hydraulic brake activation

The main problem of the cable brake was the great maintenance effort and the uneven braking effect caused by uneven friction during mechanical transmission.

This was remedied when Lockheed introduced a hydraulically actuated brake in 1919. A special brake fluid now transmitted the brake pedal force uniformly to the actuating cylinders of the wheel brakes over metal lines and hoses, without the need for levers, joints and cables.

Hydraulic brake activation also made it possible to amplify the foot pressure applied by the driver by using intake manifold depression as a source of power for a brake servo system. The principle was patented in 1919 by Hispano-Suiza.

On commercial vehicles and railway rolling stock, air brakes established themselves as the system of choice.

In 1926, the "Adler Standard" was the first car in Europe to be equipped with a hydraulic brake system. The first hydraulic braking force reinforcement in auto racing were used in the Mercedes-Benz "Silver Arrows" in 1954. This ultimately became standard equipment for many mass-production vehicles.

Because a possible failure of the brake circuit could completely disable the early single-circuit brakes, the dual-circuit brake was later prescribed by law. According to VW Golf developer Dr. Ernst Fiala, the early "Beetles" (the standard model VWs) still had a cable-operated brake for that very reason: at the time it was feared that a hose in the hydraulic brakes could explode. Later, however – if only for competitive reasons – the VW Export and VW Transporter featured hydraulic braking systems.

Disk brake

Although British automaker Lancaster had patented the disk brake in 1902, it was a long time until this type of brake was introduced. Not until some fifty years later, beginning in 1955, did the legendary Citroën DS-19 become the first mass-produced car to be fitted with disk brakes. The disk brake was derived from the multi-plate brake and was initially developed for the aircraft industry.

In the disk brake, one brake lining presses the brake disk from the inside and outside. The brake disk (which is normally made of cast iron or, less commonly, of steel) is connected to the wheel. Its advantage is its simple and easy-to-assemble structure. It also counteracts the reduction in braking effect caused by overheating and prevents misalignment of the wheels of an axle.

The first German car with disk brakes on the front wheels was the BMW 502 in 1959. The first German cars to have disk brakes on all four wheels were the Mercedes 300 SE, the Lancia Flavia and the Fiat 2300 in 1961. Today, virtually all cars have a disk braking system, at least on the front wheels. In 1974, the first Formula 1 racecars with carbon fiber composite brake disks were introduced. These disks are considered especially light and heat resistant and thus have gained widespread use in motorsports and aviation.

Brake pads and shoes

Suitable brake linings had to be developed for drum and disk brakes, for which asbestos proved to be particularly effective. Not until it became known that asbestos fibers were harmful to health was the material replaced by plastic fiber.

Driving stability systems

The age of electronic brake systems dawned in 1978 with the arrival of the antilock braking system (ABS) for cars developed by Bosch. During braking, ABS provides early detection of the incipient lock of one or more wheels and prevents wheel locking. It ensures the steerability of the vehicle and substantially reduces the danger of skidding. In 1986, it was followed by the traction control system (TCS) with which Bosch extended system capability to the control of wheel spin under acceleration.
Fig. 11 shows road tests of these systems on the Bosch proving grounds in Boxberg in southern Germany.

As a further improvement of driving safety, Bosch introduced the electronic stability program (ESP) in 1995, which integrates the functions of ABS and TCS. It not only prevents the vehicle wheels from locking and spinning, it also keeps the vehicle from pulling to the side. Alternative systems, such as four-wheel steering and rear-axle kinematics, which were developed in the 1980s and 90s and were installed in some mass-production vehicles, did not catch on because they weighed too much, cost too much or were not effective enough.

Meanwhile, the (electrohydraulic) sensotronic brake control has found its place in automobile construction. It provides all of the ESP functions and decouples the mechanical operation of the brake pedal by means of an electronic control system. For safety purposes, a hydraulic fallback system is automatically available.

11 Steep uphill drive in the Bosch proving grounds in Boxberg for testing the driving stability systems of passenger cars and commercial vehicles

Classification of car braking systems

The entirety of the braking systems on a vehicle is referred to as braking equipment. Car braking systems can be classified on the basis of

- design and
- method of operation

Designs

Based on legal requirements, the functions of the braking equipment are shared among three braking systems:

- the service brakes,
- the secondary-brake system, and
- the parking brake

On commercial vehicles, the braking equipment also includes a continuous-operation braking system (e.g. retarder) that allows the driver to keep the vehicle at a steady speed on a long descent. The braking equipment of a commercial vehicle also includes an automatic braking system that operates the brakes of a trailer if it is detached from its towing vehicle either deliberately or by accident.

Service brakes

The service-brake system ("foot brake") is used to slow down the vehicle, to keep its speed constant on a descent, or to bring it completely to a halt.

The driver can infinitely vary the braking effect by means of the pressure applied to the brake pedal.
The service-brake system applies the brakes on all four wheels.

Secondary-brake system

The secondary-brake system must be capable of providing at least some degree of braking if the service-brake system fails. It must be possible to infinitely vary the level of braking applied.

The secondary-brake system does not have to be an entirely separate third braking system (in addition to the service brakes and the parking brake) with its own separate actuation device. It can consist of the remaining intact circuit of a dual-circuit service-brake system on which one circuit has failed, or it can be a parking-brake that is capable of graduated application.

Parking-brake system

The parking brake (hand brake) performs the third function required of the braking equipment. It must prevent the vehicle from moving when stopped or parked, even on a gradient and when the vehicle is unattended.

According to the legal requirements, the parking-brake system must also have an unbroken mechanical link, e.g. a rod linkage or a cable, between the actuation device and the brakes that it operates.

The parking-brake system is generally operated by means of a hand-brake lever positioned near the driver's seat, or in some cases by a foot pedal. This means that the service and parking-brake systems of a motor vehicle have separate actuation devices and means of force transmission.

The parking-brake system is capable of graduated application and operates the brakes on one pair of wheels (front or rear) only.

Methods of operation

Depending on whether they are operated entirely or partially by human effort or by another source of energy, braking systems can be classed either as

- muscular-energy (unassisted) braking systems,
- power-assisted braking systems, or
- power-brake systems,

Muscular-energy braking systems

On this type of braking system frequently found on cars and motorcycles, the effort applied to the brake pedal or hand-brake lever is transmitted to the brakes either mechanically (by means of a rod linkage or cable) or hydraulically. The energy by which the braking force is generated is produced entirely by the physical strength of the driver.

Power-assisted braking systems

The power-assisted braking system is the type most commonly used on cars and light commercial vehicles.
It amplifies the force applied by the driver by means of a brake servo which utilises another source of energy (vacuum or hydraulic power). The amplified muscle power is transmitted hydraulically to the brakes.

Power-brake systems

Power-brake systems are generally used on commercial vehicles but are occasionally fitted on large cars in conjunction with an integral ABS facility. This type of braking system is operated entirely without muscular-energy.

The system is operated by hydraulic power (which is based on fluid pressure) transmitted by hydraulic means. The brake fluid is stored in energy accumulators (hydraulic accumulators) which also contain a compressed gas (usually nitrogen). The gas and the fluid are kept apart by a flexible diaphragm (diaphragm accumulator) or a piston with a rubber seal (piston accumulator). A hydraulic pump generates the fluid pressure, which is always in equilibrium with the gas pressure in the energy accumulator. A pressure regulator switches the hydraulic pump to idle as soon as the maximum pressure is reached.

Since brake fluid can be regarded as practically incompressible, small volumes of brake fluid can transmit large brake-system pressures.

Components of a car braking system

Figure 1 shows the schematic layout of a car braking system. It consists of the following main component groups:

- Energy supply system,
- Actuation device,
- Force transmission system, and
- Wheel brakes.

Energy supply system

The energy supply system encompasses those parts of the braking system that provide, control and (in some cases) condition the energy required to operate the brakes. It ends at the point where the force transmission system begins, i.e. where the various circuits of the braking system are isolated from the energy supply system or from each other.

Car braking systems are in the main power-assisted braking systems in which the physical effort of the driver, amplified by the vacuum in the brake servo unit, provides the energy for braking.

Actuation device

The actuation device encompasses those parts of a braking system that are used to initiate and control the action of that braking system. The control signal may be transmitted within the actuation device, and the use of an additional energy source is also possible.

The actuation device starts at the point at which the actuation force is directly applied. It may be operated in the following ways:

- by direct application of force by hand or foot by the driver,
- by indirect control of force by the driver.

The actuation device ends at the point where distribution of the braking-system energy begins or where a portion of that energy is diverted for the purpose of controlling braking. Among the essential components of the actuation device are the vacuum servo unit and the master cylinder.

Force transmission system

The force transmission system encompasses those parts of the braking system that transmit the energy introduced by the energy supply system(s) and controlled by the actuation device. It starts at the point where the actuation device and the energy supply system end. It ends at the point where it interfaces with those parts of the braking system that generate the forces that inhibit or retard vehicle motion. It may be mechanical or hydromechanical in design.

The components of the force transmission system include the transmission medium, hoses, pipes and, on some systems, a pressure regulating valve for limiting the braking force at the rear wheels.

Wheelbrakes

The wheelbrakes consist of those parts of the braking system in which the forces that inhibit or retard the movement of the vehicle are generated. On car braking systems, they are friction brakes (disc or drum brakes).

1 Layout of a car braking system

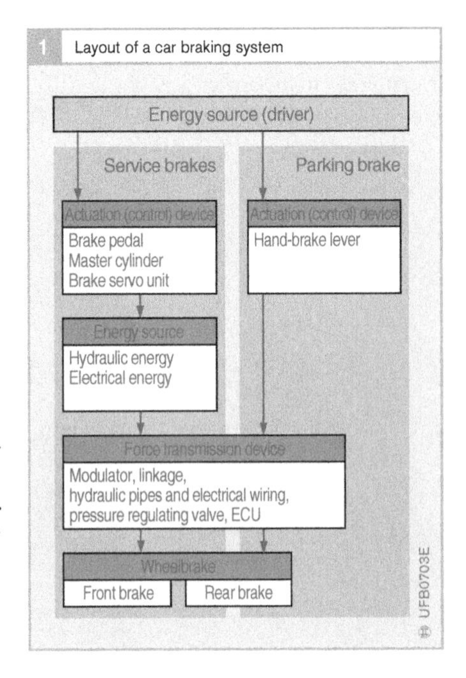

Brake-circuit configuration

Legal requirements demand that braking systems incorporate a dual-circuit forcetransmission system.

According to DIN 74 000, there are five ways in which the two brake circuits can be split (Figure 1). It uses the following combinations of letters to designate the five different configurations: II, X, HI, LL and HH. Those letters are chosen because their shapes roughly approximate to the layout of the brake lines connecting the master cylinder and the brakes.

Of those five possibilities, the II and X configurations, which involve the minimum amount of brake piping, hoses, disconnectable joints and static or dynamic seals, have become the most widely established. That characteristic means that the risk of failure of each of the individual circuits due to fluid leakage is as low as it is for a single-circuit braking system. In the event of brake-circuit failure due to overheating of one of the brakes, the HI, LL and HH configurations have a critical weakness because the connection of individual brakes to both circuits means that failure of one brake can result in total failure of the braking system as a whole.

In order to satisfy the legal requirements regarding secondary-braking effectiveness, vehicles with a forward weight-distribution bias are fitted with the X configuration. The II configuration is particularly suited to use on cars with a rearward weight-distribution bias.

II configuration

This layout involves a front-axle/rear-axle split – one circuit operates the rear brakes, the other operates the front brakes.

X configuration

This layout involves a diagonal split – each circuit operates one front brake and its diagonally opposed rear brake.

HI configuration

This layout involves a front/front-and-rear split - one brake circuit operates the front and rear brakes, the other operates only the front brakes.

LL configuration

This arrangement involves a two-front/one-rear split. Each circuit operates both front wheels and one rear wheel.

HH configuration

The circuits are split front-and-rear/front-and-rear. Each circuit operates all four wheels.

1 Brake-circuit configuration

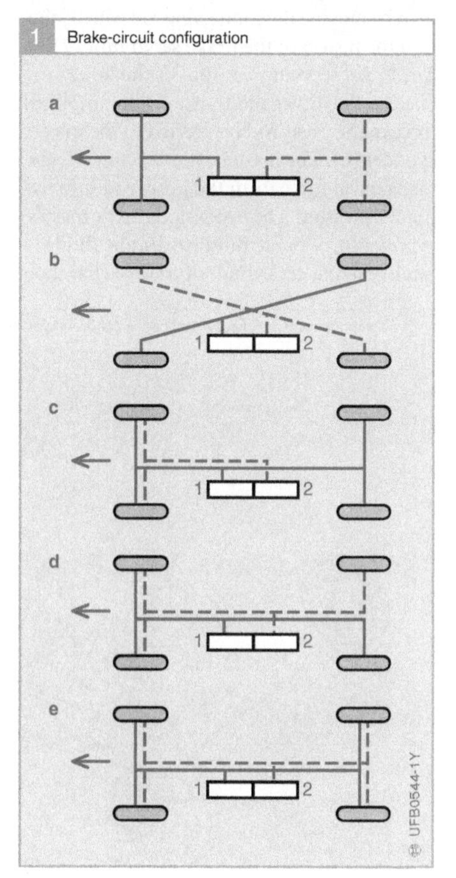

Fig. 1
a II configuration
b X configuration
c HI configuration
d LL configuration
e HH configuration
1 Brake circuit 1
2 Brake circuit 2
← Direction of travel

Car braking-system components

Along with steering and changing gear, braking is one of the most frequently performed operations when driving a car.
The components of the braking system must take account of that fact by making optimum use of the force applied to the pedal by the driver and by ensuring that the force required remains as constant as possible for the desired braking effect.

Overview

Figure 1 shows a conventional braking system of the dual-circuit type with a front/rear configuration and without electronic safety systems.

In order to operate the brakes, the driver applies foot pressure to the brake pedal (8), thereby moving the connecting rod which joins it to the piston of the brake servo unit (7). The brake servo unit amplifies the force applied by the driver and transmits the amplified force to the push-rod connected to the master cylinder (6). The master cylinder converts the mechanical force from the push-rod into hydraulic pressure. The two pistons in the master cylinder force brake fluid (hydraulic fluid) out of the master cylinder pressure chambers into the brake pipes (4) and brake hoses (2), thereby transmitting hydraulic pressure to the disk brakes (1) on the front wheels and the drum brakes (12) on the rear wheels. If one of the brake circuits fails, the other remains fully functional so that the effect of a secondary-brake system is guaranteed. The brake-fluid reservoir (6) connected to the master cylinder (5) compensates for volume fluctuations in the brake circuits.

Increasing deceleration during the braking process shifts an increasingly larger proportion of the vehicle's weight from the rear to the front wheels (dynamic axle-load shift). Accordingly, the pressure regulating valve (11) lowers the braking pressure at the rear wheels to prevent them being overbraked. This is a brake-force balancing process rather than a brake-force control process such as is effected by antilock braking systems.

The parking-brake system linked to the rear brakes (12) is operated by the hand-brake lever (9) and the hand-brake cable (10).

1 Components of a hydraulic dual-circuit car braking system

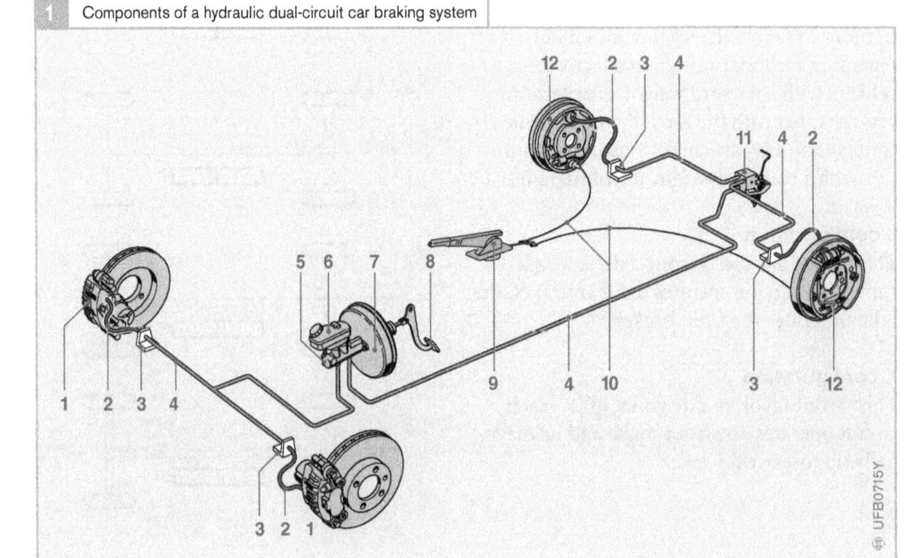

Fig. 1
1 Front brake (disk brake)
2 Brake hose
3 Connecting union between brake pipe and brake hose
4 Brake pipe
5 Master cylinder
6 Brake-fluid reservoir
7 Brake servo unit
8 Brake pedal
9 Hand-brake lever (parking brake)
10 Hand-brake cable (parking brake)
11 Pressure regulating valve
12 Rear brake (drum brake)

Brake pedal

Function

During the braking sequence, the force applied by the driver is transmitted to the braking system through the brake-pedal. This requires a sensitive response to the force applied.

Design

There are two types of brake-pedal design:

- suspended and
- floor-mounted (brake-pedal module).

Suspended design

Most cars have suspended-type brake pedals (Figure 1). The pedal (7) is attached to the pedal mount (5) through its spindle (6). A return spring (3) attached to the front bulkhead (2) holds the brake pedal in the neutral position when it is not being operated.

Brake-pedal module

On the brake-pedal module (Figure 2), the pedal (3), brake servo unit (2) and master cylinder (1) are combined to form a single unit. Design considerations dictate that the module illustrated is mounted below the vehicle floor in the area of the driver's footwell. A sealed casing (4) protects the mechanism from dirt and damp.

Method of operation

During the braking sequence, the driver applies foot pressure to the pedal (Item 7, Figure 1), thereby overcoming the retaining force of the return spring (3) and transmitting force to the brake servo unit (1) via the connecting rod (4).

1 Brake pedal (suspended design)

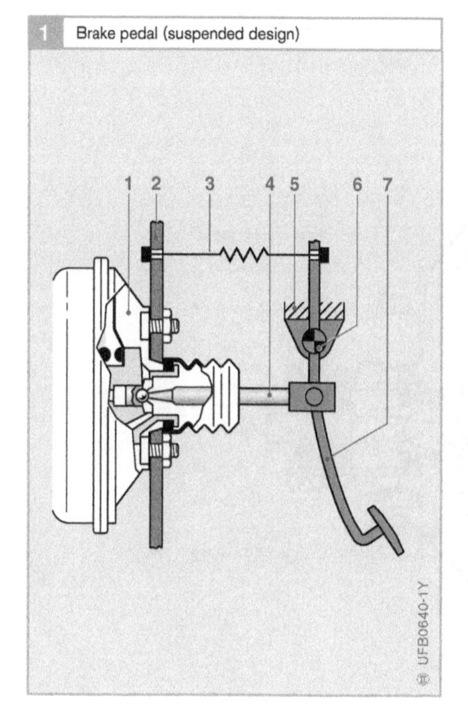

2 Brake pedal module

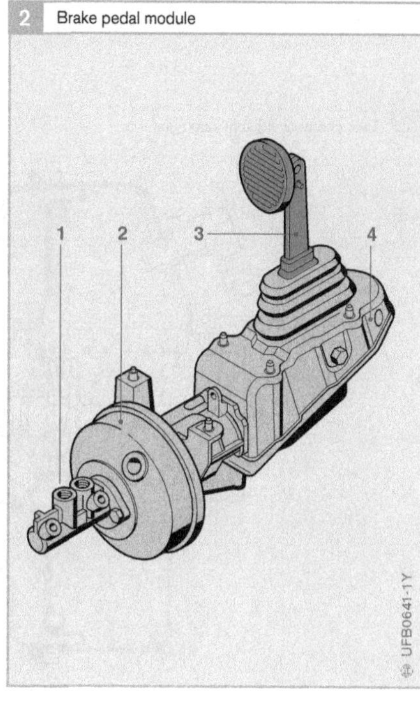

Fig. 1
1 Brake servo unit
2 Bulkhead
3 Return spring
4 Connecting rod
5 Pedal mount
6 Pedal spindle
7 Pedal

Fig. 2
1 Master cylinder
2 Brake servo unit
3 Pedal
4 Casing

Brake servo unit

The brake servo unit amplifies the force applied by the driver's foot when pressing the brake pedal and thus reduces the effort required. It is combined with the master cylinder to form a single unit and is a component of the majority of car braking systems. A fundamental technical requirement demanded of brake servo units is that they reduce the physical effort required of the driver without impairing the response sensitivity of the braking system and the feedback on the level of braking. The two common designs of brake servo unit, the vacuum servo unit and the hydraulic servo unit, utilise existing sources of energy in the vehicle – the vacuum in the intake manifold or the hydraulic pressure generated by a hydraulic pump.

Dual-chamber vacuum servo unit

Most cars are fitted with a brake servo unit of the vacuum type. There are two versions of this type of unit which both function in a similar manner:

- the dual-chamber type (Figure 1) and
- the four-chamber tandem type (Figure 2) for greater power assistance.

Function

Vacuum brake servo units utilise the negative pressure generated in the intake manifold during the induction stroke on gasoline engines or the vacuum (0.5 ... 0.9 bar) produced by a vacuum pump on diesel engines to amplify the force applied by the driver's foot. The level of power assistance increases proportionally to the foot pressure until the "saturation point" is reached. This is close to the locking pressure for the front wheels and may be between 60 and 100 bar depending on the vehicle. From that point onwards, the level of power assistance does not increase any further.

1 Twin-chamber vacuum servo unit

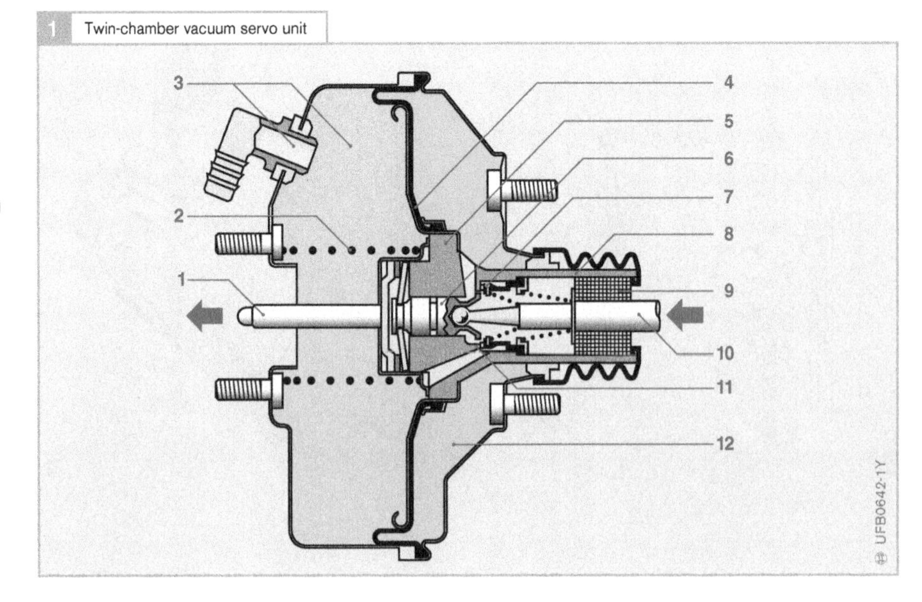

Fig. 1
1 Push rod (transmits output force to master cylinder)
2 Compression spring
3 Vacuum chamber with vacuum pipe connection
4 Diaphragm and diaphragm disk
5 Working piston
6 Sensing piston
7 Double valve
8 Valve body
9 Air filter
10 Connecting rod (transmits pedal force)
11 Valve seat
12 Working chamber

Design

A diaphragm (Item 4, Figure 1) separates the vacuum chamber (3) with its vacuum pipe connection from the working chamber (12). The connecting rod (10) transmits the applied foot pressure to the working piston (5) and the amplified force is passed to the master cylinder via the push rod (1).

Method of operation

When the brakes are not being applied, the vacuum chamber (3) and working chamber (12) are connected to one another via channels in the valve body (8). Given that the vacuum pipe connection (3) is connected to a vacuum source, this means that there is a vacuum in both chambers.

As soon as a braking sequence is initiated, the connecting rod (10) moves in the direction of the vacuum chamber (3) and presses the seal of the double valve (7) against the valve seat (11). Consequently, the vacuum chamber and working chamber are then isolated from each other. Since further movement of the connecting rod lifts the sensing piston (6) away from the double-valve seal, air at atmospheric pressure is able to flow into the working chamber. The pressure in the working chamber is then greater than in the vacuum chamber. The atmospheric pressure acts via the diaphragm (4) on the diaphragm disk with which it is in contact. Because the diaphragm disk is attached to the valve body (8), the latter moves when the disk moves, thereby assisting the foot pressure transmitted by the connecting rod. The foot pressure and the assisting force thus combine to overcome the force of the compression spring (2) and move the push rod (1) forwards, thereby transmitting the output force of the servo unit to the master cylinder.

When the brake pedal is released and braking ceases, the vacuum chamber and working chamber are connected again so that there is a vacuum in both.

Four-chamber/tandem vacuum servo unit

Function

Like the dual-chamber vacuum servo unit, the four-chamber/tandem version uses vacuum to amplify the foot pressure applied to the brake pedal by the driver. The four-chamber design allows greater power assistance to be achieved than with the dual-chamber version.

2 Four-chamber/tandem vacuum servo unit

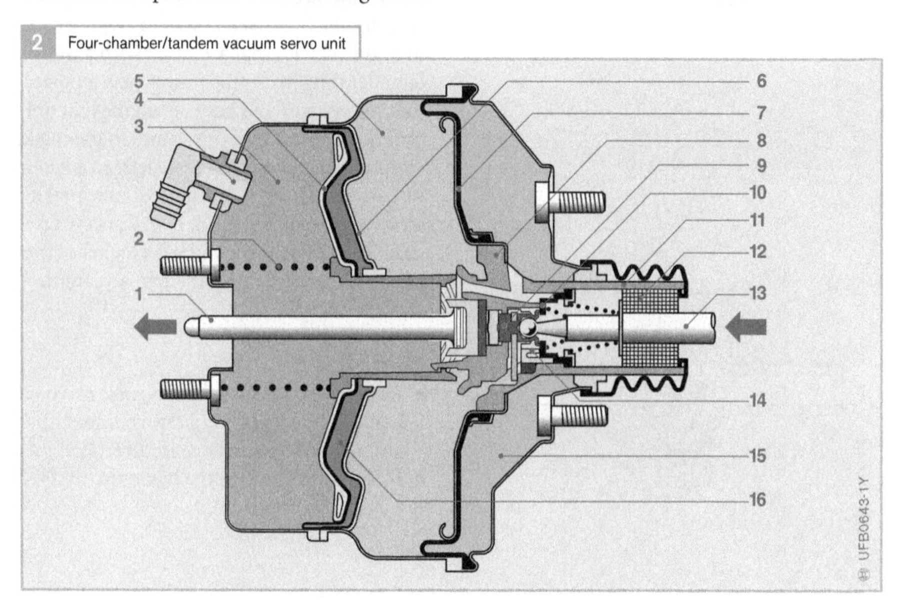

Fig. 2
1 Push rod (transmits output force to master cylinder)
2 Compression spring
3 Vacuum chamber II with vacuum pipe connection
4 Diaphragm II and diaphragm disk II
5 Vacuum chamber I
6 Partition
7 Diaphragm I and diaphragm disk I
8 Working cylinder
9 Sensing piston
10 Double valve
11 Valve body
12 Air filter
13 Connecting rod (transmits pedal force)
14 Valve seat
15 Working chamber I
16 Working chamber II

Design

There are four chambers – working chamber I (previous page, Figure 2, Item 15), vacuum chamber I (5), working chamber II (16) and vacuum chamber II (3) – arranged in series, one behind the other. Between vacuum chamber I and working chamber II there is a partition (6). The connecting rod (13) transmits the applied foot pressure to the working piston (8) and the amplified force is passed to the master cylinder via the push rod (1).

Method of operation

The method of operation is similar to that of the dual-chamber vacuum servo unit. When the brakes are not being applied, there is a vacuum in all four chambers, given that the vacuum pipe connection (3) is connected to a vacuum source. As soon as a braking sequence is initiated, the sensing piston (9) lifts away from the seal of the double valve (10). Air at atmospheric pressure is then able to enter working chambers I (15) and II (16) thus creating a pressure difference between the working chambers and the vacuum chambers. That pressure difference acts as an assisting force and amplifies the foot pressure transmitted via the connecting rod.

3 Vacuum non-return valve

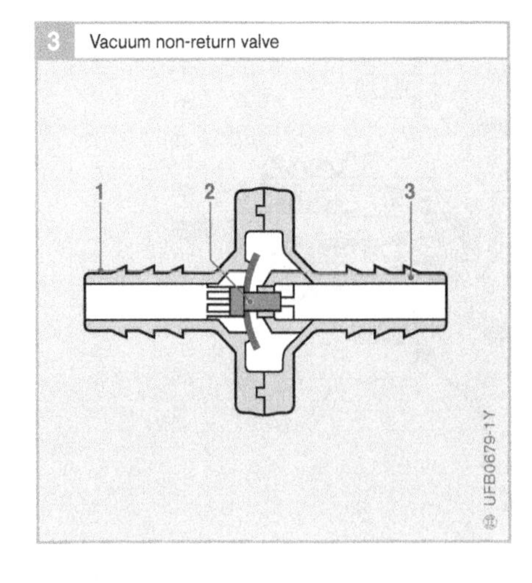

Fig. 3
1 Connection to vacuum brake servo
2 Non-return valve
3 Connection to engine

Vacuum non-return valve

All braking systems equipped with a vacuum servo unit have a non-return valve (Figure 3) in the vacuum pipe between the vacuum source (engine intake manifold or vacuum pump) and the vacuum servo unit. While there is a vacuum present, the non-return valve remains open. It closes when the vacuum source ceases to produce a vacuum (engine is switched off) so that the vacuum inside the brake servo unit is retained. On gasoline-engined cars, it also prevents fuel vapours being drawn into the brake servo unit and damaging rubber components. The throttle effect of the valve also attenuates the pulsations generated by the intake manifold.

Note on fitting

The vacuum non-return valve must be fitted in the vacuum pipe so that the arrow points towards the vacuum source (engine intake manifold or vacuum pump).
If it is fitted the wrong way round it will not function as intended. It is fitted near to the engine intake manifold but not so close that the heat from the engine impairs its function.

Hydraulic brake servo system

A hydraulic brake servo system (Figure 4) is used on vehicles that are fitted with a means of generating hydraulic power (e.g. a power steering system) and have an engine that only produces a small level of vacuum in the intake manifold (e.g. diesel or turbocharged engine). As part of such a system, the hydraulic brake servo unit (Page 33, Fig. 4, Item 7) occupies significantly less space and has a higher output pressure (approx. 160 bar) than a vacuum servo unit.

Notes on use

- Hydraulic oil is used in the brake servo circuit (A), the intake and return lines (B) and the power-steering circuit (C).
- Brake fluid is used in the brake circuits (D).

Function
Hydraulic brake servo units amplify the force applied to the brake pedal by the driver when braking. They use hydraulic power that is generated by a hydraulic pump and stored in a hydraulic accumulator.

Design
The hydraulic brake servo system consists of the following components:
- Steering pump (1)
- Reservoir (2) with filter
- Pressure-controlled flow regulator (3) with hydraulic accumulator (4)
- Master cylinder (5) with brake-fluid reservoir (6)

The steering pump supplies the brake servo unit (7) and the power steering system (8) with hydraulic pressure.

Method of operation of flow regulator and hydraulic accumulator
The flow regulator and hydraulic accumulator are identified by the numbers 3 and 4 in the diagram of the system (Figure 4).

The steering pump delivers hydraulic fluid to port C1 of the pressure-controlled flow regulator (Figure 5 overleaf, Items 1 ... 5). The flow regulator piston (4) directs the majority of the flow via port C2 to the power-steering system, while the smaller proportion is used to charge the hydraulic accumulator (6 ... 8) to a pressure of 36 ... 57 bar. When its shut-off pressure is reached, the changeover valve (5) connects the spring chamber of the flow-regulator piston (4) to the hydraulic reservoir via port B. The entire hydraulic flow is then available to the power-steering system.

4 Hydraulic brake servo system

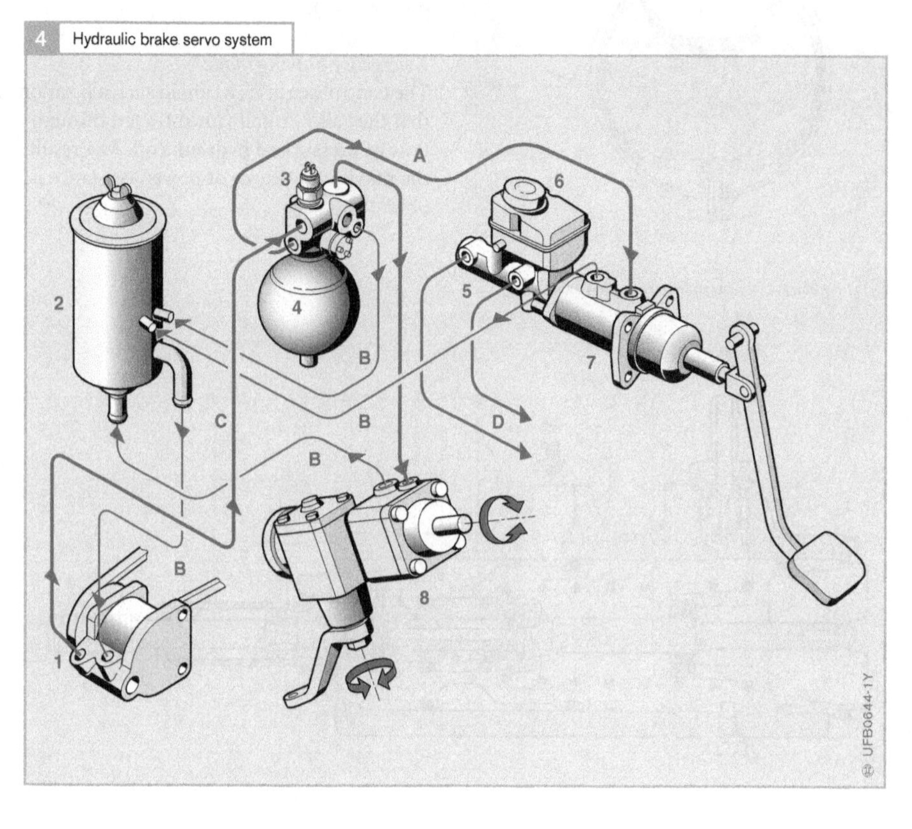

Fig. 4
1 Steering pump
2 Oil reservoir with filter
3 Pressure-controlled flow regulator
4 Hydraulic accumulator
5 Master cylinder
6 Brake-fluid reservoir
7 Hydraulic brake servo
8 Power-steering servo

Piping systems
A Brake servo circuit
B Intake and return lines
C Power-steering circuit
D Brake circuits

Method of operation of the hydraulic brake servo unit

The hydraulic brake servo unit is identified by the number 7 in the diagram of the system (Figure 4).

5 Pressure-controlled flow regulator (1...5) with hydraulic accumulator (6...8)

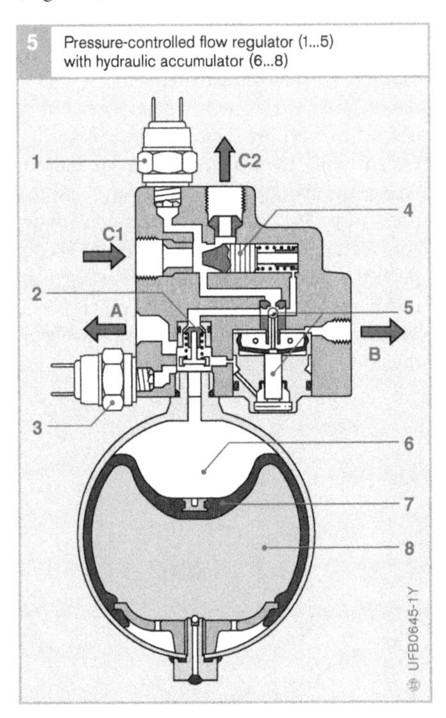

Fig. 5
1 Circulation-pressure warning switch
2 Non-return valve
3 Accumulator-pressure warning switch
4 Flow-regulator piston
5 Changeover valve
6 Hydraulic chamber
7 Diaphragm
8 Pneumatic chamber

Ports
A to brake servo unit
B to oil reservoir
C1 from power-steering pump
C2 to power-steering servo

Non-braking mode
The control edges (Figure 6, Items 4...6) block the inflow of hydraulic oil from the pressure-controlled flow regulator via port C2 while allowing unpressurised oil to flow out of port B to the hydraulic oil reservoir.

Normal braking mode
The foot pressure applied by the driver is transmitted to the actuating piston (9) which moves the control piston (7), thereby shifting its control edges. As a result, pressurised oil from the pressure-controlled flow regulator is allowed to enter the servo via port C2 and outflow of oil to the reservoir via port B is blocked. The pressurised hydraulic oil acts on the transfer piston (3) and the actuating piston (9) as an assisting force supplementing the foot pressure applied by the driver until a state of equilibrium is reached with the output force from the master cylinder acting on the push rod (1).

Emergency braking mode
The control edges (5, 6) are in such a position that they allow totally unrestricted through-flow of pressurised hydraulic oil. As a result, the maximum degree of power assistance is obtained.

6 Hydraulic brake servo (in non-braking mode)

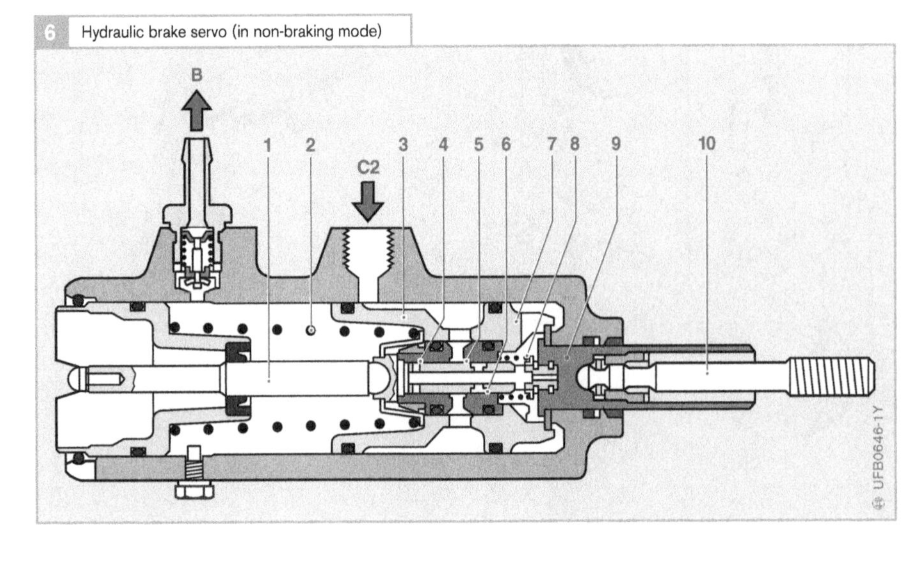

Fig. 6
1 Push rod
2 Return spring
3 Transfer piston
4...6 Control edges
7 Control piston
8 Compression spring
9 Actuating piston
10 Connecting rod

Ports
B to oil reservoir
C2 from pressure-controlled flow regulator

Master cylinder

The master cylinder (also called a tandem master cylinder) converts the mechanical force applied by the driver to the brake pedal – and amplified by the brake servo where fitted – into hydraulic brake force by forcing brake fluid into the brake circuits in proportion to the mechanical force applied and controlling it accordingly.

Statutory regulations require that cars are fitted with two separate brake circuits. That requirement is accommodated by using a tandem-type master cylinder which consists in effect of two master cylinders connected in series. If one of the brake circuits fails, full braking pressure can then still be maintained in the other circuit.

There are a number of possible variations on the master-cylinder design, as described below. Other specialised designs such as the graduated master cylinder, the multistage master cylinder or the Twintax master cylinder are only rarely used in motor vehicles.

Master cylinder with captive piston spring

Design

The "captive" piston spring (Figure 1, Item 9), a compression spring, keeps the push-rod piston (11) and the float piston (7) – also called the intermediate piston – the same distance apart when the cylinder is at rest. This prevents the piston spring (9) moving the float piston to a position where its primary seal (13) has passed beyond the balancing port (5) when the cylinder is at rest. If that were to happen, pressure equalisation in the secondary circuit via the balancing port (5) would not be possible and residual pressure in the circuit after release of the brakes would prevent the brake shoes/pads from retracting from the drum/disk.

Method of operation

When the brakes are operated, the push-rod piston (11) and the float piston (7) move to the left, pass over the balancing ports (5) and force brake fluid into the brake circuits via the outlet ports (2). As the pressure increases, the float piston ceases to be moved by the captive piston spring (9) and is moved instead by the pressure of the brake fluid.

1 Master cylinder with captive piston spring

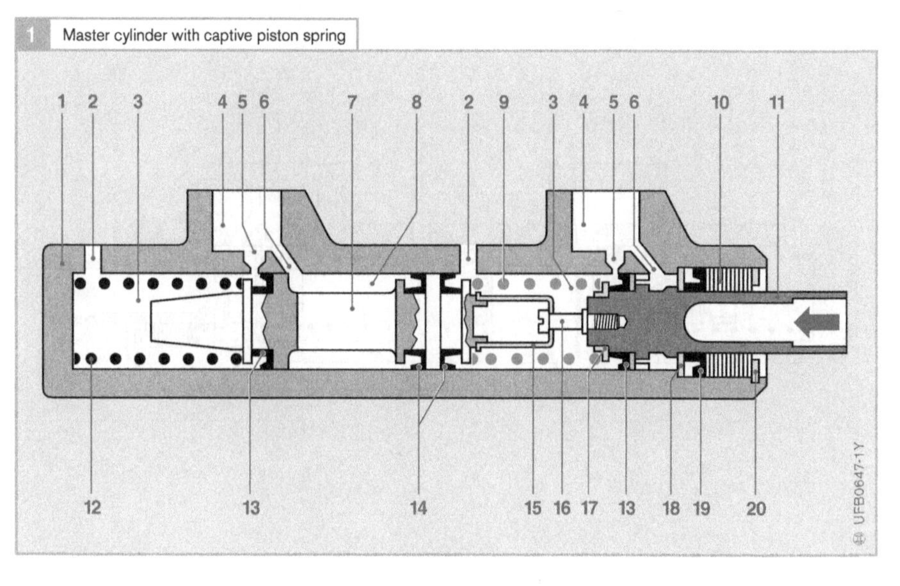

Fig. 1
1 Cylinder housing
2 Pressure outlet to brake circuit
3 Pressure chamber
4 To brake-fluid reservoir
5 Balancing port
6 Snifter bore
7 Float piston
8 Intermediate chamber
9 Captive piston spring
10 Plastic sleeve
11 Push-rod piston (transmits input force from brake servo)
12 Compression spring (secondary circuit)
13 Primary seal
14 Isolating seal
15 Captive sleeve
16 Retaining screw
17 Support ring
18 Stop disk
19 Secondary seal
20 Circlip

Master cylinder with central valve

Design

Basically, this type of master cylinder (Figure 2) is similar to the master cylinder with captive piston spring described above. It was developed for vehicles with antilock braking systems (ABS).

The particular feature of this master cylinder is a float piston with an integral central valve that allows the brake fluid to flow back through the valve-pin channel (18) when the braking system is not under pressure. The balancing port in the secondary circuit is thus dispensed with as the central valve performs the same function. The intermediate chamber (9) is permanently connected via a channel to the brake-fluid reservoir.
As there is a risk on vehicles fitted with ABS that when the piston passes over the balancing port (11), damage to the primary seal (17) can result at high pressures (leading to failure of one of the brake circuits), master cylinders on such vehicles are generally fitted with two central valves.

Method of operation

The force applied to the brake pedal acts directly against the push-rod piston (14) and pushes it to the left. This moves it past the balancing port (11) and the fluid in the pressure chamber (3) is then able to push the float piston (6) to the left as well. Once the float piston has moved about 1 mm to the left, the valve pin (18) is no longer resting against the clamping sleeve (7) and the valve seal (16) seals off the pressure chamber (3) from the intermediate chamber (9) by pressing against the float piston. As the force applied to the brake pedal increases, so the pressure in both pressure chambers (3) rises. As the foot pressure is released, the two pistons (14 and 6) move to the right until the balancing port (11) is open again or the valve pin (18) presses against the clamping sleeve (7) and the valve seal (16) lifts away from the float piston (6). The brake fluid can then flow back into the fluid reservoir and the brakes are no longer under pressure.

Fig. 2
1 Cylinder housing
2 Pressure outlet to brake circuit
3 Pressure chamber
4 Valve spring
5 To brake-fluid reservoir
6 Float piston
7 Clamping sleeve
8 Intermediate piston
9 Intermediate chamber
10 Compression spring
11 Balancing port
12 Snifter bore
13 Plastic sleeve
14 Push-rod piston (transmits input force from brake servo)
15 Compression spring (secondary circuit)
16 Valve seal
17 Primary seal
18 Valve pin
19 Isolating seal
20 Support ring
21 Stop disk
22 Secondary seal
23 Circlip

2 Master cylinder with central valve

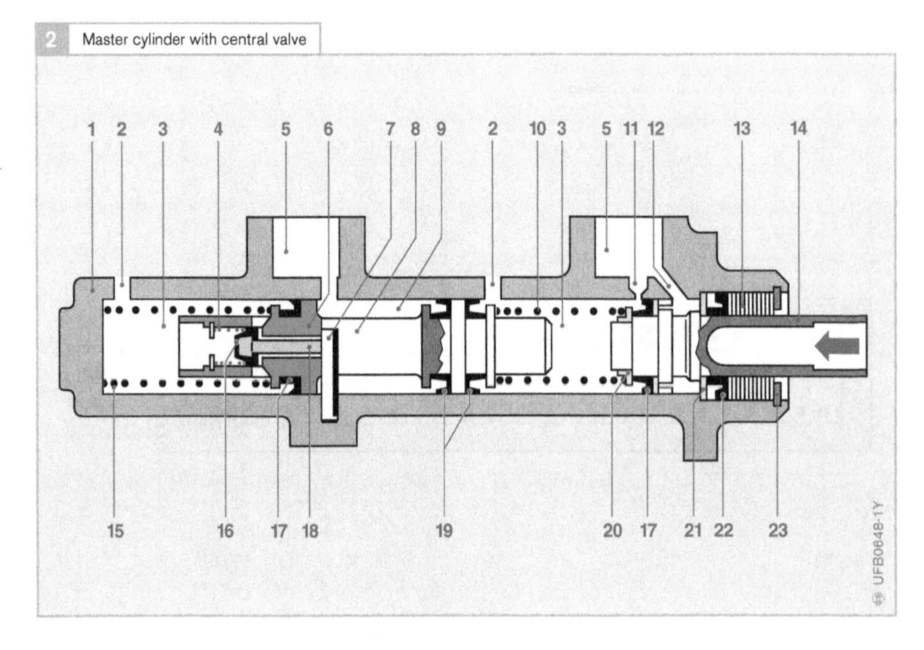

Brake-fluid reservoir

The brake-fluid reservoir is generally attached directly to the master cylinder. It acts not only as a reservoir for the brake fluid, but also as an expansion vessel which accommodates volume fluctuations in the brake circuits. These occur when the brakes are released, or as a result of brake-lining wear, temperature differences within the braking system or when the ABS or ESP systems are active. The brake-fluid reservoir is connected to the master cylinder by two outlets (Figure 3, Item 8).

The warning device for indicating when the fluid level is low operates according to the float principle. When the fluid level drops below the minimum level, the float (5) completes the electrical circuit of the warning device (1) by means of the float switch (2) so that the warning lamp (4) lights up.

Pilot-pressure valve

Pilot-pressure valves maintain a pressure of 0.4 to 1.7 bar in hydraulic brake circuits in order to ensure that the cup seals in the wheel cylinders function properly. They are used instead of bottom valves and, if there is insufficient space in the master cylinder or if the vehicle has both drum and disk brakes, are fitted external to the master cylinder.

The ball valve shuts off the connection between the master cylinder and the wheel-brake cylinders as soon as the pressure in the master cylinder drops below the set level (Figure 4).

Pilot-pressure valves are superfluous on braking systems that incorporate the latest technological advances. That is because the seals in the wheel cylinders function without having to be under pressure.

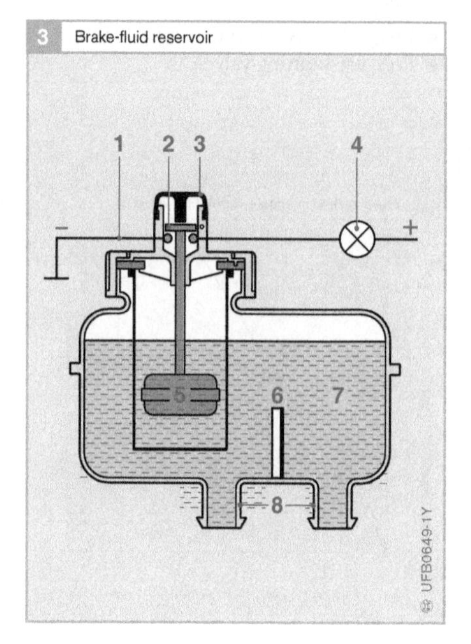

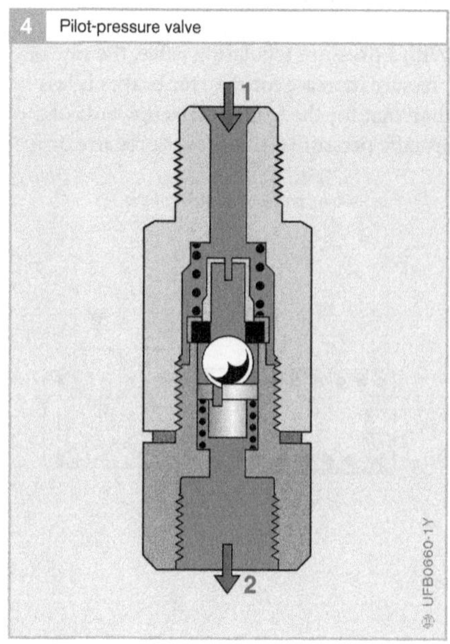

Fig. 3
1 Electrical circuit for level warning lamp
2 Float switch
3 Reservoir cap
4 Warning lamp
5 Float
6 Fluid level indicator
7 Brake fluid
8 Connection to master cylinder

Fig. 4
1 From master cylinder
2 To wheel-brake cylinders

Note on fitting:
The arrow on the hexagonal valve body must point away from the master cylinder.

Components for braking-force distribution

As a result of dynamic axle-load shift under braking, more braking force can be applied to the front wheels of a vehicle than to the rear wheels. Consequently, the front brakes are more generously dimensioned than the rear brakes. The reduction of the load on the rear axle is not a linear progression, however, but advances at a faster rate as deceleration increases. On vehicles with "invariable braking-force distribution", therefore, overbraking of either the front wheels of the rear wheels will occur at some point depending on the force-distribution setting.

Overbraking of the rear wheels has a negative effect on vehicle handling and can cause skidding. By adopting appropriate measures (fitting a rear-wheel pressure regulating valve), however, the handling characteristics of the vehicle can be positively influenced and the actual braking force made to approximate more closely to the ideal braking force (no wheel lock-up).
A distinction is made between

- static or dynamic pressure regulating valves, and
- pressure limiting valves.

With a pressure regulating valve, the rate of pressure increase for the rear brakes is less than that for the front brakes upwards of a specific pressure (changeover pressure or changeover point). Upwards of the changeover point, static pressure regulating valves regulate the brake pressure according to a fixed characteristic, while dynamic pressure regulating valves do so on the basis of a regulating ratio that depends on the vehicle load or the rate of deceleration.

Pressure regulating valves must be designed in such a way that under practical conditions, the braking force is distributed at a level well below the ideal level. The effect of variations in the frictional coefficient of the road surface, the engine braking torque and the tolerance limits of the pressure regulating valve must also be taken into account in order to prevent rear-wheel lock-up.

The pressure limiting valve (described at the end of this chapter) prevents the brake pressure to the rear wheels rising any further once a specific level (shut-off pressure) has been reached.

Depending on the type of vehicle and the braking system used by the manufacturer, there are essentially five versions employed:

- Fixed-setting pressure regulating valve,
- Load-dependent pressure regulating valve,
- Deceleration-dependent pressure regulating valve,
- Integral pressure regulating valve, and
- Pressure limiting valve

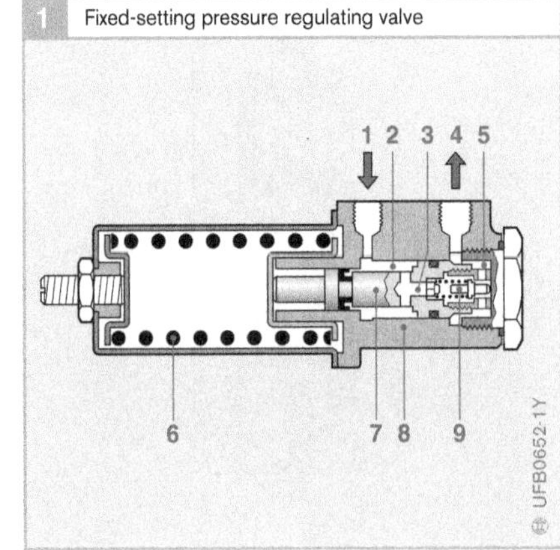

Fig. 1
1 Inlet port (from master cylinder)
2, 5 Annular chambers
3 Channel
4 Outlet port (to brakes)
6 Compression spring
7 Graduated piston
8 Valve body
9 Valve

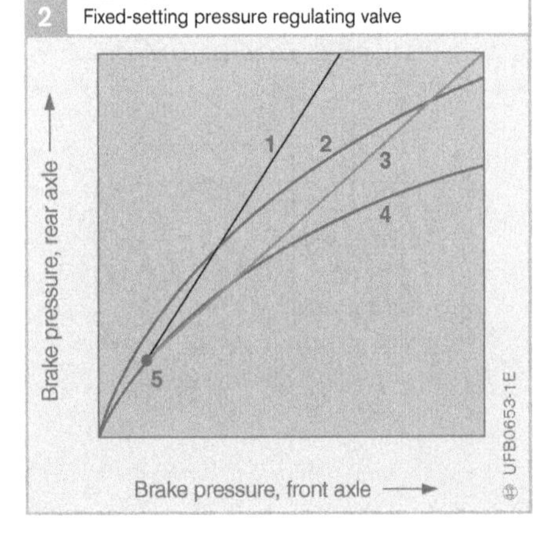

Fig. 2
1 Unregulated pressure
2 Ideal pressure curve (laden vehicle)
3 Regulated pressure
4 Ideal pressure curve (unladen vehicle)
5 Changeover point

Fixed-setting pressure regulating valve

The pressure regulating valve (Figure 1) is fitted in the rear-axle brake circuit. The valve body (8) encloses a graduated piston (7) with an integral valve (9). The output pressure is reduced relative to the input pressure in proportion to the ratio of the effective areas of the annular chambers (2, 5).

When the brakes are applied, hydraulic pressure from the master cylinder passes via the inlet port (1), the annular chamber (2), the channel (3) in the graduated piston (7) and second annular chamber (5) to the outlet port (4). Shortly before the changeover pressure is reached, the pressure acting on the annular-chamber effective area (2) pushes the graduated piston to the right as far as the stop so that the valve (9) closes off the channel to the outlet port (4). As the pressure continues to increase, the graduated piston moves rapidly back and forth, opening and closing the valve (9) accordingly, thereby regulating the output pressure in proportion to the ratio of the effective areas (2, 5). Once the braking sequence ends, the pressure at the outlet port (4) pushes the graduated piston (7) against the compression spring (6) until the excess pressure in the annular chambers (2, 5) has reduced. Figure 2 shows the pressure curves.

Load-dependent pressure regulating valve

Vehicles whose payload can alter significantly from one journey to the next require so-called load-dependent pressure regulating valves (Figure 3) so that the braking forces can be adjusted according to the weight being carried. This type of pressure regulating valve is attached to the bodywork and connected to the vehicle's rear axle (6) by means of a rod linkage (5). The relative movement between suspension and body as the springs are compressed is transmitted to the graduated piston (1). The piston then compresses the control springs (2) according to the amount of suspension travel, thereby altering the changeover point. This achieves an adaptive response of the rear-axle brake pressure relative to the weight of the vehicle payload (Figure 4).

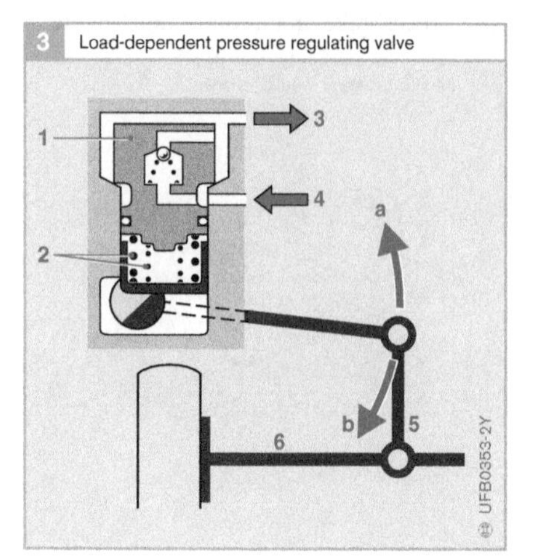

Fig. 3
a Laden vehicle
b Unladen vehicle

1 Graduated piston
2 Control springs
3 Outlet port to brakes
4 Inlet port from master cylinder
5 Linkage
6 Rear axle

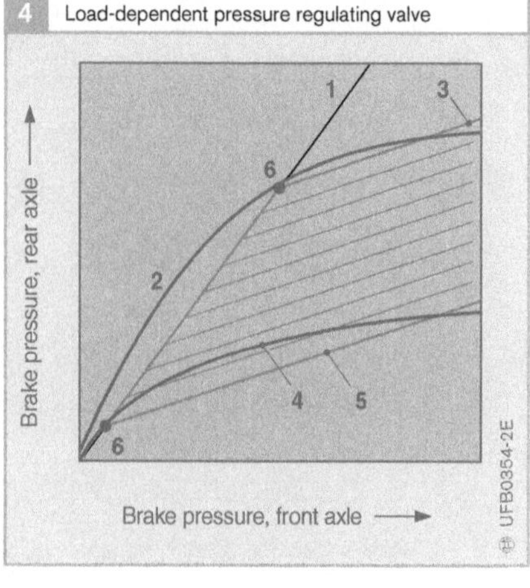

Fig. 4
1 Non-reduced pressure
2 Ideal pressure curve (laden vehicle)
3 Reduced pressure (laden vehicle)
4 Ideal pressure curve (unladen vehicle)
5 Reduced pressure (unladen vehicle)
6 Changeover points

Deceleration-dependent pressure regulating valve

Installation

This type of pressure regulating valve must be fitted in the rear-axle brake circuit and set at an angle, α, to the vehicle's horizontal axis in such a way that when the vehicle is stationary, the ball (4) rests at the back of the valve away from the stepped piston (2, Figure 5).

Design and method of operation

The main components of the valve are a stepped piston (2) and a ball (4). As the brake pressure required to obtain a given rate of deceleration depends on the weight being carried by the vehicle, this type of valve is load-dependent as well as deceleration-dependent.

5 Deceleration-dependent pressure regulating valve

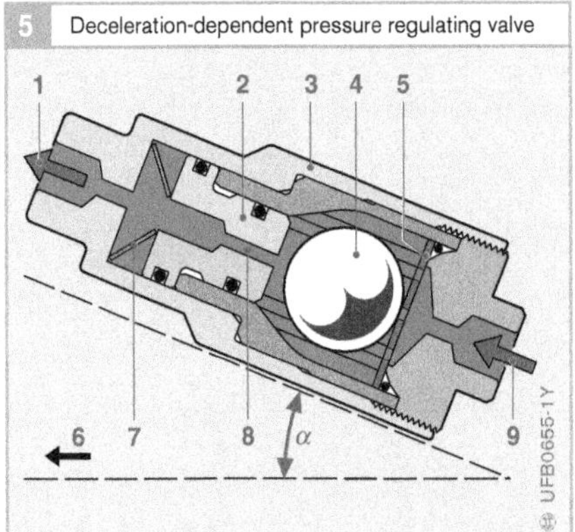
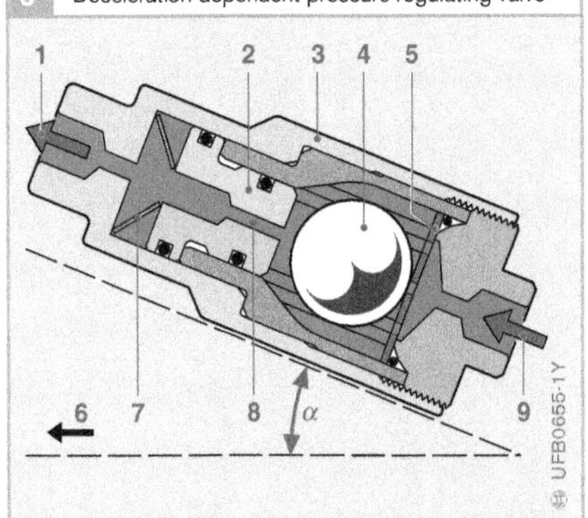

Fig. 5
1 Outlet port (to brakes)
2 Stepped piston
3 Valve body
4 Ball
5 Perforated disk
6 Front of vehicle
7 Leaf spring
8 Channel
9 Inlet port (from master cylinder)
α Angle to horizontal axis

As soon as the rate of deceleration under braking reaches a certain level, the inertia of the ball causes it to roll up the inclined plane – assisted by the pressure from the inlet port (9) – and close off the channel (8) through the stepped piston (2). This point represents the first changeover point (1 or 3, Figure 6) because further increase of pressure at the inlet port (9) cannot initially be passed through to the rear-axle brake circuit (pressure limiter function). As the pressure at the inlet port (9) continues to increase, however, the stepped piston (2) is pushed forwards against the action of the leaf spring (7) towards the outlet port (1) and away from the ball. At that stage, the second changeover point (2 or 4) has been reached. The piston channel (8) has been opened again and the passage of fluid between the two ports (9 and 1) is possible again. The pressure in the rear-axle brake circuit can now rise again at a reduced rate (pressure regulating function).

Figure 6 shows the changeover points of a deceleration-dependent pressure regulating valve for an unladen and a fully laden vehicle.

Integral pressure regulating valve

This device operates in the same way as the fixed-setting pressure regulating valve described previously. Because of its light weight and small dimensions (2), it can be integrated in the master cylinder (1) by being screwed into the rear-wheel brake circuit connection (Figure 7).

6 Deceleration-dependent pressure regulating valve

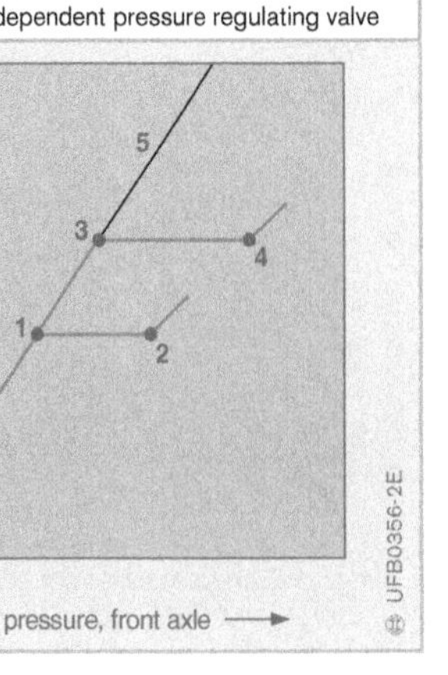

Fig. 6
1, 2 Changeover points (unladen vehicle)
3, 4 Changeover points (laden vehicle)
5 Unregulated pressure

7 Integral pressure regulating valve

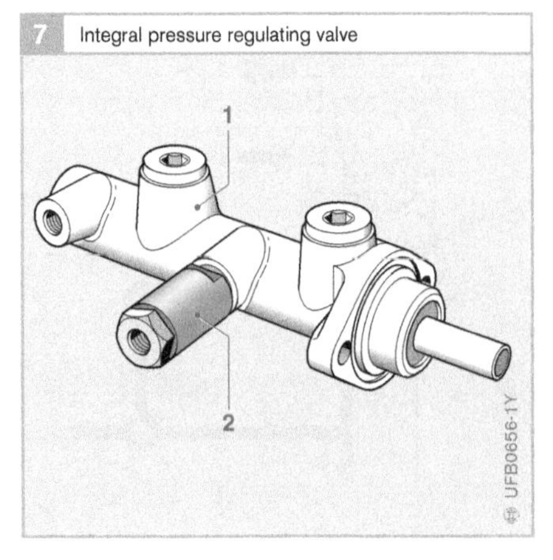

Fig. 7
1 Master cylinder
2 Integral pressure-regulating valve

Braking-force limiter

The braking-force limiter (Figure 8) is fitted in the rear-axle brake circuit and prevents the rear-axle brake pressure rising beyond its shut-off pressure. At that point, the valve piston (6) compresses the compression spring (5) and brings the valve cone (4) into contact with the valve seat (8) so that no further pressure increase at the outlet port (9) is possible. When the braking sequence comes to an end, the valve opens and releases the pressure.

8 Braking-force limiter

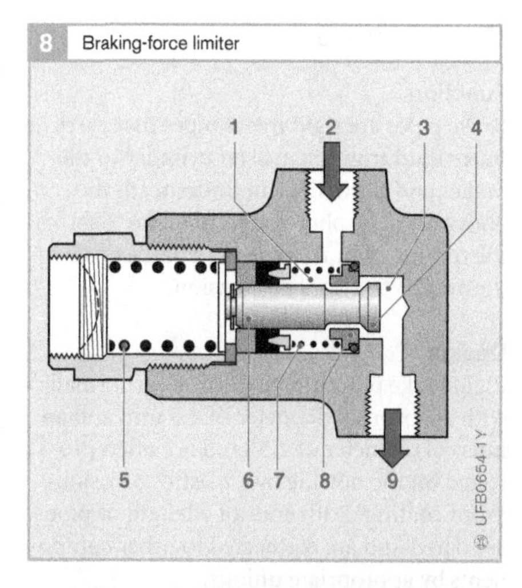

Fig. 8
1 Inlet chamber
2 Inlet port (from master cylinder)
3 Outlet chamber
4 Valve cone
5 Compression spring
6 Valve piston
7 Compression spring
8 Valve seat
9 Outlet port (to brakes)

Reaction distance and total braking distance

According to ISO 611, the total braking distance is the distance travelled during the total braking time (see chapter "Basic principles of vehicle dynamics", "Definitions" section). Thus, the point at which the driver first applies force to the actuation device is a decisive factor in determining the total braking distance. However, as far as the overall braking sequence is concerned, the distance travelled from when the driver first identifies a hazard to when the brakes are first applied is also of significance. This is the driver's reaction time and is different for every driver.

The total distance travelled from identification of a hazard to the point at which the vehicle comes to a halt is thus made up of a number of components consisting of

- the distance travelled during the reaction time and the brake-system response time at a constant velocity, v,
- the distance travelled during the brake-pressure build-up time at an increasing rate of deceleration,
- the distance travelled under fully developed deceleration at a constant rate of deceleration.

Alternatively, half the period of increasing deceleration can be taken to be under fully developed deceleration at the rate a, and the remaining period taken to be under zero deceleration. This time period is added to the other periods of zero deceleration (reaction time and brake-system response time) to form the dead time, t_{vz}. Thus the distance required for braking is defined by the formula

$$s = v \cdot t_{vz} + \frac{v^2}{2a}$$

The maximum rate of deceleration is limited by the friction between the tires and the road. Minimum rates of deceleration are defined by law.

Assuming a dead time of 1 s, the table below shows the combined reaction and total braking distance at various speeds.

	Vehicle speed in km/h prior to braking												
	10	30	50	60	70	80	90	100	120	140	160	180	200
	Distance travelled during dead time of 1 s (unbraked) in m												
	2.8	8.3	14	17	19	22	25	28	33	39	44	50	56
Deceleration a in m/s²	**Reaction and total braking distance in m**												
4.4	3.7	16	36	48	62	78	96	115	160	210	270	335	405
5.0	3.5	15	33	44	57	71	87	105	145	190	240	300	365
5.8	3.4	14	30	40	52	65	79	94	130	170	215	265	320
7.0	3.3	13	28	36	46	57	70	83	110	145	185	230	275
8.0	3.3	13	26	34	43	53	64	76	105	135	170	205	250
9.0	3.2	12	25	32	40	50	60	71	95	125	155	190	225

Brake pipes

Function

Brake pipes are rigid metal pipes that carry brake fluid from the master cylinder to the brakes and normally run underneath the bodywork. The photograph in Figure 1 shows the routing of brake pipes and brake hoses in the area of the front suspension.

Design

Rigid brake pipes are made of steel (normally with an external diameter of 4.5 mm and an internal diameter of 2.5) and are often protected on the outside by a plastic corrosion-proof coating. Both ends of a length of pipe are flared and are connected to other components by appropriate unions.

Usage

When routing brake pipes, care must be taken to ensure that, apart from at the intended fixing points, the pipes are not in contact with the body or any other components.

Brake hoses

Function

Brake hoses form a flexible link between the brake pipes that are rigidly attached to the bodywork and the brakes, which are attached to the suspension/wheel hubs and therefore subject to movement.

Design

Flexible brake hoses consist of an inner layer of rubber, two rayon reinforcing layers (for withstanding the fluid pressure), an outer rubber coating and the fittings (unions).

Usage

The lengths and usages of flexible brake hoses are set down in specifications that in some cases are specific to particular vehicles. In general, the regulations require that the brake hoses are not permitted to come into contact with suspension or body components and that the specified temperature and pressure ranges are not exceeded.

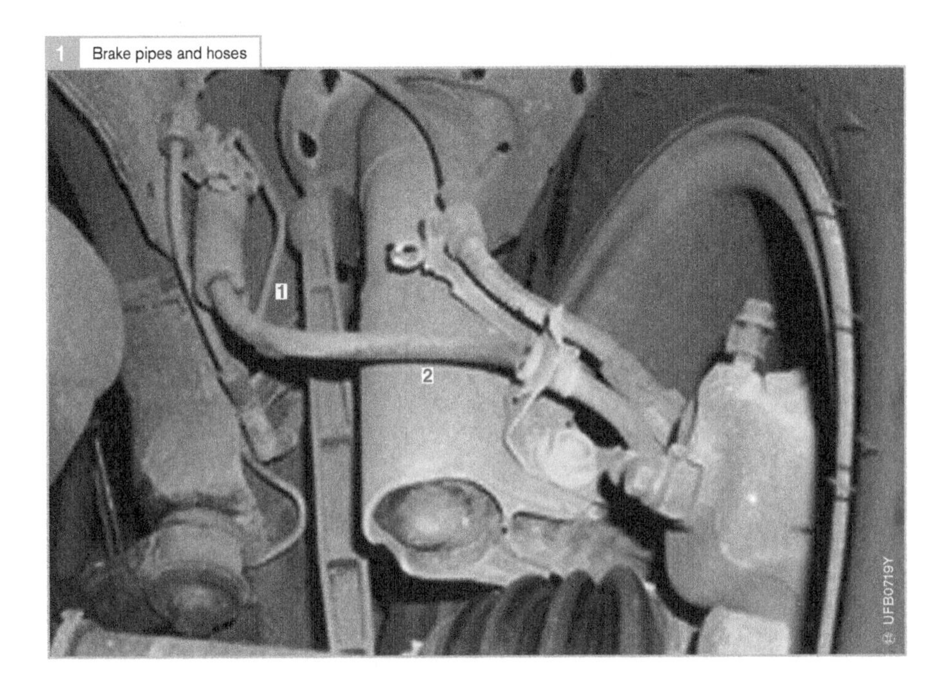

Fig. 1
1 Brake pipe
2 Brake hose

Brake fluid

Brake fluid is the hydraulic medium employed to transmit actuation forces within the braking system. The applicable quality requirements are set down in the standards SAE J 1703, FMVSS 116, ISO 4925, and Table 1 below.

Requirements

Equilibrium boiling point

The equilibrium boiling point provides an index of the brake fluid's resistance to thermal stress. Thermal stress can be particularly high in the wheel cylinders (this is where the highest temperatures in the braking system occur). At temperatures above the brake fluid's momentary boiling point, vapour bubbles are produced. If that occurs, operation of the brakes is no longer possible.

Wet boiling point

The wet boiling point is the brake fluid's equilibrium boiling point after it has absorbed water under defined conditions. On hygroscopic (glycol-based) fluids in particular, the effect is a substantial lowering of the boiling point.

The purpose of testing the wet boiling point is to determine the properties of used brake fluid, which can absorb water primarily by diffusion through the brake hoses. This is the main reason why it is necessary to replace the brake fluid in a vehicle every 1 ... 2 years. This is absolutely essential for the safety of the braking system, which must always be bled afterwards.

Viscosity

In order to ensure efficient and reliable operation of the brakes over a wide range of temperatures (-40 °C...+100 °C), the brake fluid's viscosity should have the minimum-possible degree of dependence on temperature. This is particularly important on ABS systems.

Warning:
All brake fluids apart from DOT5 (silicone-based) are harmful to the skin and corrosive to paint.

Compressibility

The compressibility of the fluid must be small and its temperature-dependence as low as possible.

Non-corrosiveness

According to FMVSS 116, brake fluid must not be corrosive to the metals normally used in braking systems. This can only be achieved by the use additives.

Elastomer swelling

The elastomers used in a braking system must be matched to the particular type of brake fluid employed. A small amount of elastomer swelling is desirable. However, it must not under any circumstances exceed 16 %, as otherwise the strength of the components will be impaired. Even small amounts of contamination of a glycol-based brake fluid with mineral oils (mineral-oil-based brake fluid, solvent) can result in damage to rubber components (such as seals) and consequently to brake failure.

Chemical composition

Although alteration of the chemical composition may improve one of the properties referred to above, it normally brings about changes to others at the same time.

[1]) FMVSS: Federal Motor Vehicle Safety Standard (USA), DOT: Department of Transportation (USA).

1 Brake fluids

Testing standard		FMVSS 116[1])			SAE J1703
Rating/version		DOT3	DOT4	DOT5, DOT5.1	11.83
Dry boiling point	min. °C	205	230	260	205
Wet boiling point	min. °C	140	155	180	140
Cold viscosity at −40 °C	mm²/s	1500	1800	900	1800

Table 1

Wheel brakes

There are two types of brake used on cars – disk brakes and drum brakes. New cars are now fitted exclusively with disk brakes at the front, and there is an increasing trend towards disk brakes for the rear wheels as well. Both types are friction brakes in which the braking energy transmitted by the braking system acts by pressing the brake pads or shoes against the brake disks/drums.

Overview

Requirements

The demands placed on the brakes are extremely exacting and include:

- short braking distance,
- fast response time,
- short pressure build-up time
- even braking effect,
- precise control,
- Insusceptibility to dirt and corrosion,
- high reliability,
- durability,
- resistance to wear,
- ease of maintenance.

In order to be able to meet those requirements while remaining acceptably economical, in Europe small cars and some medium-sized cars are fitted with disk brakes on the front wheels and drum brakes on the rear (drum brakes represent a cost saving). The more expensive mid-range cars, as well as executive/luxury and sports cars have disk brakes all round.

This is because of the fact that the heavier weights and higher speeds of those vehicles are such that only disk brakes are capable of coping with the levels of heat generated. Consequently, particular attention has to be paid to

- heat conduction,
- brake ventilation and
- stable frictional properties of the brake pads.

Assessment of brakes

The brake coefficient C* is used as an assessment criterion for brake performance, and indicates the ratio of braking force to application force. It takes account of the effect of internal force transmission (ratio of input to output force) within the brake and of the frictional coefficient, which is chiefly dependent on the parameters speed, braking pressure and temperature. Figure 1 shows the brake coefficient for various types of brake.

1 Brake coefficients C* as a function of coefficient of friction and road speed at start of braking

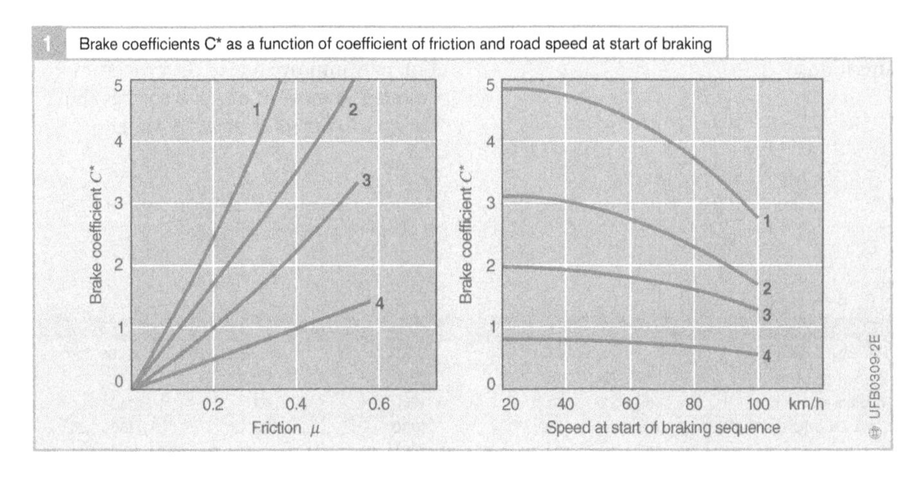

Fig. 1
1 Double-servo drum brake
2 Double-duplex drum brake
3 Simplex drum brake
4 Disk brake

Bosch test center at Boxberg

An important component of the development process for vehicle systems are the practical trials performed by the system supplier. Not all tests can be carried out on public roads. Since 1998 Bosch has performed this part of the development process at its test center near Boxberg between Heilbronn and Würzburg (south Germany). The 92-hectare site provides facilities for testing all conceivable handling, safety and convenience systems and components to the limit. Seven different sections of test track allow systems to be tested to their physical limits under all types of road conditions and driving situations – and under the safest possible conditions for the test drivers and vehicles.

The **rough-surface sections** (1) are designed for speeds of up to 50 km/h and 100 km/h respectively. The following types of surface are provided:

- pot holes
- undulations
- high-vibration surface
- cobblestones, and
- variable-surface sections.

The asphalted **gradient sections** (2) for hill-start and uphill acceleration testing with gradients of 5%, 10%, 15% and 20% include sprayable, paved sections of various widths.

There are two **ford sections** (3) with lengths of 100 meters and 30 meters respectively and depths of 0.3 and 1 meter.
There are **special sprayed sections** (4) with the following surfaces:

- "chessboard" (asphalt and paving slabs)
- asphalt
- paving slabs
- blue basalt
- concrete
- an aquaplaning section and
- a trapezium-shaped blue-basalt section

The **skid pan** (5) for testing cornering characteristics has an asphalt surface 300 meters in diameter. Parts of it can be watered to simulate ice and wet roads. It is surrounded by a safety barrier made of tires in order to protect drivers and vehicles.
The **high-speed circuit** (6) has three tracks and can be used by commercial vehicles as well as cars. This section is designed to allow speeds of up to 200 km/h.
The **handling track** (9) incorporates two sections – one for speeds of up to 50 km/h, and one for speeds up to 80 km/h. Both sections have corners of varying severity and degrees of camber. The handling track is mainly used for testing handling-stability control systems.

View of the test-track section modules

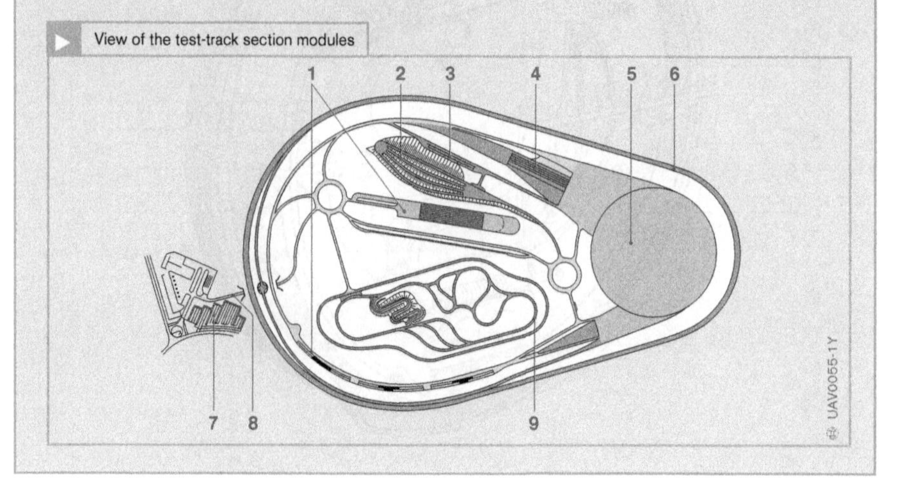

Fig. 1
1 Rough-surface sections
2 Gradient sections
3 Ford sections
4 Special watered sections
5 Skid pan
6 High-speed circuit
7 Building
- Workshops
- Offices
- Test benches
- Laboratories
- Filling station and
- Staff common rooms
8 Access road
9 Handling track

Drum brakes

Principle of operation

Drum brakes for cars generate their braking force on the inner surface of a brake drum. This principle is explained below based on the example of a simplex drum brake with integral parking-brake mechanism (Figure 1).

A double-acting wheel cylinder (1) operates the brake shoes of the drum brake. This forces the friction lining (2) of the leading brake shoe (12) and the trailing brake shoe (5) against the inside of the brake drum (6). The other ends of the brake shoes on the opposite side to the wheel cylinder are braced by a support bearing (15) that is attached to the brake anchor plate (13)

The drum brake can also be operated as a parking brake by means of the hand-brake lever (7) and hand-brake cable (8).

Automatic self-adjusting mechanism

The automatic self-adjusting mechanism (Figures 1 and 2) maintains a constant clearance between the brake shoes and the brake drum (gap between the brake shoe and drum when the brakes are not applied).

The adjusting mechanism consists of

- the pressure sleeve (18),
- the adjusting screw (16) and adjuster wheel (11),
- the return springs (4),
- the bimetal strip (10),
- the elbow lever (17) and
- the adjusting lever (20).

The adjusting lever is attached to the pressure sleeve in such a way that it is able to flex and its adjusting dog (19) engages with the adjuster wheel. This Bosch/Bendix patented automatic adjusting mechanism achieves optimum adjustment increments of approximately 0.02 mm per adjustment cycle.

Note:
Identical components in Figures 1 and 2 have the same index numbers.

1 Simplex drum brake with integral parking-brake mechanism (on right rear wheel)

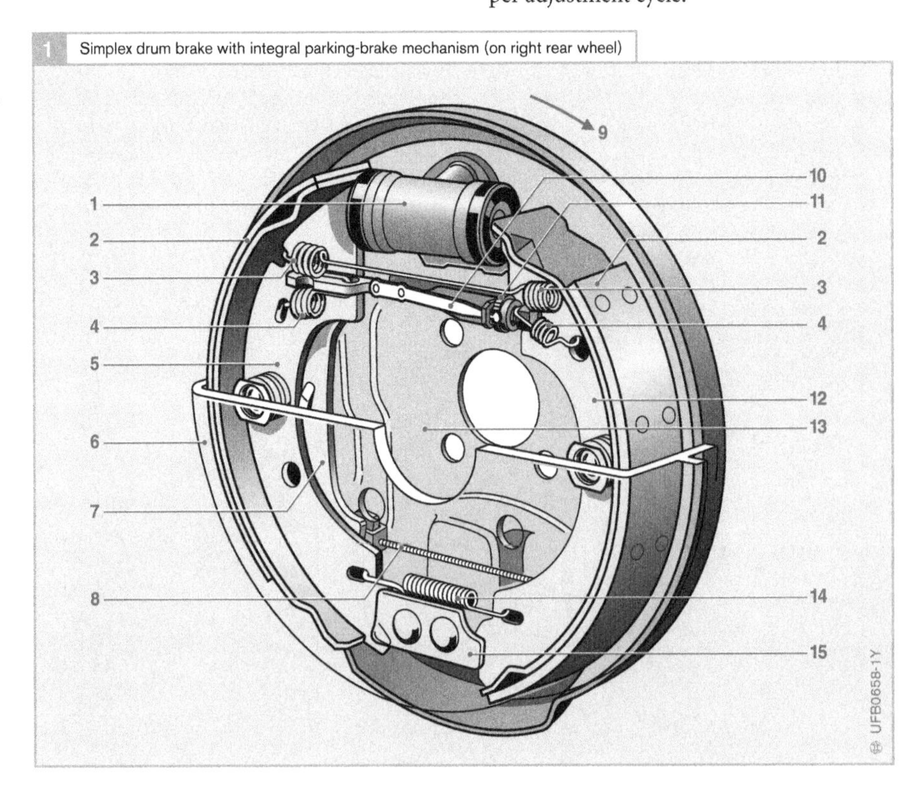

Fig. 1
1 Wheel cylinder
2 Friction lining
3 Return spring (brake shoe)
4 Return spring (self-adjusting mechanism)
5 Trailing shoe
6 Brake drum
7 Hand-brake lever
8 Hand-brake cable
9 Direction of drum rotation
10 Bimetal strip (self-adjusting mechanism)
11 Adjuster wheel (with elbow lever)
12 Leading shoe
13 Anchor plate
14 Return spring (brake shoe)
15 Support bearing

Non-braking mode

The return springs (3, 14) hold the two brake shoes (5, 12) and their attached friction linings (2) away from the brake drum (6). This presses the brake shoes against the adjusting screw (16) with its adjuster wheel (11) so that they are held against the pressure sleeve (18) with the result that the elbow lever (17) in between is prevented from moving.

Braking mode ($t < 80\,°C$)

When the brakes are applied, the pistons in the wheel cylinder (1) force the brake shoes (5, 12) and their attached friction linings (2) against the inside of the brake drum (6).
At the same time, the return springs (4) pull the adjusting screw (16) and adjuster wheel (11) away from the pressure sleeve (18). This leaves a gap in which the elbow lever (17) can move.

2 Automatic self-adjusting mechanism

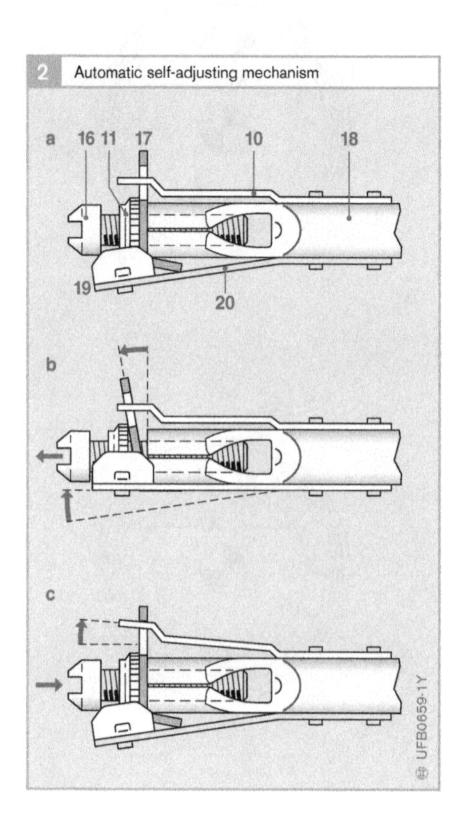

At temperatures inside the brake drum of $< 80\,°C$, the spring-action adjusting lever (20) pushes the lower arm of the elbow lever (17) upwards. This allows the angled adjusting dog (19) to engage in the teeth of the adjuster wheel (11). If, as a result of brake wear, the clearance between the brake drum and brake linings is larger than the design requirement, the adjusting lever turns the adjuster wheel by the width of one tooth, thereby unscrewing the adjusting screw (16) a small amount and increasing the overall length of the adjusting mechanism. This restores the clearance to the correct amount.

Braking mode ($t > 80\,°C$)

Temperatures of $> 80\,°C$ generated by extended or frequent braking (for example on long descents) cause the brake drum to expand. The clearance between the brake shoes and the drum then becomes larger (expansion clearance) than the design requirement. In such cases, the bimetal strip (10) prevents automatic adjustment. The bimetal strip bends upwards and holds the elbow lever (17) in position. Consequently, the adjusting lever (20) cannot move and no adjustment can take place.

Hand-brake lever

The hand-brake mechanism operates the drum brake by means of a cable (8) that is attached to the lower end of the hand-brake lever (7). The hand-brake lever is pivoted at the top in the trailing brake shoe (5) and engages in the adjusting screw (16) of the self-adjusting mechanism. When the hand brake is operated, the cable pulls the hand-brake lever to the right at the bottom. Its rounded upper right edge levers against the self-adjusting mechanism and first of all pushes the trailing shoe (5) outwards until it contacts the drum (6). It then pushes the leading shoe (12) against the brake drum by levering through the adjusting screw of the self-adjusting mechanism.

Fig. 2
10 Bimetal strip
11 Adjuster wheel
16 Adjusting screw
17 Elbow lever
18 Pressure sleeve
19 Adjusting dog
20 Adjusting lever

a Non-braking mode
b Braking mode ($t < 80\,°C$)
c Braking mode ($t > 80\,°C$)

Types of drum brake

Two different types of drum brake are distinguished according to the way in which the brake shoes are mounted and pivot:

- Drum brakes with fixed-pivot brake shoes (Figures 3a and 3b)
- Drum brakes with sliding shoes with parallel or sloping anchors (Figures 3c and 3d)

Fixed-pivot brake shoes can wear unevenly as they are not self-centring like sliding shoes.

In addition they also have a tendency for the trailing shoe to be self-inhibiting (i.e. it diminishes the braking force applied, the opposite of self-augmenting). Sliding-shoe guides are used on simplex, duplex, double-duplex, servo and double-servo brakes. Most drum brakes fitted to modern cars are of the sliding-shoe type which do not have self-inhibiting properties.

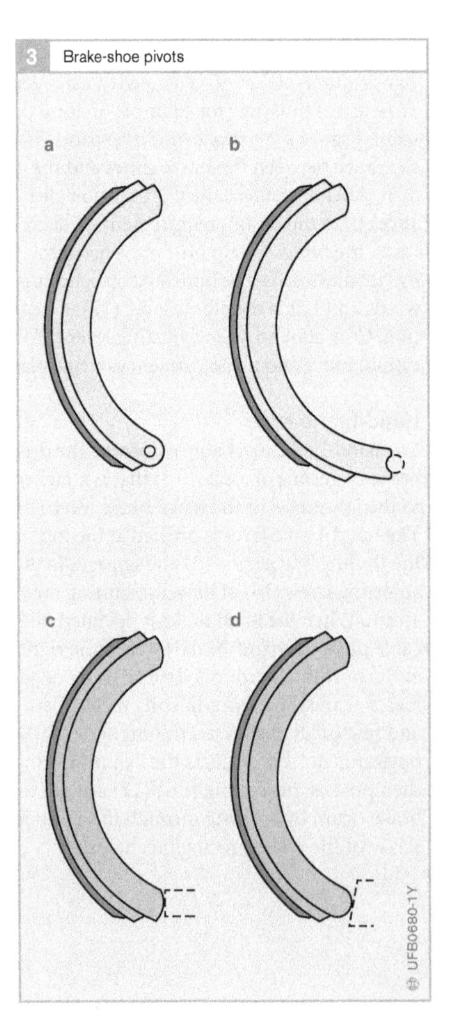

Fig. 3
a Brake shoe with fixed pivot point (single pivot)
b Brake shoe with fixed pivot point (double pivot)
c Parallel-anchor brake shoe
d Sloping-anchor brake shoe

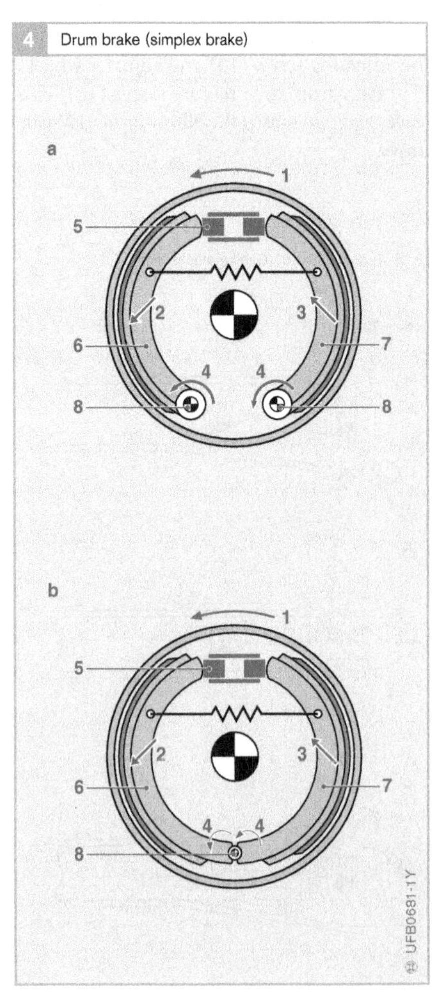

Fig. 4
a Brake shoes with 2 single pivots
a Brake shoes with 1 double pivot
One leading shoe, small degree of self-augmentation
1 Direction of drum rotation (when travelling forwards)
2 Self-augmenting effect
3 Self-inhibiting effect
4 Torque
5 Double-acting wheel cylinder
6 Leading shoe
7 Trailing shoe
8 Anchor point (pivot point)

Simplex brake
A double-acting wheel cylinder (Figure 4, Item 5) operates the brake shoes (6, 7). The anchor points (8) of the brake shoes are also pivot points (2 single pivots or 1 double pivot). When the vehicle is travelling forwards, the leading shoe (6) has a self-augmenting effect (2) while the trailing shoe (7) is self-inhibiting (3); when the vehicle is moving backwards, those effects are reversed.
Self-augmenting factor: approx. 2 to 4.

Duplex brake
Each brake shoe is operated by a single-acting wheel cylinder (Figure 5, Item 4). The sliding-type brake shoes (6) are anchored against the back of the opposing wheel cylinder in each case. The duplex brake is a single-acting brake, i.e. when the vehicle is moving forwards, it has two leading, self-augmenting (2) brake shoes. When reversing there is no self-augmenting effect. *Self-augmenting factor:* up to approx. 6.

Duo-duplex brake
Two double-acting wheel cylinders (Figure 6, Item 4) operate the sliding-type brake shoes (6) that are anchored against the opposing wheel cylinder in each case. The duo-duplex brake is a double-acting brake, i.e. it has two leading, self-augmenting (2) brake shoes when travelling forwards or in reverse. *Self-augmenting factor:* up to approx. 6.

Servo brake
A double-acting wheel cylinder (Figure 7, Item 4) operates the two sliding-type brake shoes (5, 6). Unlike simplex and duplex brakes, the brake shoes are not anchored against a fixed point but have floating ends that bear against a pressure pin (7) that can move in one direction only. When the vehicle is travelling forwards, this transmits the bracing force of the primary shoe (5) to the secondary shoe (6), imparting to it an even greater self-augmenting effect than achieved by the primary shoe. When the vehicle is reversing, the servo brake acts in the same way as a simplex brake. *Self-augmenting factor:* up to approx. 6.

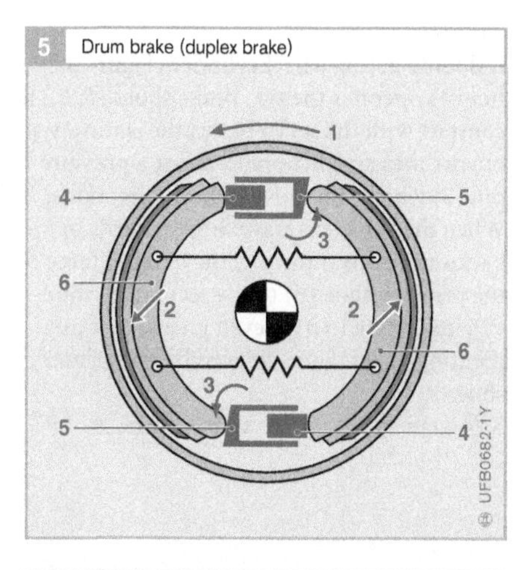

Fig. 5
Two leading shoes, large degree of self-augmentation
1 Direction of drum rotation (when travelling forwards)
2 Self-augmenting effect
3 Torques
4 Wheel cylinder
5 Anchor points
6 Brake shoes

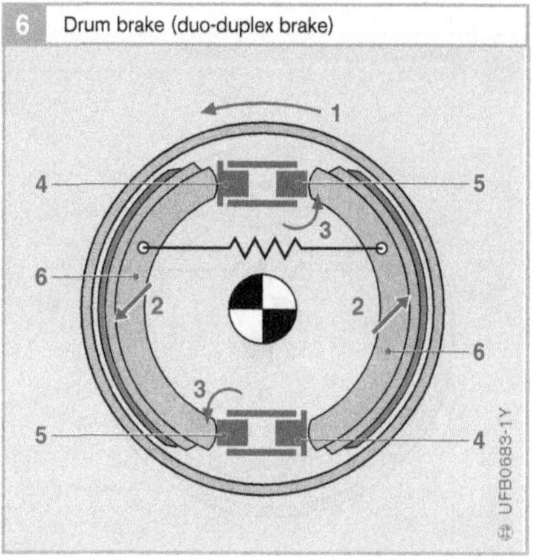

Fig. 6
Two leading shoes actuated by floating wheel cylinders, large degree of self-augmentation
1 Direction of drum rotation (when travelling forwards)
2 Self-augmenting effect
3 Torques
4 Wheel cylinder
5 Anchor points
6 Brake shoes

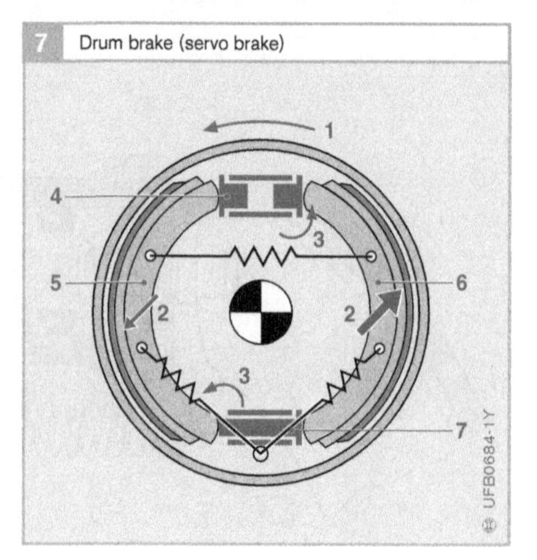

Fig. 7
Two leading shoes with floating anchor, unidirectional pressure pin, large degree of self-augmentation
1 Direction of drum rotation (when travelling forwards)
2 Self-augmenting effect
3 Torques
4 Wheel cylinder
5 Primary shoe
6 Secondary shoe
7 Pressure pin

Duo-servo brake

A double-acting wheel cylinder (Figure 8, Item 4) operates the two brake shoes (5, 6). In contrast with the servo brake, the sliding-type brake shoes are anchored against a pressure pin (7) which can move in both directions. When the vehicle is travelling forwards or backwards, this transmits the bracing force of the primary shoe (6) to the secondary shoe (5), imparting to it an even greater self-augmenting effect than achieved by the primary shoe.

Self-augmenting factor: up to approx. 6.

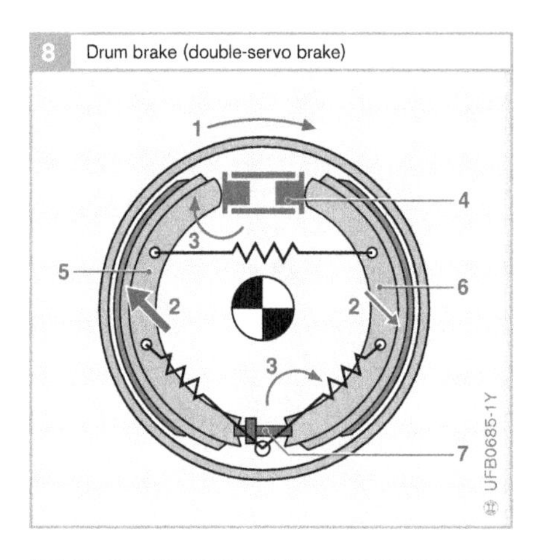

Fig. 8
As servo brake except that pressure pin moves in both directions; self-augmenting when travelling forwards or in reverse
1 Direction of rotation (when reversing)
2 Self-augmenting effect
3 Torques
4 Wheel cylinder
5 Secondary shoe
6 Primary shoe
7 Pressure pin (anchor bearing)

Self-augmenting effect

The extent of the self-augmenting effect is an important property of drum brakes. Self-augmentation is an effect whereby the effective braking force is greater than the force that would result directly from the application force generated by the master cylinder. It is brought about by the fact that the friction between the leading brake shoe and the drum creates a turning force around the brake-shoe pivot which forces the shoe against the drum, thereby supplementing the application force of the braking system. Only on the simplex brake does the trailing brake shoe also create a turning force around the shoe pivot that diminishes the force applied by the hydraulic system, in other words a self-inhibiting or self-diminishing effect.

Adjusting mechanisms

As brake friction linings are subject to wear, and wear enlarges the clearance between the brake linings and the drum, drum brakes are fitted with adjusting mechanisms for adjusting the shoes. There are several types of adjusting mechanism:

- manually operated adjusting mechanisms on the wheel cylinder
- manually operated adjusting mechanisms on the anchor bearings
- automatic adjusting mechanisms (see above, Figures 1 and 2)

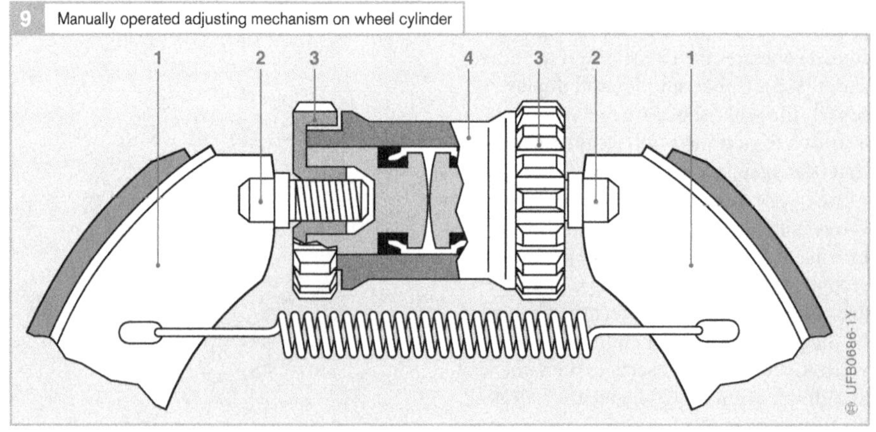

Fig. 9
1 Brake shoes
2 Adjuster pin
3 Combined cylinder cap/adjuster wheel
4 Wheel cylinder

Figure 9 shows the method of operation of the adjusting mechanism on a double-acting wheel cylinder (4) for simplex, duplex and servo brakes. The combined cylinder cap and adjuster wheel (3) is tapped in the centre for the threaded adjuster pin (2) which has a slotted head into which the brake shoe (1) locates. By passing a screwdriver through a hole in the brake backplate, the adjuster wheel (3) can be turned to adjust the brake-shoe clearance.

Figure 10 shows the method of operation of an adjusting mechanism integrated in the anchor bearing of a servo brake. In order to adjust the brake-shoe clearance, two adjusting screws (5) with slotted heads into which the brake shoes (1) locate, are screwed in or out by turning the adjuster wheels (2, 4). These are integral with the adjusting nuts (6).

The desire on the part of vehicle manufacturers for an automatic self-adjusting mechanism has produced the following designs:

- friction adjustment on the brake shoe
- adjustment incorporated in the wheel cylinder
- incremental adjustment

The incremental system is the only one which is of any significance and has been described previously in the section dealing with the basic principle of a drum brake (Figures 1 and 2).

Wheel cylinder

On drum brakes, the brake pressure generated in the master cylinder is transmitted by the wheel cylinder via its cup seal (Figure 11, Item 4), piston (5) and the pressure pins (1, 7) to the brake shoes and forces them against the brake drum. The piston spring (2) ensures that the pressure pins are always held against the brake shoe. A rubber dust cap (6) protects the cylinder against dirt and damp.

There are single-acting and double-acting wheel cylinders, as well as designs with an integral pressure regulating valve (patented by Bosch/Bendix).

11 Wheel cylinder (single-acting)

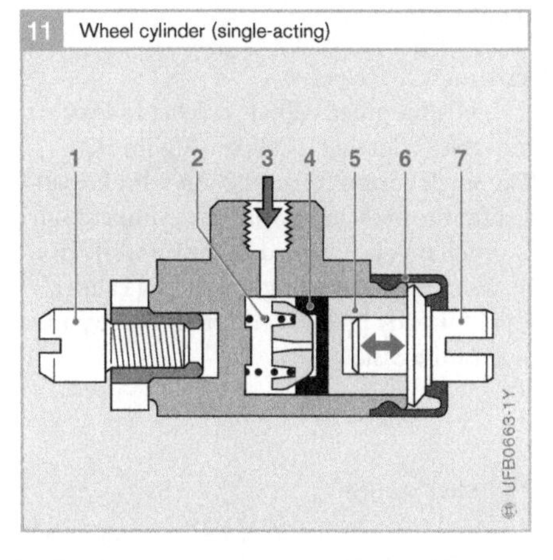

Fig. 11
1 Fixed pressure pin (threaded for adjustment)
2 Compression spring
3 Inlet port from master cylinder
4 Cup seal
5 Piston
6 Rubber dust cap
7 Movable pressure pin

10 Manually operated adjustment mechanism on support bearing

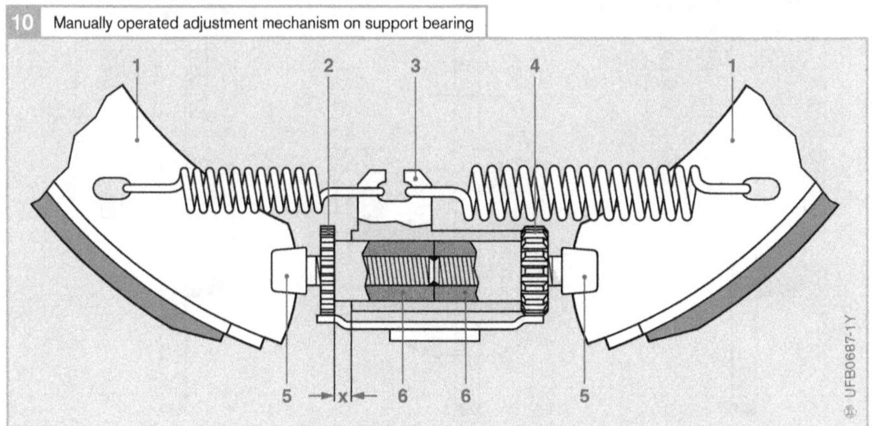

Fig. 10
1 Brake shoes
2 Adjuster wheel
3 Support bearing
4 Adjuster wheel
5 Adjusting screws
6 Adjusting nuts

x Play between adjuster wheel (2) and anchor bearing (3)

Disk brakes

Principle of operation

Disk brakes generate the braking forces on the surface of a brake disk that rotates with the wheel. The U-shaped brake caliper is attached to non-rotating suspension components.

Designs (Summary)

There are three types of disk brake that can be distinguished. Their basic principles are briefly explained below and described in more detail on the following pages.

Fixed-caliper brake

Two pistons in a rigidly mounted caliper press the brake pads against the disk from both sides (Figure 1a).

Floating-caliper brake

A rigidly mounted caliper bracket holds a movable ("floating") caliper (Figure 1b). The single piston forces the inner brake pad against the brake disk while the cylinder body is simultaneously forced in the opposite direction, thereby moving the sliding caliper, and indirectly pressing the outer brake pad against the disk.

Sliding-caliper brake

The sliding-caliper brake is a variation of the floating-caliper brake (Figure 1c). The single piston inside the sliding caliper presses the inner brake pad directly against the brake disk. The resulting reaction force simultaneously shifts the caliper body in the opposite direction, thereby indirectly pressing the outer pad against the disk.

Components of a disk brake

Piston seal

A rubber seal with a rectangular cross section sits in a groove around the inside of the cylinder and forms a seal around the piston while also automatically adjusting the clearance between brake pad and disk (Figure 2). The internal diameter of the seal is slightly smaller than the piston diameter, so that the seal is under tension and grips the piston. When the brakes are applied, the piston moves towards the brake disk and in so doing stretches the seal, which is designed so that its static friction prevents it from sliding over the piston until (as a result of pad wear) the distance travelled by the piston in order to close the gap between itself and the disk is greater than the design clearance. Because the seal is elastic, it stores energy that returns it to its original shape and position so that it pulls the piston back when the brake is

1 Disk brake types

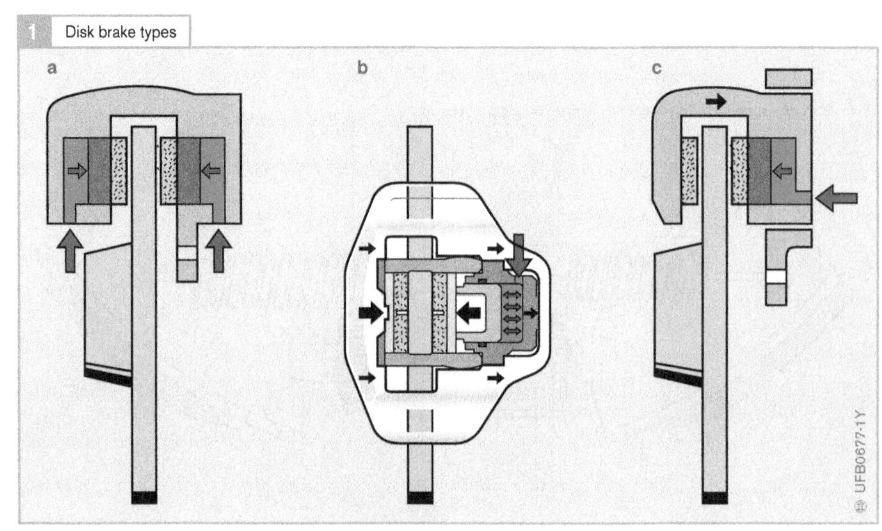

Fig. 1
a Fixed-caliper brake (front view)
b Floating-caliper brake (top view)
c Sliding-caliper brake (front view)

released, i.e. when the hydraulic pressure is removed. This is only possible when the pressure in the brake pipes feeding the disk brake has completely dissipated. The pad clearance on a disk brake is around 0.15 mm and is therefore in the vicinity of the maximum permissible *static* disk runout.

As the brake pad wears and the piston travel increases, the seal slides over it, thus allowing it to protrude further and thereby effecting automatic infinitely variable adjustment of the brake-pad clearance. Consequently, the pad clearance is kept constant and the disk can rotate freely when the pad is not under pressure.

Expander spring

The expander spring used on fixed-caliper brakes is in the shape of a cross and has the job of pressing the brake pads against the pistons and assisting release of the brakes (Figure 3).

Brake piston (positioning)

If the pistons of fixed-caliper and floating-caliper disk brakes have heels (Figure 4, Item 4) with a depth of approx. 3 mm on the pad leading-edge side, those heels must be set an angle of 20° to the horizontal using a piston gauge (3). The piston heel acting on the brake pad reduces the pressure of the brake pad against the disk on the pad leading edge and therefore helps to produce more even pad wear. A similar effect is produced by piston offset, whereby the piston does not act centrally on the brake pad but is offset between 2 and 6 mm towards the trailing edge of the pad. The more evenly distributed pad pressure that this produces leads to more even pad wear. As an additional bonus, this feature also has a noise-reducing effect.

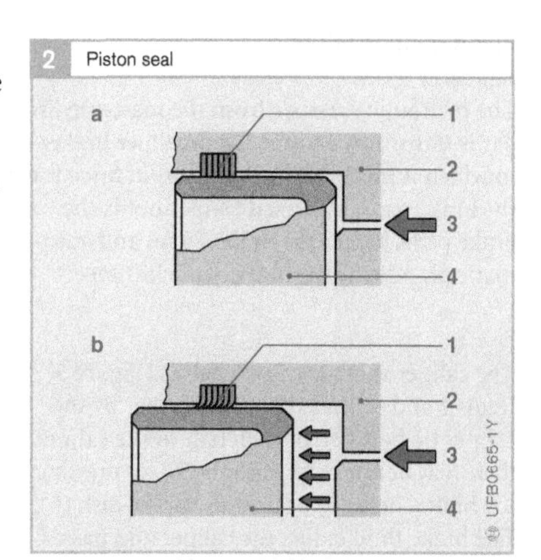

Fig. 2
a Non-braking mode
b Braking mode

1 Piston seal
2 Caliper body
3 Inlet port from master cylinder
4 Piston

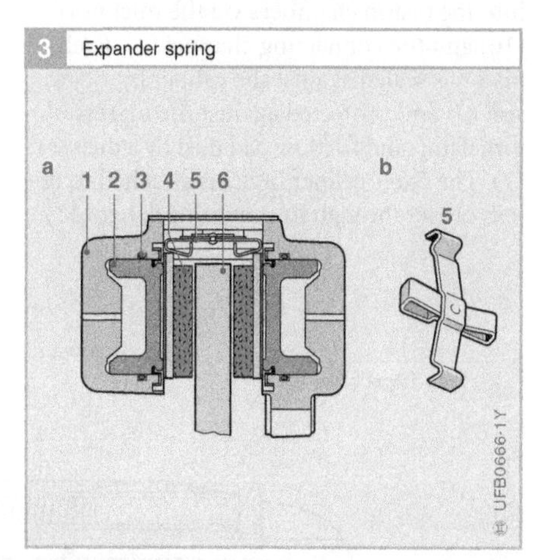

Fig. 3
a Expander spring in fitted position (top view)
b Expander spring

1 Caliper body
2 Piston
3 Piston seal
4 Brake pad
5 Expander spring
6 Brake disk

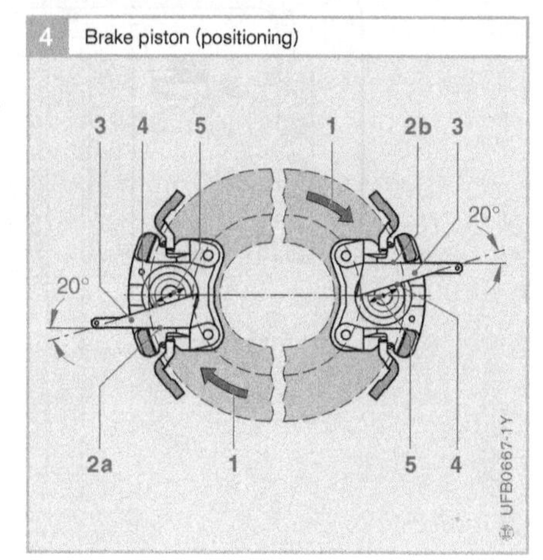

Fig. 4
1 Direction of disk rotation when travelling forwards
2a Upper contact surface in caliper opening
2b Lower contact surface in caliper opening
3 Piston gauge
4 Piston heel
5 Piston bore

Fixed-caliper brake

Function

The hydraulic pressure from the master cylinder is transmitted to the fixed-caliper brake, in which it produces the application force for the brake pads. The fixed caliper holds the brake pads, braces the brake forces and automatically adjusts the brake-pad clearance.

Design

The caliper is made of two halves (Figure 5, Items 1 and 9) that are held together by the caliper tie bolt (2). In each half of the caliper there may be one or two pistons (8) for pressing the brake shoes (5) against the brake disk (6). The brake fluid enters the caliper and passes into the piston chambers via the inlet port (10) and the connecting channel (4). Each piston is sealed against the caliper by a ring seal (3) and protected against the ingress of dirt, damp and friction-pad dust by a dust seal (7). The fixed-caliper brake is attached to the hub carrier through its mounting flange (11).

Method of operation

When the brakes are applied, hydraulic pressure from the master cylinder acts via the inlet port (10) on both pistons (8), thereby producing the actuating force by which the brake pads (5) are pressed against the friction surfaces of the brake disk (6). The size of that controllable actuating force is determined by the foot pressure applied to the brake pedal. When the brake is released, i.e. when the foot is taken off the brake pedal, the master-cylinder piston is returned to its original position by the force of its compression spring and the pressure transmitted to the brake caliper through the brake pipe is released. The pistons (8) are then drawn back to their original positions by the elastic piston seals (3). Having been released by the brake pads, the brake disk (6) is then free to rotate again. If the piston travel is greater than the design clearance between brake pad and disk due to pad wear, the piston slides through the seal when the brakes are applied and the clearance is reset to its correct amount.

5 Fixed-caliper brake

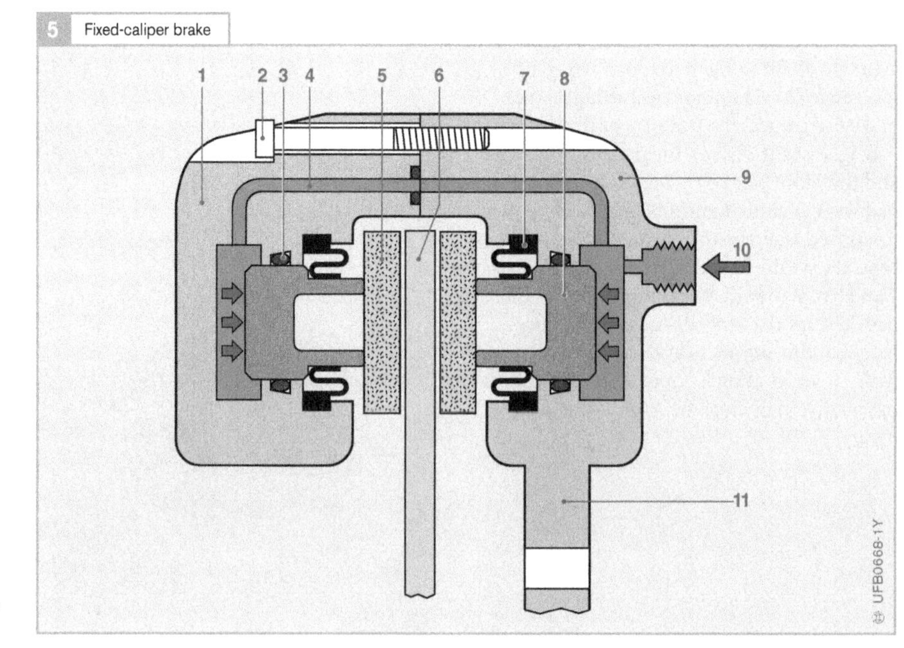

Fig. 5
1 Caliper half (cap)
2 Caliper tie bolt
3 Piston seal
4 Hydraulic-fluid connecting channel
5 Brake pad
6 Brake disk
7 Dust seal
8 Piston
9 Caliper half (flanged)
10 Inlet port from master cylinder
11 Mounting flange

Notes
Adjustment or readjustment of the fixed-caliper brake is unnecessary due to the self-adjusting effect of the elastic piston seals. Because of their high mechanical strength, fixed-caliper brakes are used on heavy and high-speed cars. Their disadvantage is their sensitivity to heat under extended periods of braking (e.g. on long descents). That sensitivity is reduced on high-performance brakes by replacing the internal hydraulic channel between the two halves of the caliper with an external brake pipe. Fixed-caliper brakes require more space inside the wheel rim so that on vehicles with a negative kingpin offset floating-caliper or sliding caliper brakes are generally preferred.

Floating-caliper brake

Function
The floating-caliper brake generates the actuating force for the brake pads from the hydraulic pressure created in the master cylinder, holds the brake pads, braces the brake forces and automatically adjusts the brake-pad clearance.

Design
The floating-caliper brake consists of two main components (Figure 6):

- the caliper bracket (3), which holds the cylinder (8) and piston (7) assembly and the brake pads (4, 5), and is rigidly attached to the hub carrier, and
- the floating caliper (2) which slides in the curved guideways of the caliper bracket.

A guide spring helps the caliper bracket and floating caliper to slide smoothly and quietly over each other. The brake fluid enters the chamber between the cylinder body and the piston via the inlet port (6).

Method of operation
When the brakes are applied, hydraulic pressure from the master cylinder acts via the inlet port (6) on the piston (7) which moves out of the cylinder, closing the clearance gap and pressing the inner brake pad (5) against the brake disk (1). The hydraulic pressure simultaneously acts against the cylinder body (8), moving the floating caliper (2) in the opposite direction to the piston, and indirectly pressing the outer brake pad (4) against the disk after closing the clearance gap on that side. When the brake pedal is released, the hydraulic pressure at the inlet port (6) is removed. As on the fixed-caliper brake, the elastic piston seal (9) draws the piston (7) back by an amount equivalent to the clearance gap and the disk is then free to rotate again.

Notes
Like the fixed-caliper brake, the floating-caliper design does not require adjustment or readjustment. Because it is more compact, this brake is particularly suited to use on vehicles where space is restricted or which have a negative kingpin offset. Because air can circulate freely around the hydraulic components, the hydraulic fluid is effectively cooled. Floating-caliper brakes can incorporate a parking-brake mechanism.

6 Floating-caliper brake

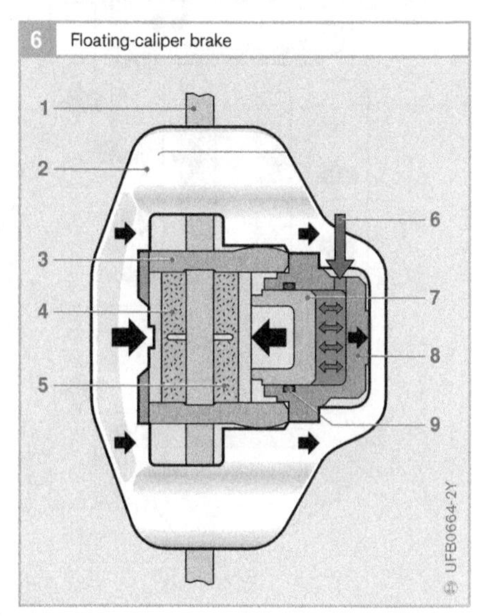

Fig. 6
1 Brake disk
2 Floating caliper
3 Bracket
4 Outer brake pad
5 Inner brake pad
6 Inlet port (from master cylinder)
7 Piston
8 Cylinder
9 Piston seal

Sliding-caliper brake

Function

The sliding-caliper brake (Figure 7) produces the application force for the brake pads by using the hydraulic pressure from the master cylinder. The sliding caliper holds the brake pads, braces the brake forces and automatically adjusts the brake-pad clearance.

Comparison with floating-caliper brake

The sliding-caliper brake is easier to service than the floating-caliper brake from which it was developed. Like the floating-caliper brake, it too has a movable caliper (3, 10) and a single piston (9). Similar too is the way in which the hydraulic pressure moves the piston against the inner pad while simultaneously pushing the caliper body (3) in the opposite direction so as to indirectly press the outer pad (4) against the disk (5). The caliper, however, slides on two guide pins (2) instead of on a caliper bracket.

Design

The sliding caliper (3, 10) is mounted on two guide pins (2) on which it is able to slide in and out. A bracket (1) attached to the hub carrier holds the two guide pins. The piston (9) acts directly on the inner brake pad (6) and indirectly on the outer brake pad (4).
The inlet port (8) connects the caliper to the master cylinder.

Method of operation

When the brakes are applied, hydraulic pressure from the master cylinder acts via the inlet port (8) on the piston (9) which moves out of the caliper and directly presses the inner brake pad (6) against the brake disk (5). As the brake fluid pressure acts with equal force against the piston and the caliper (10), the sliding caliper is pushed in the opposite direction to the piston. The caliper then slides on the guide pins (2) and draws the outer brake pad (4) against the disk. Both brake pads are then pressed against the disk with equal force.
When the brake is released, the elastic piston seal (7) pulls the piston back to its original position.

7 Sliding-caliper brake

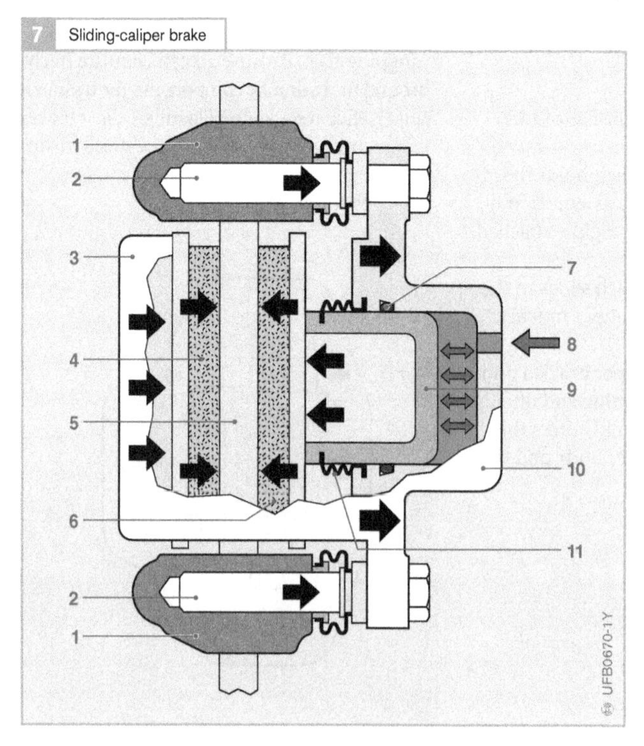

Fig. 7
1 Bracket
2 Guide pin
3 Caliper body
4 Outer brake pad
5 Brake disk
6 Inner brake pad
7 Piston seal
8 Inlet port from master cylinder
9 Piston
10 Caliper body
11 Dust seal

Sliding-caliper brake with integral parking-brake mechanism

Function

This sliding-caliper brake functions both as a service brake and a parking brake. It produces the application force for the brake pads from the hydraulic pressure from the master cylinder or the tension applied by the handbrake lever. The sliding caliper holds the brake pads, braces the brake forces and automatically adjusts the brake-pad clearance.

Design

The sliding-caliper body (Figure 8, Item 8) is mounted on two guide pins (2) on which it is able to slide in and out. A bracket attached to the hub carrier holds the two guide pins. The piston (6) presses the inner brake pad (5) and the outer brake pad (3) against the disk (4) directly and indirectly respectively. The piston is hydraulically operated by the brake fluid entering the inlet port (11).
A metal casing (10) and a sealing disk (13) isolate the hydraulic system from the parking-brake mechanism which is operated by the hand-brake lever (17).

Method of operation

When the service brakes are applied, hydraulic pressure from the master cylinder acts via the inlet port (11) on the piston (6) which moves out of the caliper and directly presses the inner brake pad (5) against the brake disk (4). At the same time, the hydraulic pressure acts with equal force against the sliding caliper-body base. The caliper body (8) then slides on the guide pins (2) and draws the outer brake pad (3) against the disk. The brake-pad pressure is thus equal on both sides of the disk. When the brake is released, the elastic piston seal (18) draws the piston back to its original position and the disk is then free to rotate again. When the parking brake is applied, a cable pulls the hand-brake lever (17) so that the cam (15) turns and presses the tappet (16) and the push-rod (12) against the piston, which then directly presses the inner brake pad against the disk. The outer brake pad is pressed against the other side of the disk by the reactive force.

In addition to the parking-brake mechanism illustrated in Figure 8, there is also the BIR (Ball in Ramp) mechanism. In that case, the inner brake pad is not moved by means of a tappet, but by balls instead. When the parking brake is applied, the parking-brake mechanism is caused to rotate, thereby moving three balls, each of which runs in a ramp-shaped groove. Those ramps convert the rotation into a linear movement by means of which the piston presses the brake pads against the disk.

8 Sliding-caliper brake with integral parking-brake mechanism

UFB0672-1Y

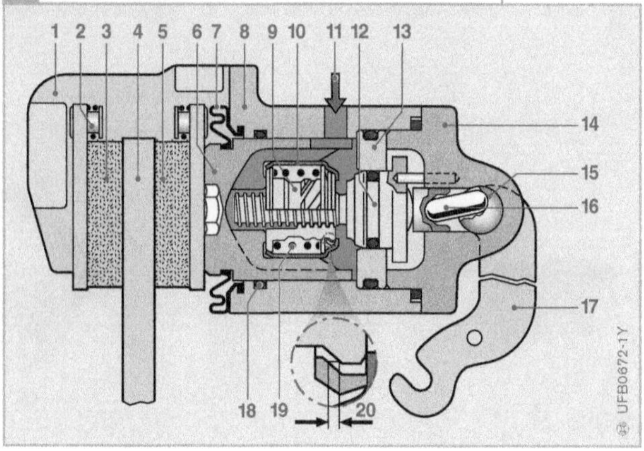

Fig. 8
1 Caliper-body base
2 Guide pin (rear guide pin concealed)
3 Outer brake pad
4 Brake disk
5 Inner brake pad
6 Piston
7 Dust seal
8 Caliper body
9 Self-adjusting mechanism
10 Metal casing
11 Inlet port from master cylinder
12 Push-rod
13 Sealing disk
14 Caliper-body cap
15 Cam
16 Tappet
17 Hand-brake lever
18 Piston seal
19 Compression spring
20 Play

Brake pads, shoes and disks

Braking sequence

During braking, the brake pads or shoes are pressed against a surface which is rotating with the wheel. This friction pairing generates a frictional or braking force that converts the kinetic energy of the vehicle into heat. On disk brakes, the braking force is generated by the friction pairing of brake-pad friction material and disk, and on drum brakes by the friction pairing of brake-shoe friction lining and drum. The kinetic friction coefficient between brake-pad/shoe friction material/lining and disk/drum determines, among other things, the amount of force that has to be applied to the brake pedal to achieve a given braking effect. It also has a fundamental effect on brake balance and vehicle handling stability under braking.

Brake-shoe friction linings are riveted or glued to the brake shoe. The brake shoes have a T-shaped cross-section which gives them the required rigidity. To make them easier to fit, brake shoes can be supplied as a pre-assembled drum-brake kit comprising shoes, wheel cylinder and accessories.

Disk-brake friction pads consist of a friction material and an intermediate layer that are glued to the metal backing plate.

Brake-type usage

Nowadays, all cars are fitted with disk brakes on the front wheels. Drum brakes are fitted to the rear wheels of smaller/medium-sized cars.

Composition of brake friction material

The friction material from which brake linings/pads are made is basically made up of four raw materials. The relative proportions of those materials depends on the specifics of the application and the required frictional coefficient (static and kinetic). Thus, the friction material used on the disk-brake pads for a large executive saloon is different from that used in the brake-shoe linings for a small "runabout". The precise details of the friction material compositions are well-guarded secrets known only to the manufacturers (Table 1 lists examples of the main constituents).

Table 1

1 Composition of disk-pad friction material (example)

Raw material group	Raw materials	% by volume
Metals	Steel wool Copper powder	14
Fillers	Aluminum oxide Mica powder Barite Iron oxide	23
Friction adjusters	Antimony sulphide Graphite Powdered coke	35
Organic components	Aramide fibre Resin filler powder Binding resin	28

1 Drum-brake friction lining attached to brake shoe

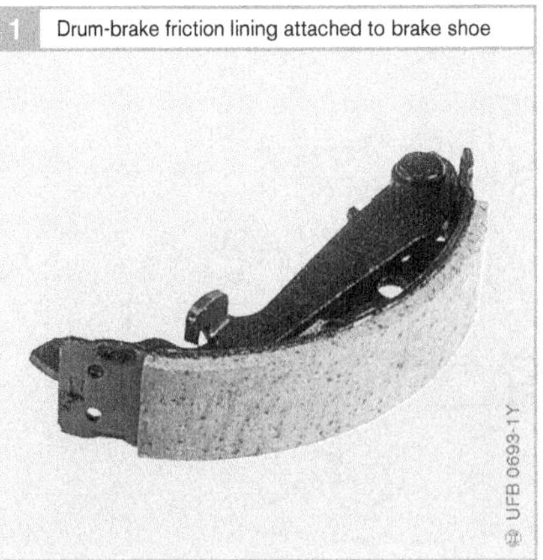

2 Disk-brake friction pad attached to backing plate

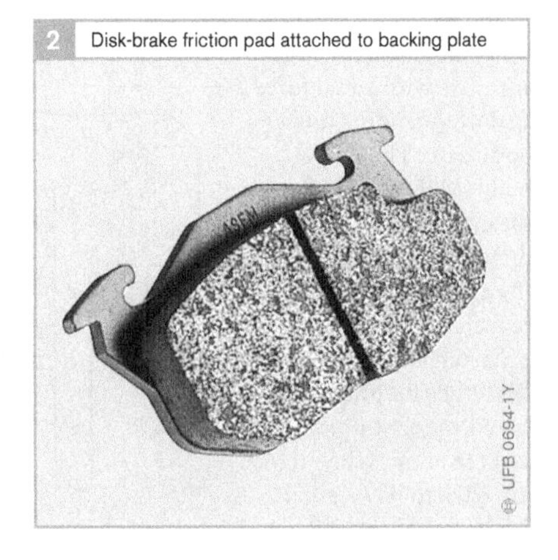

Replacement intervals for brake shoes/pads

As a rule drum-brake shoes have to be replaced twice as frequently as the drums, while disk-brake pads normally need to be renewed five times as often as the disks.
For the spare-parts trade, this obviously means that brake pads for disk brakes are sold in much larger numbers than brake shoes for drum brakes.

On some vehicles, a wear sensor integrated in the brake pad comes into contact with the disk when the friction material has been worn down to a thickness of 3.5 mm. In so doing, it completes an electrical circuit connected to an indicator lamp on the dashboard, thus warning the driver that the pads need replacing.

Note
If the brake pads/shoes or the disk/drum on one wheel require replacement, the pads/shoes or disk/drum on the opposite wheel should always be replaced at the same time in order to ensure that even braking efficiency is maintained on both (front or rear) wheels. When replacing brake shoes, pads, drums or disks, it is important to use only those components approved by the brake manufacturer.

Quality requirements for brake friction materials

The quality requirements for brake friction materials can be divided into the three categories of safety, comfort/convenience and durability, and must be counterbalanced against one another to suit each particular application.

Safety
- Stability of frictional coefficient
- Shear resistance
- Compressibility
- Dimensional stability
- Thermal conductivity
- Flammability
- Corrosion resistance
- Running-in characteristics

Comfort/convenience
- Noise-generating characteristics
- Vibration absorption
- Response characteristics

Durability
- Wear characteristics

▶ Assessment criteria for car disk-brake pads

Performance	
• *Speed-dependent coefficient of friction:*	Friction measurements under braking from 40 km/h → 5 km/h and from 180 km/h → 150 km/h at 40 bar, 100 °C; decisive criterion for overall assessment.
• *Difference in coefficient of friction:*	Friction measurements under braking from 40 km/h → 5 km/h and from 180 km/h → 150 km/h at 40 bar, 100 °.
• *Motorway braking:*	Friction measurements under braking from 180 km/h → 100 km/h.
• *Fade:*	Reduction in braking effect at high temperatures resulting from reduced friction caused by chemical reactions; measured under braking from 0.9 v_{max} → 0.5 v_{max}.
• *Running-in characteristics:*	Measurement of number of braking operations required to stabilise frictional coefficient of new brake pads.
• *General friction coefficient levels*	Averaged figure for comparable brake pads made by different manufacturers
Wear	
• *Brake pad:*	Wear in millimetres of friction-pad thickness at 150 °C, 300 °C and 400 °C.
• *Brake disk:*	Wear in grammes
Visual appearance after testing	
• *Brake pad:*	subjective assessment
• *Brake disk:*	subjective assessment
Physical data	
• *Shear-strength levels:*	Rejection criterion; measurement of minimum and maximum levels in kN
• *Compressibility:*	Comfort feature; measurement in μm.

Approval of disk-brake pads

A requirements profile detailing vehicle specifications, operating conditions and specific customer requirements forms the basis for selection of a suitable base material. A bench test checks the brake friction material for performance, noise/vibration characteristics and wear. Once that test has been passed, the road test follows.

The road test encompasses a performance test, an endurance test to check for wear, an extreme-load test including a drive over a mountain pass, and a test for perceptible vibrations and noise. If all the tests are successfully completed, the friction material is approved. If only the noise characteristics are unsatisfactory, secondary modifications are made to the friction pad such as application of a rubber coating or attachment of damper plates. On completion of those modifications, the brake pads are put through the bench test and road test again.

Approved disk-brake pads are identified by a number on the pad surface (Figure 3).

Braking noises produced by disk-brake pads

Uneven frictional processes between the brake pad and the disk generate vibrations, whose sound waves can be discernible inside the vehicle to the driver, depending on their frequency. The major determinants of brake noise are brake pressure, brake-disk temperature, vehicle speed and climatic conditions.

Brake noises that occur during the braking sequence are distinguished according to whether they are produced at the point when the brakes are applied, continuously while the brakes are on, or when they are released. Low-frequency noises between 0 and 500 Hz are imperceptible from inside the car. Noises with a frequency of 500 ... 1,500 Hz are not distinguishable by the driver as brake noises. High-frequency noises of between 1,500 and 15,000 Hz are recognisable by the driver as brake noises.

Table 2 contains a trouble-shooting chart for noisy brakes detailing fault causes and remedies.

Table 2

2 Trouble-shooting chart for noisy disk pads

Cause	Remedy
Worn or incorrect brake pads	Replace pads on both (front/rear) wheels
New pads not yet run-in	Take vehicle on test run to run-in pads
Friction pads contaminated with oil	Locate oil leak and remedy; fit new brake pads
Worn accessories	Replace accessories
Incorrect positioning of brake-caliper piston	Check caliper-piston position and adjust as necessary
Excessive wheel-bearing play	Adjust wheel-bearing play
Uneven disk runout	Replace brake disk
Wet or dirty brakes	Clean brakes and lubricate with suitable lubricant

3 Disk-brake pad numbering

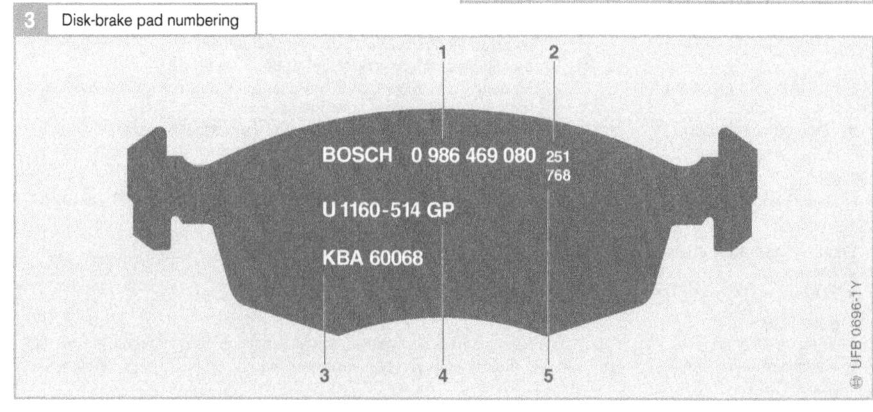

Fig. 3
1 Ten-digit Bosch part number
2 Bosch production-plant number
3 KBA (German vehicle-registration authority) number
4 Friction-material manufacturer number
5 Bosch manufacturing date

Brake disks

Like brake drums, brake disks are attached to the wheel hubs and for most applications are made of cast iron or steel. Compared with the shoes in drum brakes, the pressure with which the pads are applied to a disk is greater as their surface area is smaller. This results in the generation of more heat and a faster rate of pad wear as compared with drum-brake shoes. A brake disk is exposed to the air and therefore effectively cooled by the air flow when the vehicle is in motion.

Disks can be of three types: unventilated, ventilated from the inside, or ventilated from the outside (Figure 4). Due to their greater mass, ventilated disks have a greater heat-storage capacity and also cool down more quickly as a result of their radial ventilation channels that have a fan effect. Consequently, ventilated disks are usually preferred for the front wheels.

Outlook

For a number of years, research has been conducted into the use of carbon fibre as a material for brake disks in order to reduce vehicle weight, particularly on racing cars. Ventilated disks made entirely of carbon fibre have been tested under race conditions.

The latest developments have made it possible to produce composite brake disks made of ceramic materials. In combination with new friction-pad materials, these new disks are expected to last as long as the vehicle itself. Since ceramic disks are only half the weight of conventional cast-iron disks, the overall weight of the vehicle can be reduced and greater fuel economy achieved. At the same time, the response characteristics of the shock absorbers are improved due to the lower unsprung mass.

4 Brake-disk types

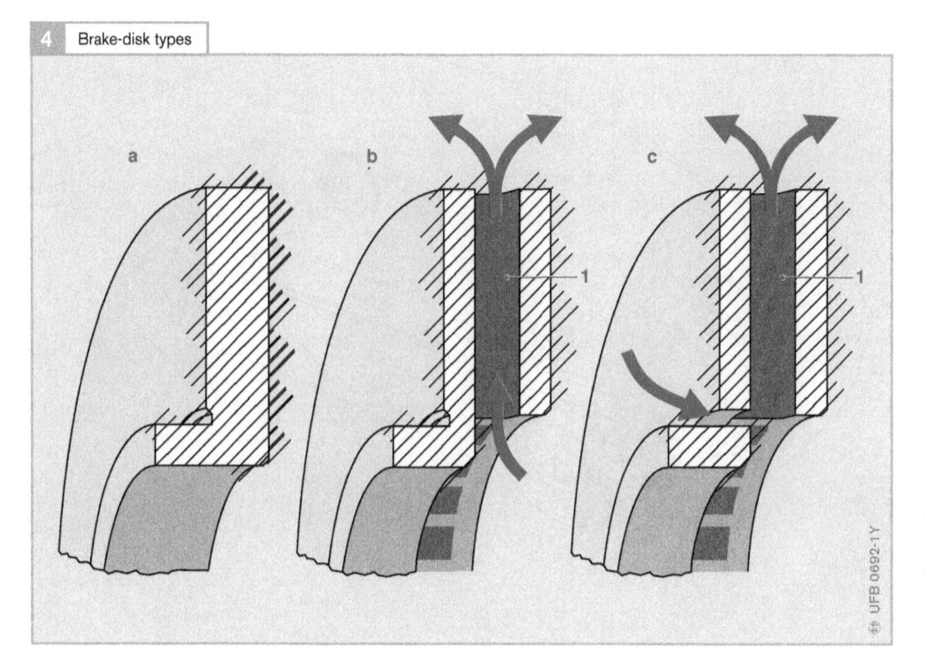

Fig. 4
a Unventilated
b Ventilated from inside
c Ventilated from outside

1 Cooling channel

Antilock braking system (ABS)

In hazardous driving conditions, it is possible for the wheels of a vehicle to lock up under braking. The possible causes include wet or slippery road surfaces, and abrupt reaction on the part of the driver (unexpected hazard). The vehicle can become uncontrollable as a result, and may go into a slip and/or leave the road. The antilock braking system (ABS) detects if one or more wheels are about to lock up under braking and if so makes sure that the brake pressure remains constant or is reduced. By so doing, it prevents the wheels from locking up and the vehicle remains steerable. As a consequence the vehicle can be braked or stopped quickly and safely.

System overview

The ABS braking system is based on the components of the conventional system. Those are

- the brake pedal (Fig. 1, 1),
- the brake booster (2),
- the master cylinder (3),
- the reservoir (4),
- the brake lines (5) and hoses (6), and
- the brakes and wheel-brake cylinders (7).

In addition there are also the following components:

- the wheel-speed sensors (8),
- the hydraulic modulator (9), and
- the ABS control unit (10).

The warning lamp (11) lights up if the ABS is switched off.

1 Braking system with ABS

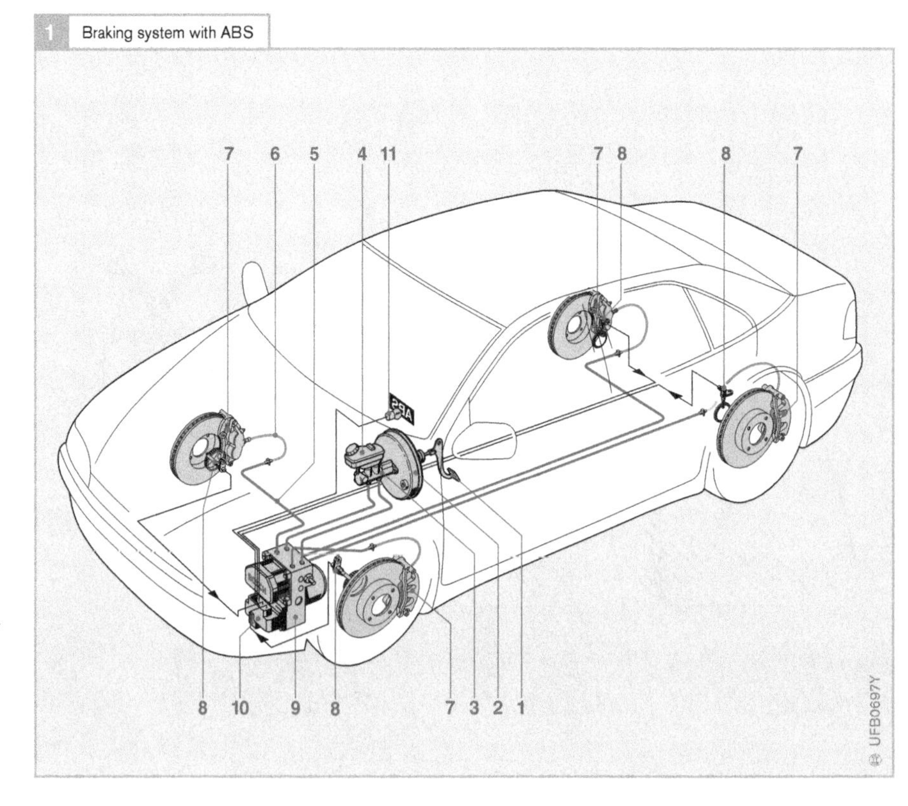

Fig. 1
1 Brake pedal
2 Brake booster
3 Master cylinder
4 Reservoir
5 Brake line
6 Brake hose
7 Wheel brake with wheel-brake cylinder
8 Wheel-speed sensor
9 Hydraulic modulator
10 ABS control unit (in this case, attached unit fixed onto hydraulic modulator)
11 ABS warning lamp

Wheel-speed sensors
The speed of rotation of the wheels is an important input variable for the ABS control system. Wheel-speed sensors detect the speed of rotation of the wheels and pass the electrical signals to the control unit.

A car may have three or four wheel-speed sensors depending on which version of the ABS system is fitted (ABS system versions). The speed signals are used to calculate the degree of slip between the wheels and the road surface and therefore detect whether any of the wheels is about to lock up.

Electronic control unit (ECU)
The ECU processes the information received from the sensors according to defined mathematical procedures (control algorithms). The results of those calculations form the basis for the control signals sent to the hydraulic modulator.

Hydraulic modulator
The hydraulic modulator incorporates a series of solenoid valves that can open or close the hydraulic circuits between the master cylinder (Fig. 2, 1) and the brakes (4). In addition, it can connect the brakes to the return pump (6). Solenoid valves with two hydraulic connections and two valve positions are used (2/2 solenoid valves). The inlet valve (7) between the master cylinder and the brake controls pressure application, while the outlet valve (8) between the brake and the return pump controls pressure release. There is one such pair of solenoid valves for each brake.

Under normal conditions, the solenoid valves in the hydraulic modulator are at the "pressure application" setting. That means the inlet valve is open. The hydraulic modulator then forms a straight-through connection between the master cylinder and the brakes. Consequently, the brake pressure generated in the master cylinder when the brakes are applied is transmitted directly to the brakes at each wheel.

As the degree of brake slip increases due to braking on a slippery surface or heavy braking, the risk of the wheels locking up also increases. The solenoid valves are then switched to the "maintain pressure" setting. The connection between the master cylinder and the brakes is shut off (inlet valve is closed) so that any increase of pressure in the master cylinder does not lead to a pressure increase at the brakes.

If the degree of slip of any of the wheels increases further despite this action, the pressure in the brake(s) concerned must be reduced. To achieve this, the solenoid valves are switched to the "pressure release" setting. The inlet valve is still closed, and in addition, the outlet valve opens to allow the return pump integrated in the hydraulic modulator to draw brake fluid from the brake(s) concerned in a controlled manner. The pressure in the relevant brake(s) is thus reduced so that wheel lock-up does not occur.

2 Principle of hydraulic modulator with 2/2 solenoid valves (schematic)

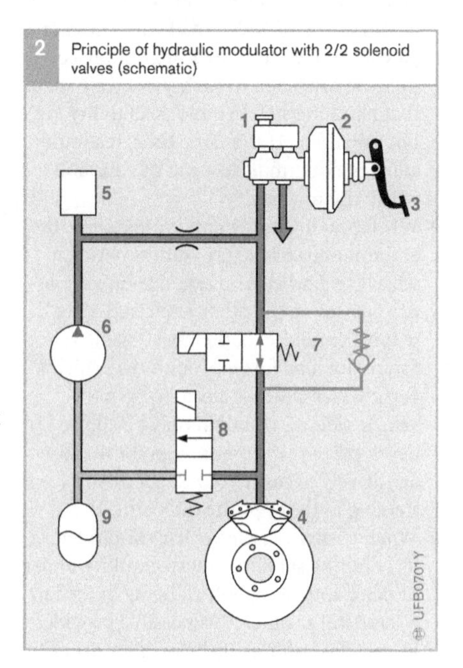

Fig. 2
1 Master cylinder with reservoir
2 Brake booster
3 Brake pedal
4 Wheel brake with wheel-brake cylinder

Hydraulic modulator with
5 Damping chamber
6 Return pump
7 Inlet valve
8 Outlet valve
9 Brake-fluid accumulator

Inlet valve: shown in open setting
Outlet valve: shown in closed setting

Requirements placed on ABS

An ABS system must meet a comprehensive range of requirements, in particular all the safety requirements associated with dynamic braking response and the braking-system technology.

Handling stability and steerability

- The braking control system should be capable of ensuring that the car retains its handling stability and steerability on all types of road surface (from dry roads with good adhesion to black ice).
- An ABS system should utilize the available adhesion between the tires and the road surface under braking to the maximum possible degree, giving handling stability and steerability precedence over minimizing braking distance. It should not make any difference to the system whether the driver applies the brakes violently or gradually increases the braking force to the point at which the wheels would lock.
- The control system should be capable of adapting rapidly to changes in road-surface grip, e.g. on a dry road with occasional patches of ice, wheel lock-up on the ice must be restricted to such short time spans that handling stability and steerability are not impaired. At the same time, it should allow maximum utilization of adhesion where the road is dry.
- When braking under conditions where the amount of available grip differs between wheels (e.g. wheels on one side on ice, on dry tarmac on the other – referred to as " μ-split" conditions), the unavoidable yaw forces (turning forces around the vehicle's vertical axis which attempt to turn the vehicle sideways) should only be allowed to develop at a rate slow enough for the "average driver" to easily counteract them by steering in the opposite direction.
- When cornering, the vehicle should retain its handling stability and steerability under braking and be capable of being braked to a standstill as quickly as possible provided its speed is sufficiently below the corner's limit speed (that is the absolute maximum speed at which a vehicle can successfully negotiate a bend of a given radius with the drive disengaged).
- The system should also be able to ensure that handling stability and steerability are maintained and the best possible braking effect obtained on a bumpy or uneven road surface, regardless of the force with which the driver applies the brakes.
- Finally, the braking control system should be able to detect aquaplaning (when the wheels "float" on a film of water) and respond appropriately to it. In so doing it must be able to maintain vehicle controllability and course.

Effective range

- The braking control system must be effective across the entire speed range of the vehicle right down to crawling speed (minimum speed limit approximately 2.5 km/h). If the wheels lock up below the minimum limit, the distance traveled before the vehicle comes to a halt is not critical.

Timing characteristics

- Adjustments to take account of braking-system hysteresis (delayed reaction to release of brakes) and the effects of the engine (when braking with the drive engaged) must take as little time as possible.
- Pitching of the vehicle in response to suspension vibration must be prevented.

Reliability

- There must be a monitoring circuit which continuously checks that the ABS is functioning correctly. If it detects a fault which could impair braking characteristics, the ABS should be switched off. A warning lamp must then indicate to the driver that the ABS is not functioning and only the basic braking system is available.

Dynamics of a braked wheel

Figs. 1 and 2 show the physical interdependencies during a braking sequence with ABS. The areas in which the ABS is active are shaded blue.

Curves 1, 2 and 4 in Fig. 1 relate to road conditions in which the level of adhesion and therefore the braking force increases up to a maximum limit as the brake pressure rises.

Increasing the brake pressure above that maximum adhesion limit on a car without ABS would constitute overbraking. When that happens, the "sliding" proportion of the tire footprint (the area of the tire in contact with the road) increases so much as the tire deforms that the static friction diminishes and the kinetic friction increases.

The brake slip λ is a measure of the proportion of kinetic friction, whereby $\lambda = 100\%$ is the level at which the wheel locks and only kinetic friction is present.

The degree of brake slip,

$$\lambda = \frac{(v_F - v_R)}{v_F} \cdot 100\%$$

indicates the degree to which the wheel's circumferential speed, v_R, lags behind the vehicle's linear speed (road speed), v_F.

From the progression shown in Fig. 1 of curves 1 (dry conditions), 2 (wet conditions) and 4 (black ice), it is evident that shorter braking distances are achieved with ABS than if the wheels are overbraked and lock up (brake slip $\lambda = 100\%$). Curve 3 (snow) shows a different pattern whereby a wedge of snow forms in front of the wheels when they lock up and helps to slow the vehicle down; in this scenario, the advantage of ABS is in its ability to maintain handling stability and steerability.

As the two curves for coefficient of friction, μ_{HF}, and lateral-force coefficient, μ_S, in Fig. 2 illustrate, the active range of the ABS has to be extended for the large lateral slip angle of $\alpha = 10°$ (i.e. high lateral force due to rapid

1 Coefficient of friction, μ_{HF}, relative to brake slip, λ

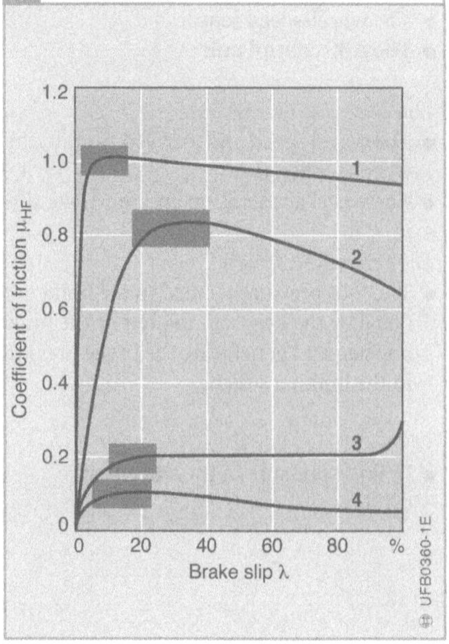

Fig. 1
1 Radial tire on dry concrete
2 Cross-ply tire on wet tarmac
3 Radial tire on loose snow
4 Radial tire on wet black ice
Blue shaded areas: ABS active zones

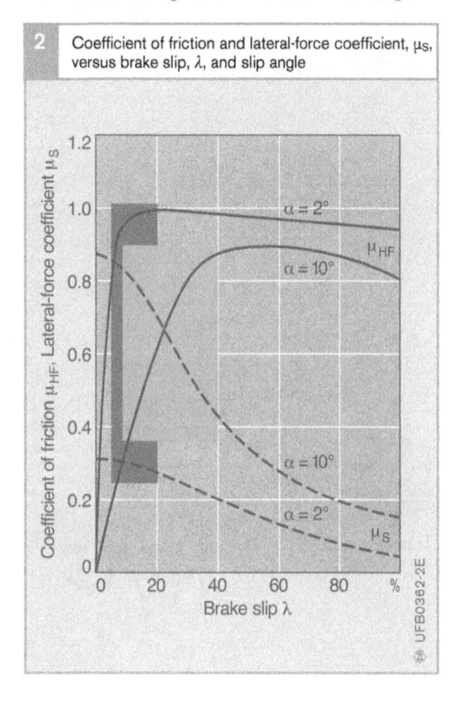

2 Coefficient of friction and lateral-force coefficient, μ_S, versus brake slip, λ, and slip angle

Fig. 2
μ_{HF} Coefficient of friction
μ_S Lateral-force coefficient
α Slip angle
Blue shaded areas: ABS active zones

lateral acceleration of the vehicle) as compared with the smaller lateral slip angle of $\alpha = 2°$, that is to say, if the vehicle is braked hard in a corner when the lateral acceleration is high, the ABS cuts in sooner and allows an initial level of brake slip of, say, 10%. At $\alpha = 10°$, an initial coefficient of friction of only $\mu_{HF} = 0.35$ is obtained, while the lateral-force coefficient is almost at its maximum at $\mu_S = 0.80$.

As the vehicle's speed and therefore its lateral acceleration are reduced by braking through the corner, the ABS is able to allow increasingly higher levels of brake slip so that the deceleration increases, while the lateral-force coefficient diminishes as the lateral acceleration reduces.

Braking while cornering causes the braking forces to rise so quickly that the overall braking distance is only marginally longer than when braking in a straight line under identical conditions.

1 ABS control loop

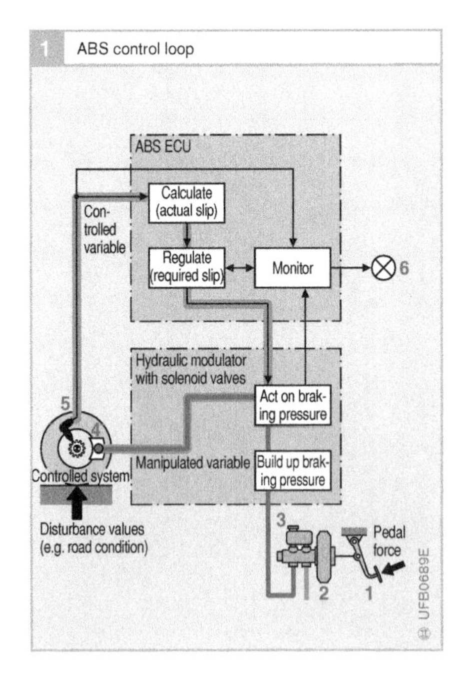

Fig. 1
1 Brake pedal
2 Brake booster
3 Master cylinder with reservoir
4 Wheel-brake cylinder
5 Wheel-speed sensor
6 Warning lamp

ABS control loop

Overview

The ABS control loop (Fig. 1) consists of the following:

The controlled system

- The vehicle and its brakes
- The wheels and the friction pairing of tire and road surface

The external variables affecting the control loop:

- Changes in the adhesion between the tires and the road surface caused by different types of road surface and changes in the wheel loadings, e.g. when cornering
- Irregularities in the road surface causing the wheels and suspension to vibrate
- Lack of circularity of the tires, low tire pressure, worn tire tread, differences in circumference between wheels, (e.g. spare wheel)
- Brake hysteresis and fade
- Differences in master-cylinder pressure between the two brake circuits

The controllers

- The wheel-speed sensors
- The ABS control unit

The controlled variables

- The wheel speed and, derived from it, the wheel deceleration,
- The wheel acceleration and the brake slip

The reference variable

- The foot pressure applied to the brake pedal by the driver, amplified by the brake booster, and generating the brake pressure in the braking system

The correcting variable

- Braking pressure in the wheel-brake cylinder.

Controlled system

The data-processing operations performed by the ABS control unit are based on the following simplified controlled system:

- a non-driven wheel,
- a quarter of the vehicle's mass apportioned to that wheel,
- the brake on that wheel and, representing the friction pairing of tire and road surface,
- a theoretical curve for coefficient of friction versus brake slip (Fig. 2).

That curve is divided into a stable zone with a linear gradient and an unstable zone with a constant progression (μ_{HFmax}). As an additional simplification, there is also an assumed initial straight-line braking response that is equivalent to a panic braking reaction.

Fig. 3 shows the relationships between braking torque, M_B (the torque that can be generated by the brake through the tire), or road-surface frictional torque, M_R (torque that acts against the wheel through the friction pairing of tire and road surface), and time, t, as well as the relationships between the wheel deceleration (a) and time, t, whereby the braking torque increases in linear fashion over time. The road-surface frictional torque lags slightly behind the braking torque by the time delay, T, as long as the braking sequence is within the stable zone of the curve for friction coefficient versus brake slip. After about 130 ms, the maximum level (μ_{HFmax}) – and therefore the unstable zone – of the curve for friction coefficient versus brake slip is reached. From that point on, the curve for friction coefficient versus brake slip states that while the braking torque, M_B, continues to rise at an undiminished rate, the road-surface frictional torque, M_R, cannot increase any further and remains constant. In the period between 130 and 240 ms (this is when the wheel locks up), the minimal torque difference, $M_B - M_R$, that was present in the stable zone rises rapidly to a high figure. That torque difference is a precise measure of the wheel deceleration ($-a$) of the braked wheel (Fig. 3, bottom). In the stable zone, the wheel deceleration is limited to a small rate, whereas in the unstable zone it increases rapidly. As a consequence, the curve for friction coefficient versus brake slip reveals opposite characteristics in the stable and unstable zones. The ABS exploits those opposing characteristics.

2 Ideal curve for friction coefficient versus slip

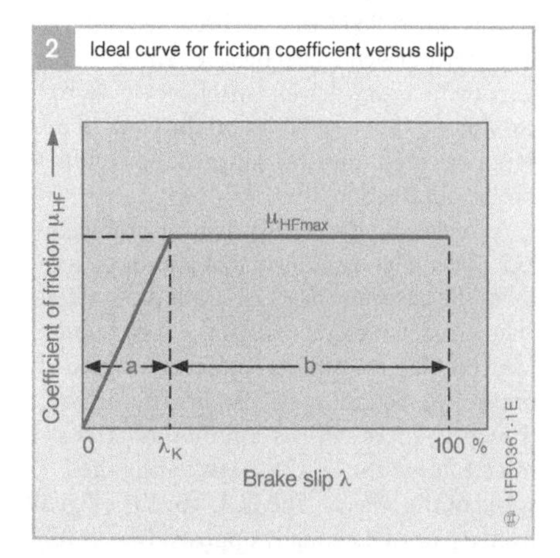

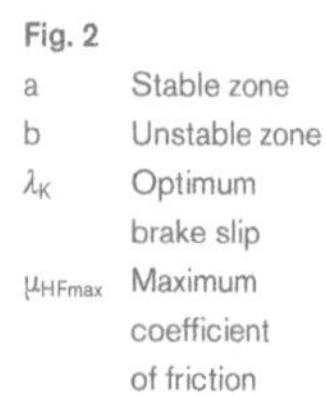

Fig. 2
a Stable zone
b Unstable zone
λ_K Optimum brake slip
μ_{HFmax} Maximum coefficient of friction

3 Initial braking phase, simplified

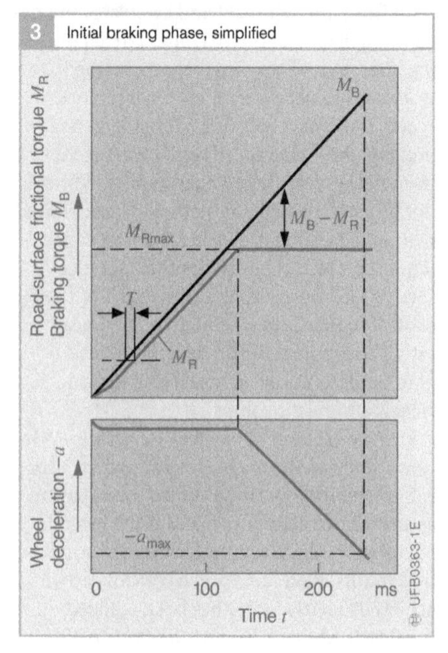

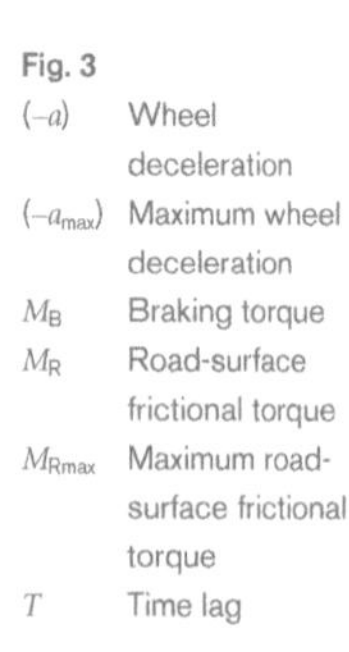

Fig. 3
$(-a)$ Wheel deceleration
$(-a_{max})$ Maximum wheel deceleration
M_B Braking torque
M_R Road-surface frictional torque
M_{Rmax} Maximum road-surface frictional torque
T Time lag

Controlled variables

An essential factor in determining the effectiveness of an ABS control system is the choice of controlled variables. The basis for that choice are the wheel-speed sensor signals from which the ECU calculates the deceleration/acceleration of the wheel, brake slip, the reference speed and the vehicle deceleration. On their own, neither the wheel deceleration/acceleration nor the brake slip are suitable as controlled variables because, under braking, a driven wheel behaves entirely differently to a non-driven wheel. However, by combining those variables on the basis of appropriate logical relationships, good results can be obtained.

As brake slip is not directly measurable, the ECU calculates a quantity that approximates to it. The basis for the calculation is the reference speed, which represents the speed under ideal braking conditions (optimum degree of brake slip). So that speed can be determined, the wheel-speed sensors continuously transmit signals to the ECU for calculating the speed of the wheels. The ECU takes the signals from a pair of diagonally opposed wheels (e.g. right front and left rear) and calculates the reference speed from them. Under partial braking, the faster of the two diagonally opposite wheels generally determines the reference speed. If the ABS cuts in under emergency braking, the wheel speeds will be different from the vehicle's linear speed and can thus not be used for calculating the reference speed without adjustment. During the ABS control sequence, the ECU provides the reference speed based on the speed at the start of the control sequence and reduces it at a linear rate. The gradient of the reference-speed graph is determined by analyzing logical signals and relationships.

If, in addition to the wheel acceleration/ deceleration and the brake slip, the vehicle's linear deceleration is brought into the equation as an additional quantity, and if the logical circuit in the ECU is modulated by computation results, then ideal brake control can be achieved. That concept has been realized in the Bosch Antilock Braking System (ABS).

Controlled variables for non-driven wheels

The wheel acceleration and deceleration are generally suitable as controlled variables for driven and non-driven wheels provided the driver brakes with the clutch disengaged. The reason can be found in the opposing characteristics of the controlled system in the stable and unstable zones of the curve for friction coefficient versus brake slip.

In the stable zone, the wheel deceleration is limited to relatively low rates so that when the driver presses harder on the brake pedal, the car brakes harder without the wheels locking up.

In the unstable zone, on the other hand, the driver only needs to apply slightly more pressure to the brake pedal to induce instantaneous wheel lock-up. This characteristic means that very often the wheel deceleration and acceleration can be used to determine the degree of brake slip for optimum braking.

A fixed wheel deceleration threshold for initiation of the ABS control sequence should only be fractionally above the maximum possible vehicle linear deceleration. This is particularly important if the driver initially only applies the brakes lightly but then increasingly applies more pressure to the pedal. If the threshold is set too high, the wheels could then progress too far along the curve for friction coefficient versus slip into the unstable zone before the ABS detects the imminent loss of control.

When the fixed threshold is initially reached under heavy braking, the brake pressure at the wheel(s) concerned should not automatically be reduced, because with modern tires, valuable braking distance would be lost on a surface with good grip, especially in cases where the initial speed is high.

Controlled variables for driven wheels

If first or second gear is engaged when the brakes are applied, the engine acts on the driven wheels and substantially increases their effective mass moment of inertia Θ_R, i.e. the wheels behave as if they were considerably heavier. The sensitivity with which the wheel deceleration responds to changes in the braking torque in the unstable zone of the curve for friction coefficient versus brake slip diminishes to an equal extent.

The starkly opposing characteristics displayed by non-driven wheels in the stable and unstable zone of the curve for friction coefficient versus brake slip are evened out to a substantial degree, so that in this situation the wheel deceleration is often insufficient as a controlled variable for identifying the degree of brake slip offering the greatest possible friction. It is necessary instead to introduce as an additional controlled variable a quantity that approximates to brake slip and to combine it in a suitable manner with the wheel deceleration.

Fig. 4 compares an initial braking sequence on a non-driven wheel and on a driving wheel that is connected to the drivetrain.
In this example, the engine's inertia increases the effective wheel inertia by a factor of four. On the non-driven wheel, a specific threshold for deceleration $(-a)_1$ is exceeded very soon after leaving the stable zone of the curve for friction coefficient versus brake slip. Because it has a moment of inertia that is four times greater, the driven wheel requires a torque difference four times as big

$$\Delta M_2 = 4 \cdot \Delta M_1$$

to exceed the threshold $(-a)_2$. Consequently, the driven wheel may by then have progressed a long way into the unstable zone of the curve for friction coefficient versus brake slip, resulting in impaired vehicle handling stability.

Effectiveness of control

An efficient antilock braking system must meet the criteria listed below for the standard of control.

- Maintenance of handling stability by provision of adequate lateral forces at the rear wheels
- Maintenance of steerability by provision of adequate lateral forces at the front wheels
- Reduction of braking distances as compared with locked-wheel braking by optimum utilization of the available adhesion between tires and road surface
- Rapid adjustment of braking force to different friction coefficients, for instance when driving through puddles or over patches of ice or compacted snow.
- Ensuring application of low braking-torque control amplitudes to prevent suspension vibration
- High degree of user-friendliness due to minimal pedal feedback ("pedal judder") and low levels of noise from the actuators (solenoid valves and return pump in the hydraulic modulator).

The criteria listed can only be optimized collectively rather than individually. Nevertheless, vehicle handling stability and steerability are always among the top priorities.

4 Initial braking phase for a non-driven wheel and a driving wheel connected to the drivetrain

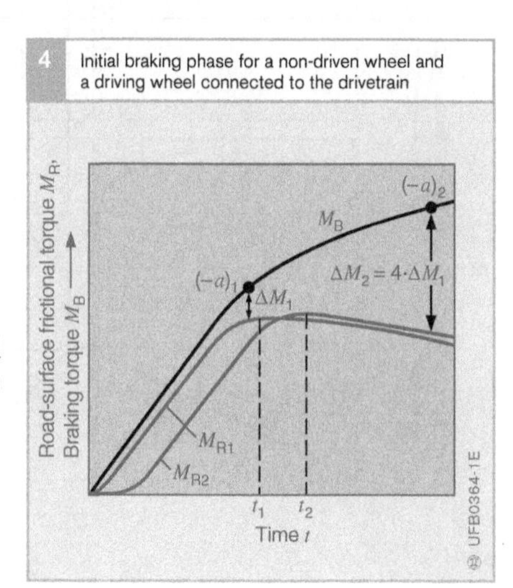

Fig. 4
Index 1: non-driven wheel
Index 2: driven wheel (in this example, the wheel moment of inertia is increased by a factor of 4)
$(-a)$ Threshold for wheel deceleration
M Torque difference $M_B - M_R$

Typical control cycles

Control cycle on surfaces with good grip (High coefficient of friction)

If the ABS sequence is activated on a road surface with good grip (high coefficient of friction), the subsequent pressure rise must be 5 to 10 times slower than in the initial braking phase in order to prevent undesirable suspension vibration. That requirement produces the control-cycle progression for high coefficients of friction illustrated in Fig. 1.

During the initial phase of braking, the brake pressure at the wheel and the rate of wheel deceleration (negative acceleration) rise. At the end of phase 1, the wheel deceleration passes the set threshold level ($-a$).

As a result, the relevant solenoid valve switches to the "maintain pressure" setting. At this point the brake pressure must not be reduced, because the threshold ($-a$) might be exceeded within the stable zone of the curve for friction coefficient versus brake slip and then potential braking distance would be "wasted". At the same time, the reference speed, v_{Ref}, reduces according to a defined linear gradient. The reference speed is used as the basis for determining the slip switching threshold, λ_1.

At the end of phase 2, the wheel speed, v_R, drops below the λ_1 threshold. At that point, the solenoid valves switch to the "reduce pressure" setting so that the pressure drops, and they remain at that setting as long as

1 Braking control cycle for high-adhesion conditions

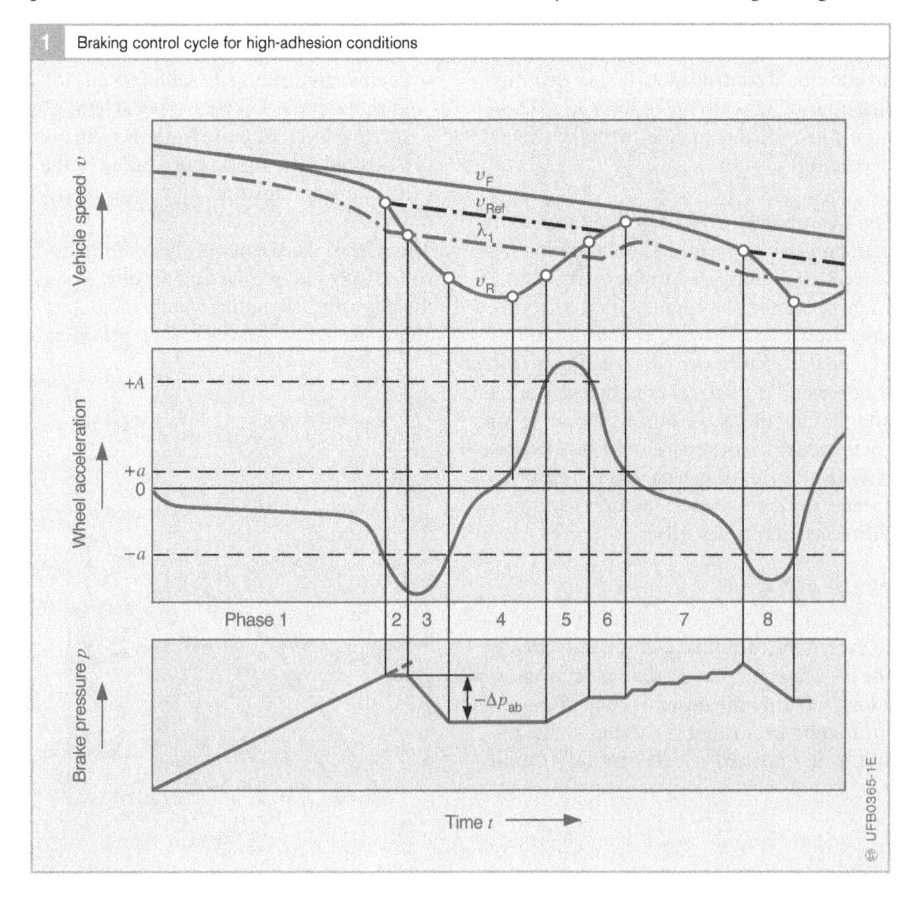

Fig. 1
v_F Vehicle speed
v_{Ref} Reference speed
v_R Wheel speed
λ_1 Slip switching threshold
Switching signals:
$+A$, $+a$ Thresholds for wheel acceleration
$-a$ Threshold for wheel deceleration
$-\Delta p_{ab}$ Brake-pressure drop

2 Braking sequence without ABS

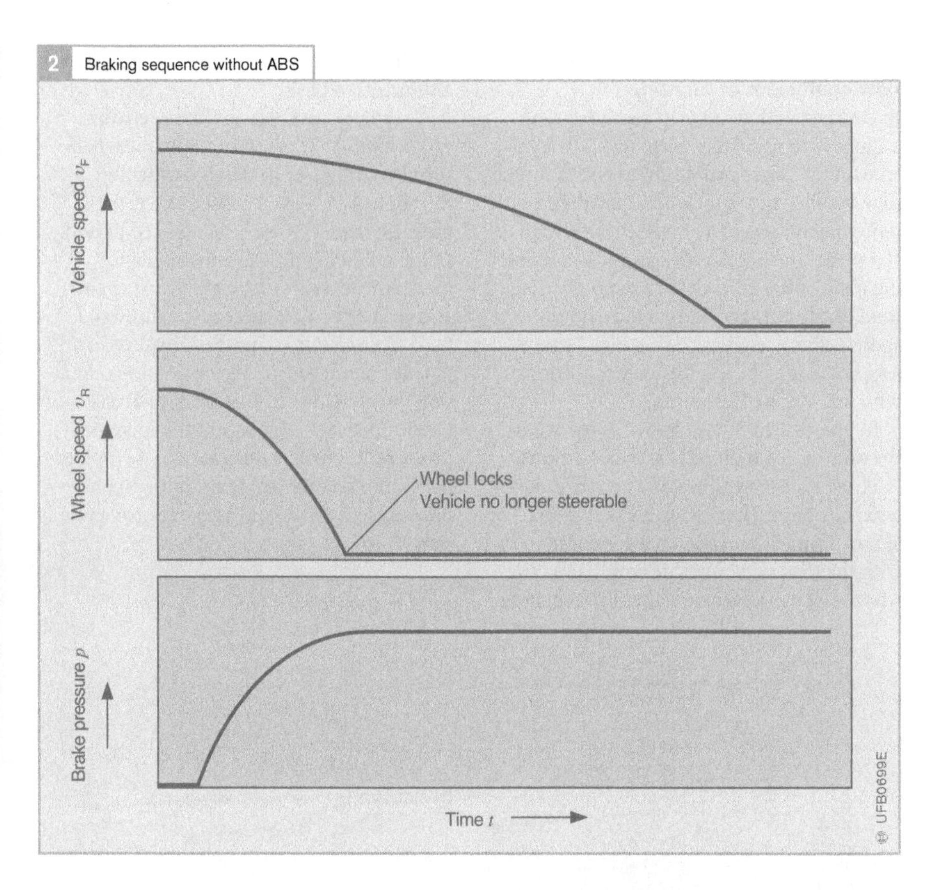

the wheel deceleration is above the threshold (–a).

At the end of phase 3, the deceleration rate falls below the threshold (–a) again and a pressure-maintenance phase of a certain length follows. During that period, the wheel acceleration has increased so much that the threshold (+a) is passed. The pressure continues to be maintained at a constant level.

At the end of phase 4, the wheel acceleration exceeds the relatively high threshold level (+A). The brake pressure then increases for as long as the acceleration remains above the threshold (+A).

In phase 6, the brake pressure is once again held constant because the acceleration is above the threshold (+a). At the end of this phase, the wheel acceleration falls below the threshold (+a). This is an indication that the wheel has returned to the stable zone of the curve for friction coefficient versus brake slip and is now slightly underbraked.

The brake pressure is now increased in stages (phase 7) until the wheel deceleration passes the threshold (–a) (end of phase 7). This time, the brake pressure is reduced immediately without a λ_1 signal being generated.

By comparison, Fig. 2 shows the progressions for an emergency braking sequence without ABS.

Control cycle on slippery surfaces (low coefficient of friction)

In contrast with good grip conditions, on a slippery road surface even very light pressure on the brake pedal is frequently enough to make the wheels lock up. They then require much longer to emerge from a high-slip phase and accelerate again. The processing logic of the ECU detects the prevailing road conditions and adapts the ABS response characteristics to suit. Fig. 3 shows a typical control cycle for road conditions with low levels of adhesion.

In phases 1 to 3, the control sequence is the same as for high-adhesion conditions.

Phase 4 starts with a short pressure-maintenance phase. Then, within a very short space of time, the wheel speed is compared with the slip switching threshold λ_1. As the wheel speed is lower than the slip switching threshold, the brake pressure is reduced over a short, fixed period.

A further short pressure-maintenance phase follows. Then, once again, the wheel speed is compared to the slip switching threshold λ_1 and, as a consequence, the pressure reduced over a short, fixed period. In the following pressure-maintenance phase, the wheel starts to accelerate again and its acceleration exceeds the threshold ($+a$). This results in another pressure-maintenance phase which lasts until the acceleration drops below the threshold ($+a$) again (end of phase 5). In phase 6, the incremental pressure-increase pattern seen in the preceding section takes place again until, in phase 7, pressure is released and a new control cycle starts.

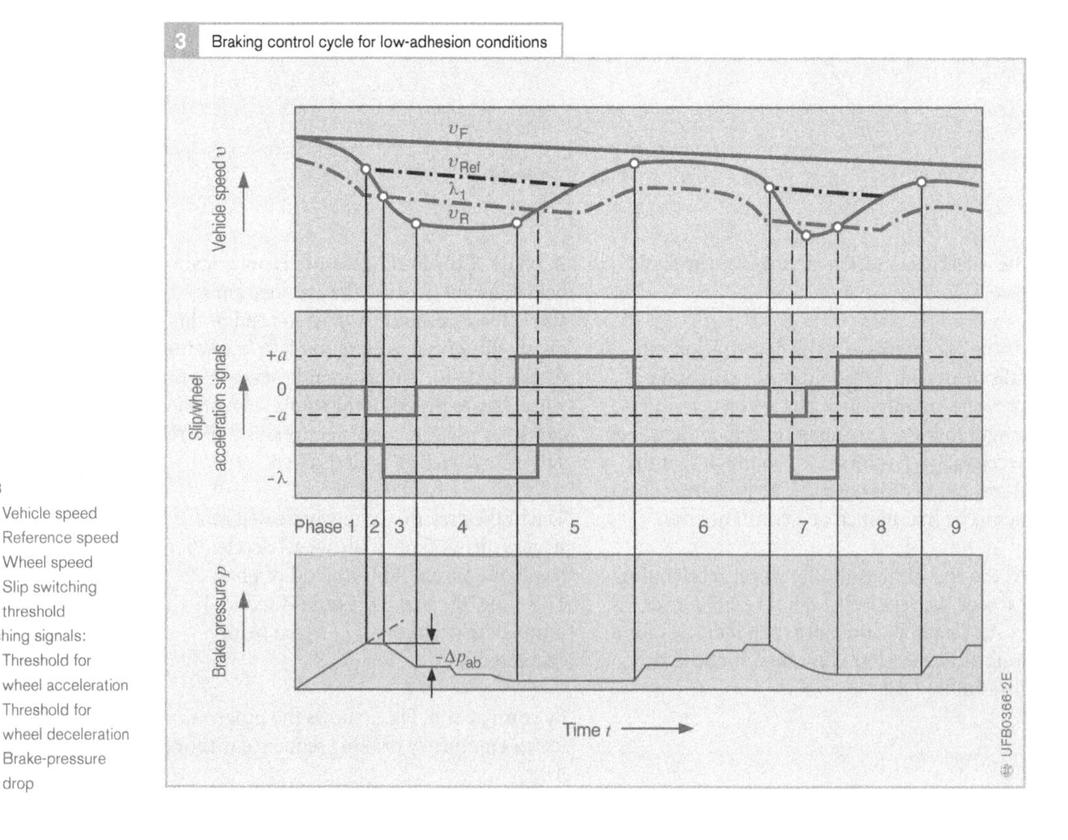

Fig. 3
v_F Vehicle speed
v_{Ref} Reference speed
v_R Wheel speed
λ_1 Slip switching threshold
Switching signals:
$+a$ Threshold for wheel acceleration
$-a$ Threshold for wheel deceleration
$-p_{ab}$ Brake-pressure drop

In the cycle described above, the control logic detected that following pressure release – triggered by the signal $(-a)$ – two more pressure-reduction stages were necessary to induce the wheel to accelerate again. The wheel remains in the higher-slip zone for a relatively long period, which is not ideal for handling stability and steerability.

In order to improve those two characteristics, this next control cycle and those that follow incorporate continual comparison of wheel speed with the slip switching threshold λ_1. As a consequence, the brake pressure is continuously reduced in phase 6 until, in phase 7, the wheel acceleration exceeds the threshold $(+a)$. Because of that continuous pressure release, the wheel retains a high level of slip for only a short period so that vehicle handling and steerability are improved in comparison with the first control cycle.

4 Yaw-moment build-up induced by areas of widely differing adhesion

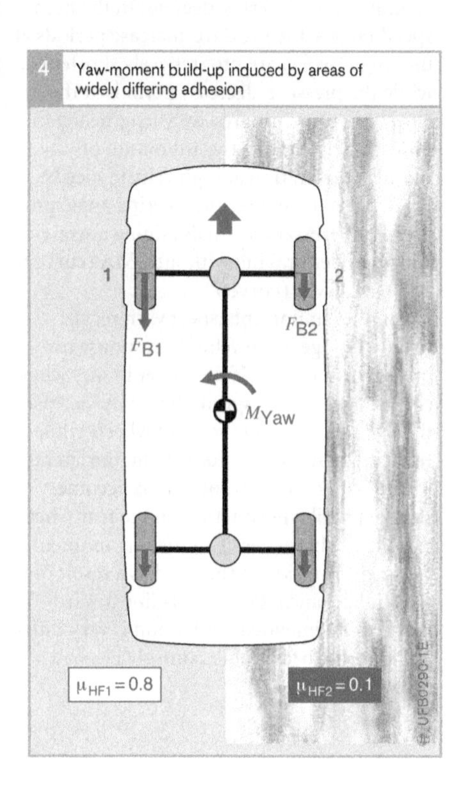

Control cycle with yaw-moment buildup delay

When the brakes are applied in situations where the grip conditions differ significantly between individual wheels ("µ-split" conditions) – for example, if the wheels on one side of the car are on dry tarmac while those on the other side are on ice – vastly different braking forces will be produced at the front wheels (Fig. 4). That difference in braking force creates a turning force (yaw moment) around the vehicle's vertical axis. It also generates steering feedback effects of varying types dependent on the vehicle's kingpin offset. With a positive kingpin offset, corrective steering is made more difficult, while a negative kingpin offset has a stabilizing effect.

Heavy cars tend to have a relatively long wheelbase and a high level of inertia around the vertical axis. With vehicles of this type, the yaw effect develops slowly enough for the driver to react and take corrective steering action during ABS braking. Smaller cars with short wheelbases and lower levels of inertia, on the other hand, require an ABS system supplemented by a yaw-moment buildup delay (GMA system) to make them equally controllable under emergency braking in conditions where there are wide differences in grip between individual wheels. Development of the yaw moment can be inhibited by delayed pressure increase in the brake on the front wheel that is on the part of the road offering the higher level of adhesion (the "high" wheel).

Fig. 5 (overleaf) illustrates the principle of the yaw-moment buildup delay:
Curve 1 shows the brake pressure, p, in the master cylinder. Without yaw-moment buildup delay, the wheel on the tarmac quickly reaches the pressure p_{high} (curve 2) and the wheel on the ice, the pressure p_{low} (curve 5). Each wheel is braked with the specific maximum possible deceleration (individually controlled).

Fig. 4
M_{yaw} Yaw moment
F_B Braking force
1 "High" wheel
2 "Low" wheel

GMA 1 system

On vehicles with less extreme handling characteristics, the GMA 1 system is used. With this system, during the initial phases of braking (curve 3), as soon as the pressure is reduced for the first time at the "low" wheel because this shows a tendency to lock up, the brake pressure at the "high" wheel is increased in stages. Once the brake pressure at the "high" wheel reaches its lock-up point, it is no longer affected by the signals from the "low" wheel, and is individually controlled so that it is able to utilize the maximum possible braking force. This method gives the type of vehicle referred to satisfactory steering characteristics under emergency braking on surfaces offering unequal grip to individual wheels. As the maximum brake pressure at the "high" wheel is reached within a relatively short time (750 ms), the braking distance is only marginally longer than for vehicles without a yaw-moment buildup delay facility.

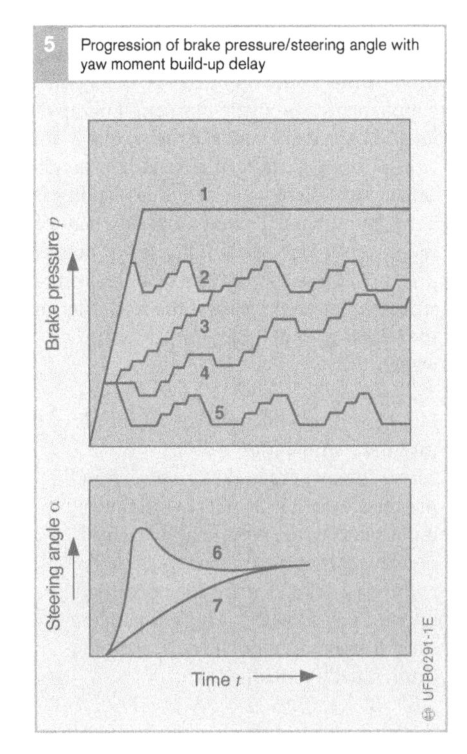

Fig. 5
1 Pressure, p_{Hz}, in the master cylinder
2 Brake pressure, p_{high}, without GMA
3 Brake pressure, p_{high}, with GMA 1
4 Brake pressure, p_{high}, with GMA 2
5 Brake pressure, p_{low}
6 Steering angle, α, without GMA
7 Steering angle, α, with GMA

GMA 2 system

The GMA 2 system is used on vehicles with more extreme handling characteristics. With this system, as soon as the brake pressure at the "low" wheel is reduced, the ABS solenoid valves for the "high" wheel are directed to maintain and then reduce the pressure for a specific period (Fig. 5, curve 4). Renewed pressure increase at the "low" wheel then triggers incremental increase of pressure at the "high" wheel, though with pressure-increase periods a certain amount longer than for the "low" wheel. This pressure metering takes place not only in the first control cycle but throughout the braking sequence.

The effect of the yaw moment on steering characteristics is all the more exaggerated the greater the speed of the vehicle when the brakes are first applied. The GMA 2 system divides the vehicle speed into four ranges. In each of those ranges, the yaw moment is inhibited to differing degrees. In the high speed ranges, the pressure-increase periods at the "high" wheel are progressively shortened, while the pressure-decrease periods at the "low" wheel are progressively lengthened in order to achieve effective inhibition of yaw-moment generation at high vehicle speeds. Fig. 5 below compares the steering angle progression necessary for maintaining a straight course under braking without GMA (curve 6) and with GMA (curve 7).

Another important aspect with regard to GMA usage is the vehicle's response to braking in a corner. If the driver brakes when cornering at a high speed, the GMA increases the dynamic load on the front wheels while reducing it at the rear. As a result, the lateral forces acting on the front wheels become stronger while those acting on the rear wheels diminish. This generates a turning moment towards the inside of the bend as a result of which the vehicle slews off its line towards the inside of its intended course and is very difficult to bring back under control (Fig. 6a).

In order to avoid this critical response to braking, the GMA also takes the lateral acceleration into account. The GMA is deactivated at high lateral acceleration rates. As a result, a high braking force is generated at the outside front wheel during the initial phase of braking in a corner and creates a turning moment towards the outside of the bend. That turning moment balances out the turning moment acting in the opposite direction that is produced by the lateral forces, so that the vehicle slightly understeers and thus remains easily controllable (Fig. 6).

The ideal method of inhibiting yaw-moment buildup involves a compromise between good steering characteristics and suitably short braking distance and is developed by Bosch individually for a specific vehicle model through consultation with the manufacturer.

6 Response to braking when cornering at critical speeds with/without GMA

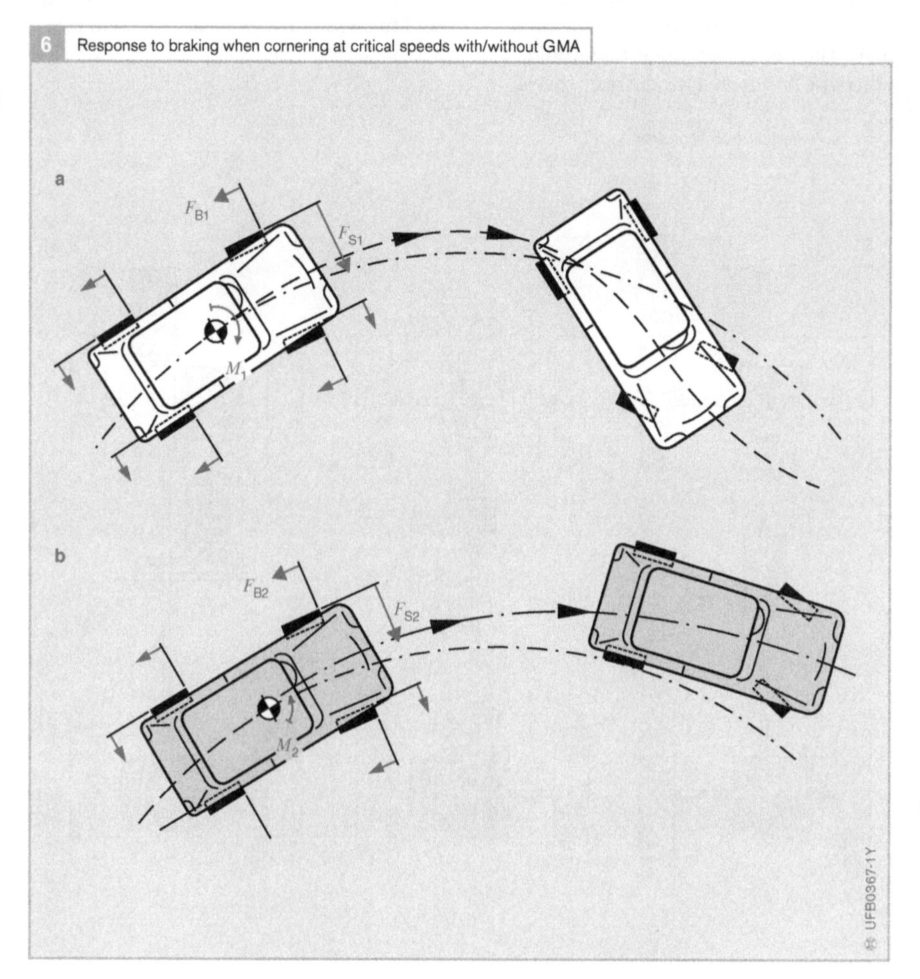

Fig. 6
a GMA activated (no individual control): vehicle oversteers
b GMA deactivated (individual control): vehicle slightly understeers
F_B Braking force
F_S Lateral force
M Torque

Control cycle for four-wheel-drive vehicles

The most important criteria for assessing the various types of four-wheel-drive configuration (Fig. 7) are traction, dynamic handling, and braking characteristics. As soon as differential locks are engaged, conditions are created that demand a different response from the ABS system.

When a rear-axle differential is locked, the rear wheels are rigidly interconnected, i.e. they always rotate at the same speed and respond to the braking forces (at each wheel) and friction levels (between each tire and the road surface) as if they were a single rigid body. The "select low" mode that would otherwise be adopted for the rear wheels (whereby the wheel with the lower degree of adhesion, μ_{HF}, determines the brake pressure for both rear wheels) is thus canceled, and both rear wheels utilize the maximum braking force. As soon as the inter-axle lock is engaged, the system forces the front wheels to assume the same average speed as the rear wheels. All four wheels are then dynamically interlinked and the engine drag (engine braking effect when backing off the throttle) and inertia act on all four wheels.

In order to ensure optimum ABS effectiveness under those conditions, additional features have to be incorporated according to the type of four-wheel-drive system (Fig. 7) in use.

7 Four-wheel-drive configurations

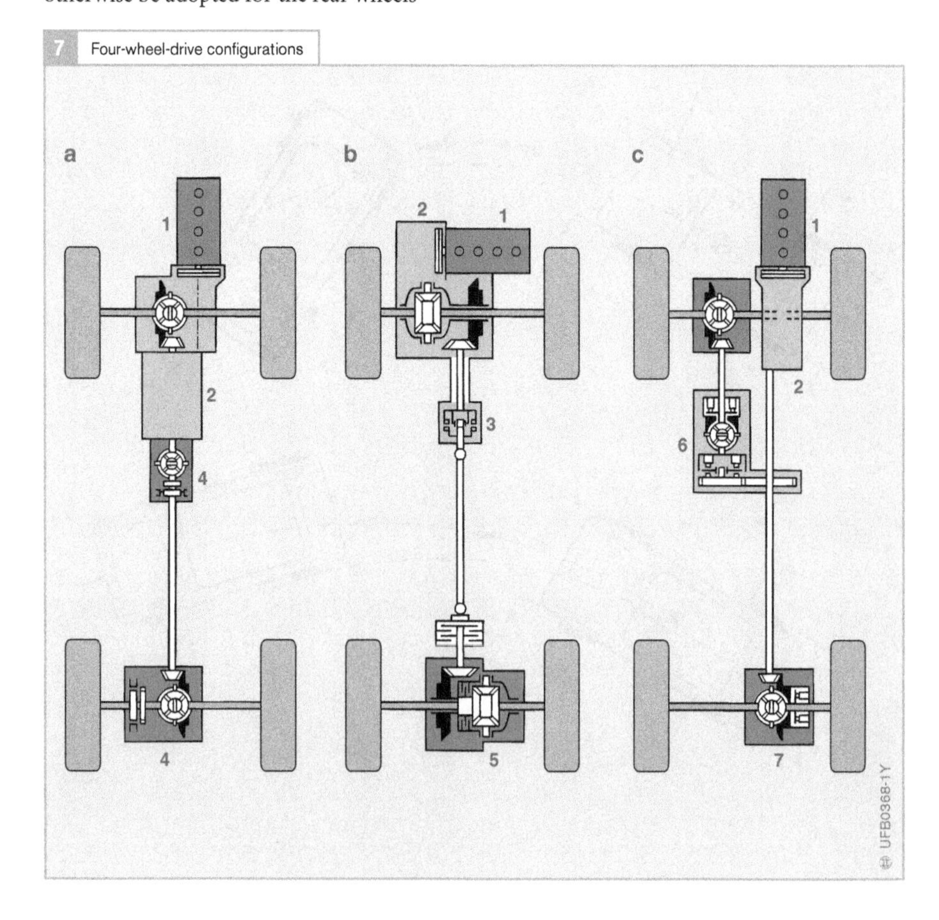

Fig. 7
a Four-wheel-drive system 1
b Four-wheel-drive system 2
c Four-wheel-drive system 3

1 Engine
2 Transmission
3 Freewheel and viscous clutch

Differential with
4 Manual lock or viscous lock
5 Proportional lock
6 Automatic clutch and lock
7 Automatic lock

Four-wheel-drive system 1

On four-wheel-drive system 1 with manual locks or permanently active viscous locks in the propeller shaft and the rear-axle differential, the rear wheels are rigidly interconnected and the average speed of the front wheels is the same as that of the rear wheels. As already mentioned, the rear differential lock results in deactivation of the "select low" mode for the rear wheels and utilization of the maximum braking force at each rear wheel. When braking on road surfaces with unequal levels of grip at the two rear wheels, this can generate a yaw moment with a potentially critical effect on vehicle handling stability. If the maximum braking force difference were also applied very quickly at the front wheels, it would not be possible to keep the vehicle on a stable course.

This type of four-wheel-drive arrangement therefore requires a GMA system for the front wheels in order to ensure handling stability and steerability are maintained in road conditions where there is a significant divergence between the levels of grip at the left and right wheels. In order to maintain ABS effectiveness on slippery surfaces, the engine drag effect – which of course acts on all four wheels on a four-wheel-drive vehicle – has to be reduced. This is done by using an engine drag-torque control system that applies just enough throttle to counteract the unwanted engine braking effect.

Another factor that demands refinement of the ABS control cycle in order to prevent wheel lock-up is the reduced sensitivity of the wheels to changes in road-surface adhesion on slippery surfaces that is caused by the effect of the engine inertia. The fact that all four wheels are dynamically linked to the engine's inertial mass therefore requires additional analysis and processing operations on the part of the ABS control unit. The vehicle longitudinal deceleration is calculated to detect smooth or slippery road surfaces where μ_{HF} is less than 0.3.

When analyzing braking on such surfaces, the response threshold ($-a$) for wheel deceleration is halved and the diminishing rise in the reference speed is limited to specific, relatively low levels. As a result, imminent wheel lock-up can be detected early and "sensitively".

On four-wheel-drive vehicles, heavy application of the throttle on slippery road surfaces can cause all four wheels to spin. In such situations, special signal-processing methods ensure that the reference speed can only increase in response to the spinning wheels within the limits of the maximum possible vehicle acceleration. In a subsequent braking sequence, the initial ABS pressure reduction is triggered by a signal ($-a$) and a specific, minimal difference in wheel speed.

Four-wheel-drive system 2

Because of the possibility of all four wheels spinning with four-wheel-drive system 2 (viscous lock with freewheel in the propeller shaft, proportional rear-axle differential lock), the same special procedures must be adopted for signal processing.

Other modifications for ensuring ABS effectiveness are not necessary as the freewheel disengages the wheels under braking. However, the system can be improved by the use of an engine drag-torque control system.

Four-wheel-drive system 3

As with the first two systems, four-wheel-drive system 3 (automatic differential/inter-axle locks) requires adoption of the signal processing procedures described above in the event of wheel-spin at all four wheels. In addition, automatic release of the differential locks whenever the brakes are applied is necessary. Other modifications for ensuring ABS effectiveness are not necessary.

Wheel-speed sensors

Application

Wheel-speed sensors are used to measure the rotational speed of the vehicle wheels (wheel speed). The speed signals are transmitted via cables to the ABS, TCS or ESP control unit of the vehicle which controls the braking force individually at each wheel. This control loop prevents the wheels from locking up (with ABS) or from spinning (with TCS or ESP) so that the vehicle's stability and steerability are maintained.

Navigation systems also use the wheel speed signals to calculate the distance traveled (e. g. in tunnels or if satellite signals are unavailable).

Design and method of operation

The signals for the wheel-speed sensor are generated by a steel pulse generator that is fixed to the wheel hub (for passive sensors) or by a multipole magnetic pulse generator (for active sensors). This pulse generator has the same rotational speed as the wheel and moves past the sensitive area of the sensor head without touching it. The sensor "reads" without direct contact via an air gap of up to 2 mm (Fig. 2).

The air gap (with strict tolerances) ensures interference-free signal acquisition. Possible interference caused for instance by oscillation patterns in the vicinity of the brakes, vibrations, temperature, moisture, installation conditions at the wheel, etc. is therefore eliminated.

Since 1998 active wheel-speed sensors have been used almost exclusively with new developments instead of passive (inductive) wheel-speed sensors.

Passive (inductive) wheel-speed sensors

A passive (inductive) speed sensor consists of a permanent magnet (Fig. 2, 1) with a soft-magnetic pole pin (3) connected to it, which is inserted into a coil (2) with several thousand windings. This setup generates a constant magnetic field.

The pole pin is installed directly above the pulse wheel (4), a gear wheel attached to the wheel hub. As the pulse wheel turns, the continuously alternating sequence of teeth and gaps induces corresponding fluctuations in the constant magnetic field. This changes the magnetic flux through the pole pin and therefore also through the coil winding. These fluctuations induce an alternating current in the coil suitable for monitoring at the ends of its winding.

The frequency and amplitude of this alternating current are proportional to wheel speed (Fig. 3) and when the wheel is not rotating, the induced voltage is zero.

Tooth shape, air gap, rate of voltage rise, and the ECU input sensitivity define the smallest still measurable vehicle speed and thus, for ABS applications, the minimum response sensitivity and switching speed.

1 Passive (inductive) wheel-speed sensors

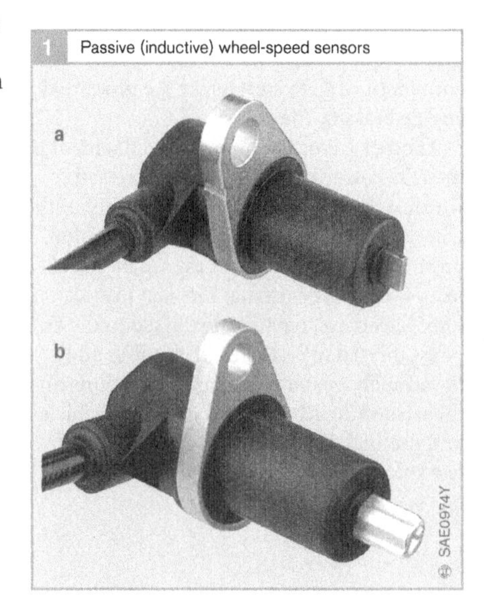

Fig. 1
a Chisel-type pole pin (flat pole pin)
b Rhombus-type pole pin (lozenge-shaped pole pin)

2 Figure illustrating the principle of the passive wheel-speed sensor

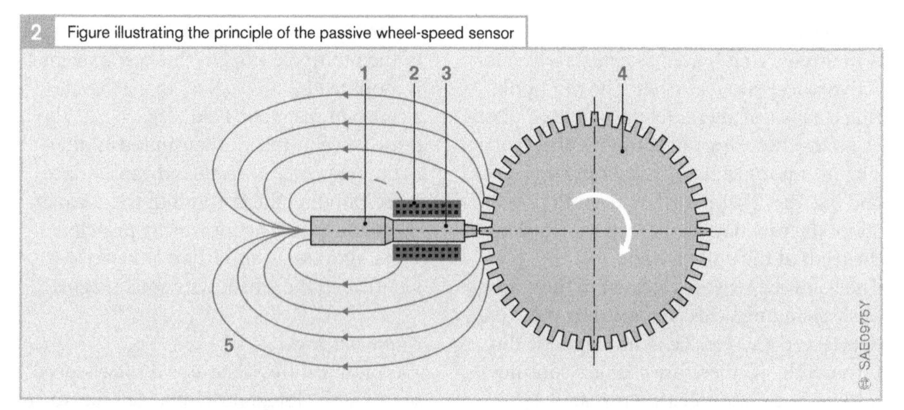

Fig. 2
1 Permanent magnet
2 Solenoid coil
3 Pole pin
4 Steel pulse wheel
5 Magnetic field lines

Various pole-pin configurations and installation options are available to adapt the system to the different installation conditions encountered with various wheels. The most common variants are the chisel-type pole pin (Fig. 1a, also called a flat pole pin) and the rhombus-type pole pin (Fig. 1b, also called a lozenge-shaped pole pin). Both pole-pin designs necessitate precise alignment to the pulse wheel during installation.

Active wheel-speed sensors

Sensor elements

Active wheel-speed sensors are used almost exclusively in today's modern brake systems (Fig. 4). These sensors usually consist of a hermetic, plastic-cast silicon IC that sits in the sensor head.

In addition to magnetoresistive ICs (the electrical resistance changes as the magnetic field changes) Bosch now uses Hall sensor elements almost exclusively. These sensors react to the smallest changes in the magnetic field and therefore allow greater air gaps compared to passive wheel-speed sensors.

3 Signal output voltage of passive wheel-speed sensor

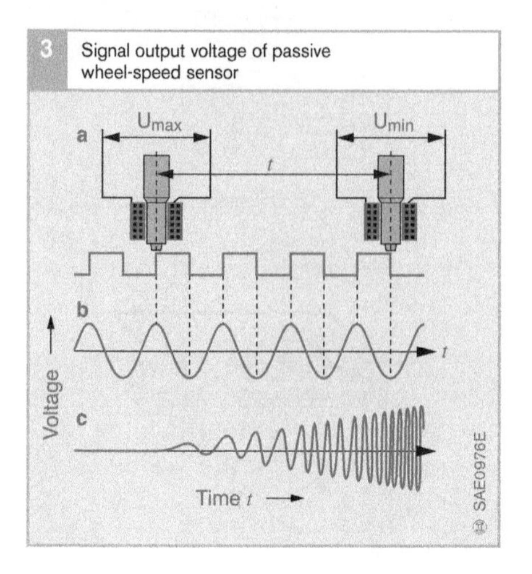

4 Active wheel-speed sensor

Fig. 3
a Passive wheel-speed sensor with pulse wheel
b Sensor signal at constant wheel speed
c Sensor signal at increasing wheel speed

Pulse wheels
A multipole ring is used as a pulse wheel for active wheel-speed sensors. The multipole ring consists of alternately magnetized plastic elements that are arranged in the shape of a ring on a nonmagnetic metal carrier (Fig. 6 and Fig. 7a). These north and south poles adopt the function formerly performed by the teeth of the pulse wheel.
The IC of the sensor is located in the continuously changing fields generated by these magnets (Fig. 6 and Fig. 7a). The magnetic flux through the IC therefore changes continuously as the multipole ring turns.

A steel pulse wheel can also be used instead of the multipole ring. In this case a magnet is mounted on the Hall IC that generates a constant magnetic field (Fig. 7b). As the pulse wheel turns, the continuously alternating sequence of teeth and gaps induces corresponding fluctuations in the constant magnetic field. The measuring principle, signal processing and IC are otherwise identical to the sensor without a magnet.

Characteristics
A typical feature of the active wheel-speed sensor is the integration of a Hall measuring element, signal amplifier and signal conditioning in an IC (Fig. 8). The wheel-speed data is transferred as an impressed current in the form of square-wave pulses (Fig. 9). The frequency of the pulses is proportional to the wheel speed and the speed can be detected until the wheel is practically stationary (0.1 km/h).

The supply voltage is between 4.5 and 20 volts. The square-wave output signal level is 7 mA (low) and 14 mA (high).

5 Explosion diagram with multipole pulse generator

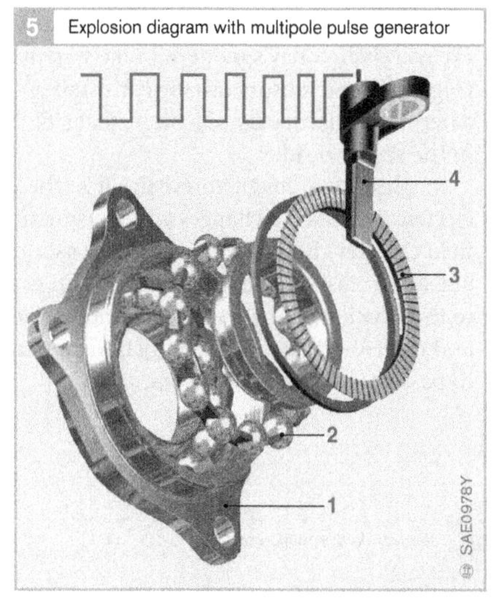

Fig. 5
1 Wheel hub
2 Roller bearing
3 Multipole ring
4 Wheel-speed sensor

6 Sectional drawing of active wheel-speed sensor

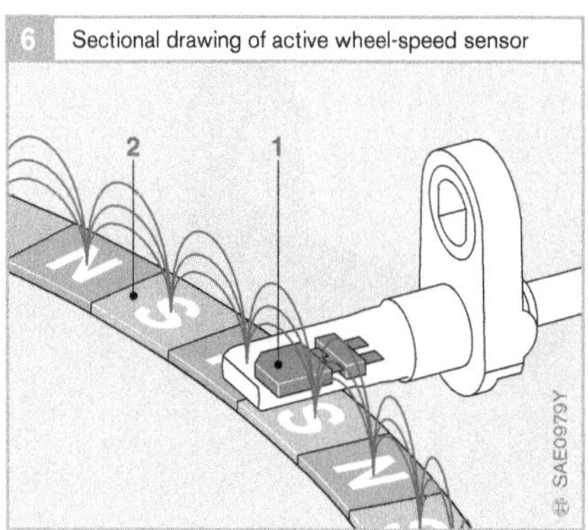

Fig. 6
1 Sensor element
2 Multipole ring with alternating north and south magnetization

7 Figure illustrating principle for measuring wheel speed

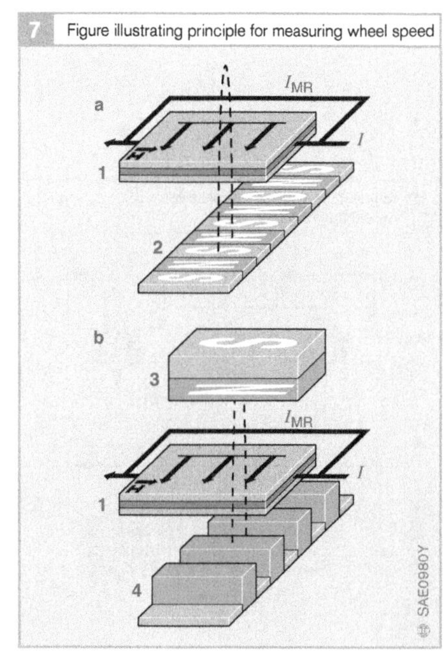

Fig. 7
a Hall IC with multipole pulse generator
b Hall IC with steel pulse generator and magnet in sensor

1 Sensor element
2 Multipole ring
3 Magnet
4 Steel pulse wheel

This type of data-transmission using digital signals is less sensitive to interference than the signals from passive inductive sensors. The sensor is connected to the ECU by a two-conductor wire.

Compact dimensions combine with low weight to make the active wheel-speed sensor suitable for installation on and even within the vehicle's wheel-bearing assemblies (Fig. 10). Various standard sensor head shapes are suitable for this.

Digital signal conditioning makes it possible to transfer coded additional information using a pulse-width-modulated output signal (Fig. 11):

- Direction of wheel rotation recognition: This is especially significant for the "hill hold control" feature, which relies on selective braking to prevent the vehicle from rolling backwards when starting off on a hill. The direction of rotation recognition is also used in vehicle navigation systems.
- Standstill recognition: This information can also be evaluated by the "hill hold control" function. The information is also used for self-diagnosis.
- Signal quality of the sensor: Information about the signal quality of the sensor can be relayed in the signal. If a fault occurs the driver can be advised that service is required.

8 Block diagram of Hall IC

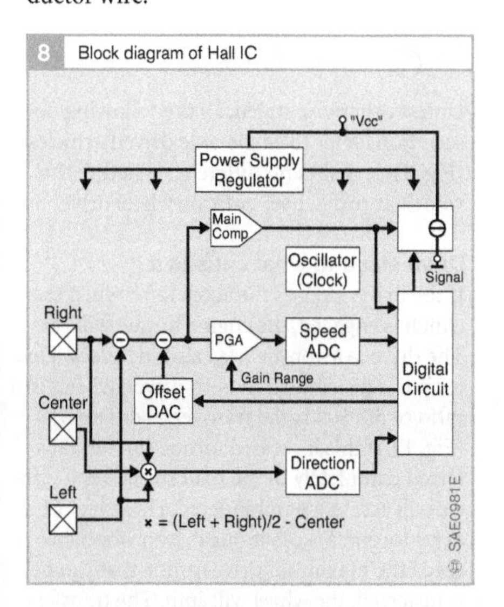

9 Signal conversion in Hall IC

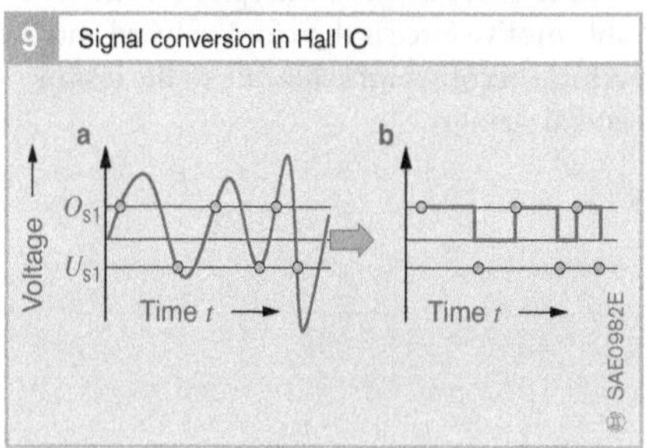

10 Wheel bearing with wheel-speed sensor

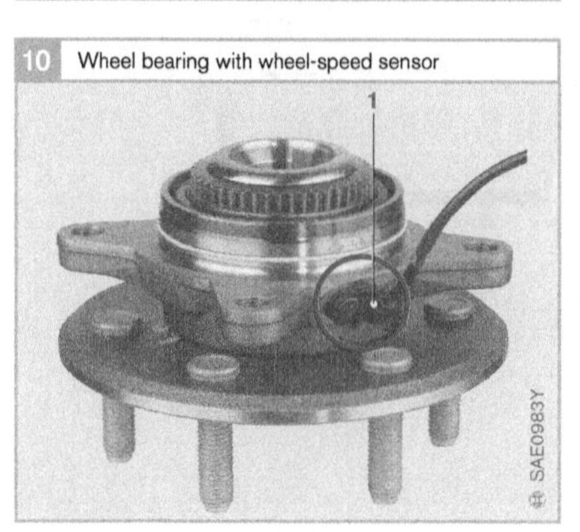

11 Coded information transfer with pulse-width-modulated signals

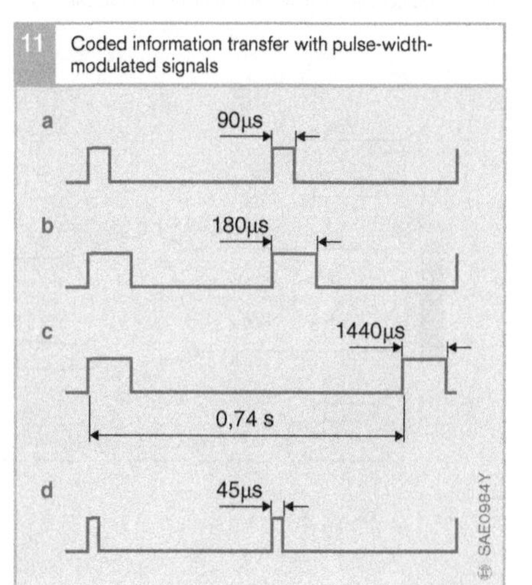

Fig. 9
a Raw signal
b Output signal

O_{S1} Upper switching threshold
U_{S1} Lower switching threshold

Fig. 10
1 Wheel-speed sensor

Fig. 11
a Speed signal when reversing
b Speed signal when driving forwards
c Signal when vehicle is stationary
d Signal quality of sensor, self-diagnosis

Traction control system (TCS)

Critical driving situations can occur not only while braking, but also whenever strong longitudinal forces should be transferred at the contact area between the tire and the ground. This is because the transferable lateral forces are reduced by this process. Critical situations can also occur when starting off and accelerating, particularly on a slippery road surface, on hills, and when cornering. These kinds of situations can overtax the driver not only causing him/her to react incorrectly but also causing the vehicle to become unstable. The traction control system (TCS) solves these problems, providing the vehicle remains within the physical limits.

Tasks

The antilock braking system (ABS) prevents the wheels from locking up when the brakes are applied by lowering the wheel brake pressures. The traction control system (TCS) prevents the wheels from spinning by reducing the drive torque at each driven wheel. TCS therefore provides a logical extension of ABS during acceleration.

In addition to this safety-relevant task of ensuring the stability and steerability of the vehicle when accelerating, TCS also improves the traction of the vehicle by regulating the optimum slip (see μ-slip curve in "Basic principles of vehicle dynamics"). The upper limit here is, of course, set by the traction requirement stipulated by the driver.

Function description

Unless otherwise stated, all the following descriptions refer to single-axle driven vehicles (Fig. 1). It makes no difference whether the vehicle is rear-wheel or front-wheel drive.

Drive slip and what causes it

If the driver presses the accelerator when the clutch is engaged, the engine torque will rise. The drive axle torque M_{Kar} also increases. This torque is distributed to both driven wheels in a ratio of 50 : 50 via the transversal differential (Fig. 1). If this increased torque can be transferred completely to the road surface, the vehicle will accelerate unhindered. However, if the drive torque $M_{Kar}/2$ at one driven wheel exceeds the maximum drive torque that can be transferred, the wheel will spin. The transferable motive force is therefore reduced and the vehicle becomes unstable due to the loss of lateral stability.

Fig. 1
1 Engine with transmission
2 Wheel
3 Wheel brake
4 Transversal differential
5 Control unit with TCS functionality

Engine, transmission, gear ratio of differential and losses are combined in one unit

M_{Kar} Drive axle torque
v_{Kar} Drive axle speed
M_{Br} Braking torque
M_{Str} Torque transferred to the road
v Wheel speed
V Front
H Rear
R Right
L Left

1 Drive concept of a single-axle driven vehicle with TCS

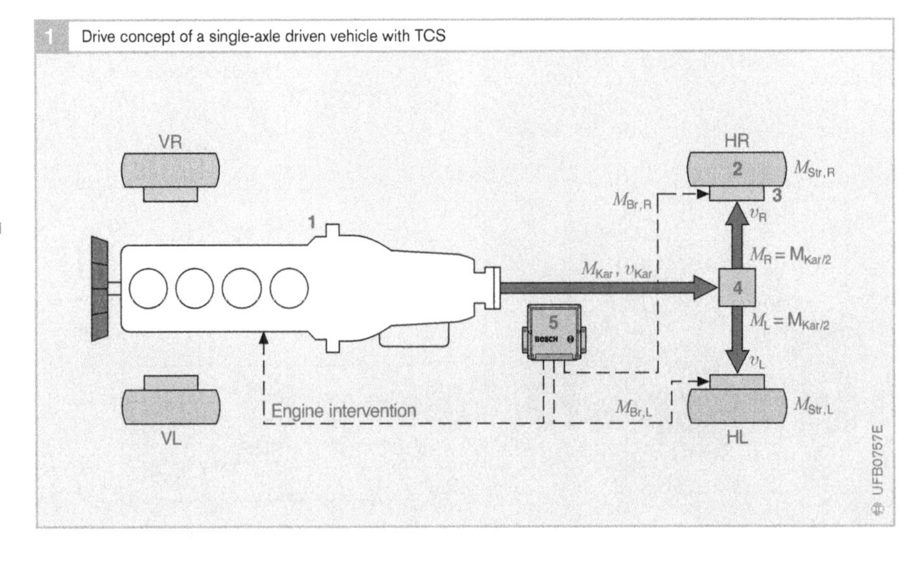

The TCS regulates the slip of the driven wheels as quickly as possible to the optimum level. To do this the system first determines a reference value for the slip. This value depends on a number of factors which are intended to represent the current driving situation as closely as possible. These factors include:

- the basic characteristic for TCS reference slip (based on the slip requirement of a tire during acceleration),
- effective coefficient of friction,
- external tractive resistance (deep snow, rough road, etc.),
- yaw velocity, lateral acceleration, and steering angle of the vehicle.

TCS interventions

The measured wheel speeds and the respective drive slip can be influenced by changing the torque balance M_{Ges} at each driven wheel. The torque balance M_{Ges} at each wheel results from the drive torque $M_{Kar}/2$ at this wheel, the respective braking torque M_{Br} and the road torque M_{Str} (Fig. 1).

$$M_{Ges} = M_{Kar}/2 + M_{Br} + M_{Str}$$

(M_{Br} and M_{Str} are negative here.)

This balance can obviously by influenced by the drive torque M_{Kar} provided by the engine as well as by the braking torque M_{Br}. Both these parameters are therefore correcting variables of the TCS which can be used to regulate the slip at each wheel to the reference slip level.

In gasoline-engine vehicles, the drive torque M_{Kar} can be controlled using the following engine interventions:

- Throttle valve (throttle valve adjustment),
- Ignition system (ignition-timing advance),
- Fuel-injection system (phasing out individual injection pulses).

The last two interventions are rapid interventions, the first a slower means of intervention (Fig. 2). The availability of these interventions depends on the vehicle manufacturer and engine version.

In diesel-engine vehicles, the drive torque M_{Kar} is influenced by the electronic diesel control system (EDC) (reduction of the quantity of fuel injected).

The braking torque M_{Br} can be regulated for each wheel via the braking system. The TCS function requires the original ABS hydraulic system to be expanded because of the need for active pressure build-up (see also "Hydraulic modulator").

Fig. 2 compares the response times with various TCS interventions. The figure shows that exclusive drive torque regulation by means of the throttle valve can be unsatisfactory due to the relatively long response time.

2 Comparison of response times with various TCS interventions

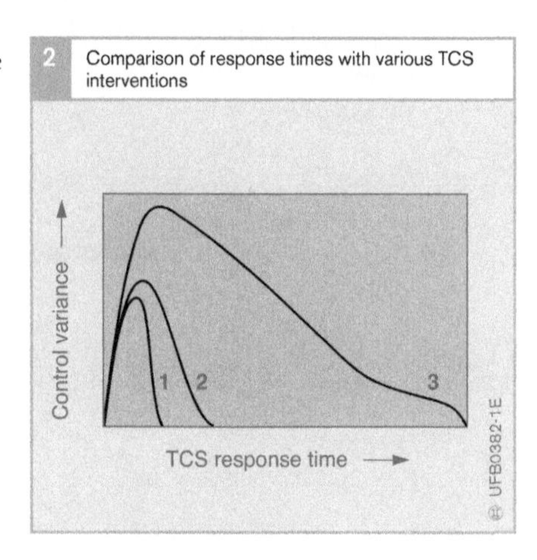

Fig. 2
1 Throttle-valve/wheel brake intervention
2 Throttle-valve intervention/ ignition adjustment
3 Throttle-valve intervention

Structure of traction control system (TCS)

The expanded ABS hydraulic system allows both symmetrical brake application (i.e. brake application at both driven wheels) as well as individual brake application. This is the key to further structuring of the TCS, i.e. structuring according to controlled variable rather than according to the actuator (engine/brake).

Drive axle speed controller

The drive axle speed v_{Kar} or the drive torque M_{Kar} can be influenced by means of engine interventions. Symmetric brake applications also influence the drive axle speed v_{Kar} and effect torque balance between the individual wheels in the same way as reducing the drive torque M_{Kar}. The *drive axle speed controller* is used to regulate the drive axle speed in this way.

Transversal differential lock controller

Asymmetric brake application (brake application at just one driven wheel) is used primarily to regulate the differential speed at the driven axle $v_{Dif} = v_L - v_R$. This task is carried out by the *differential speed controller*. Asymmetric brake application at just one driven wheel is only noticeable at first in the torque balance of this wheel. The brake application has basically the same effect as an asymmetric distribution ratio of the transversal differential (but applied to a drive torque M_{Kar} that is reduced by the asymmetric braking torque). The differential speed controller is also referred to as the *transversal differential lock controller* because it can be used to influence to a certain extent the distribution ratio of the transversal differential, i.e. to mimic the effect of a differential lock.

Together the drive axle speed controller and transversal differential lock controller form the TCS system (Fig. 3). The drive axle speed controller uses the drive axle speed v_{Kar} to regulate the drive torque M_{Kar} provided by the engine. The transversal differential lock controller functions primarily like a controller that uses the differential speed v_{Dif} to regulate the distribution ratio M_L to M_R of the transversal differential and therefore the distribution of the drive torque M_{Kar} to the driven wheels.

3 TCS controller concept for a single-axle-driven vehicle (rear-wheel drive)

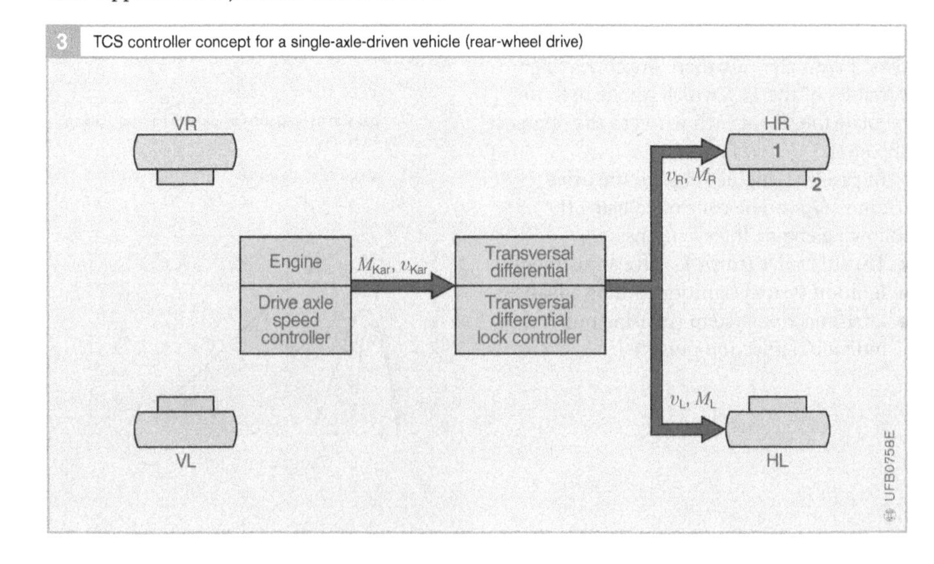

Fig. 3
1 Wheel
2 Wheel brake

v_R, v_L Wheel speeds
v_{Kar} Drive axle speed
M_{Kar} Drive axle torque
V Front
H Rear
R Right
L Left

Typical control situations

-Split: Transversal differential lock controller

Fig. 4 shows a typical situation ("μ-split") whereby the transversal differential lock controller of the TCS becomes active when the vehicle pulls away after being stationary. The left side of the vehicle is on a slippery road surface with a low coefficient of friction μ_l ("l" for low) and the right side of the vehicle is on dry asphalt with a considerably higher coefficient of friction μ_h ("h" for high).

Without brake application by the transversal differential lock controller only the drive force F_l could be transferred on both sides since the differential distributes the drive torque equally on both sides. A greater drive torque M_{Kar} causes the wheel on the side with μ_l to spin and leads to a differential speed of $v_{Dif} > 0$ (see also Fig. 5). In this case, the excess drive torque is lost as lost torque in the differential, engine, and transmission.

To prevent the wheel on the side of the vehicle with μ_l from spinning if the drive torque is too high, the braking force F_{Br} is applied (Fig. 4, see also Fig. 5). The differential can then transfer the force $F_{Br} + F_l$ to this side (or a torque corresponding to this force), whereby F_{Br} is diffused by the brake action. The drive force F_l remains as before. On the side of the vehicle with μ_h the force $F_{Br} + F_l$ is also transferred (characteristic of the differential). Since the brakes are not applied on this side, the entire force can be used as drive force $F_{Br}^* + F_l$ (F_{Br}^* results from F_{Br} taking the different effective radii into account). Overall the drive force transferred is increased by F_{Br}^* (the drive torque M_{Kar} must of course also be increased accordingly). This ability of the transversal differential lock controller to increase the traction is part of the traction control system (TCS).

The drive torque can be regulated to a maximum possible drive force. The value of μ_h represents a physical upper limit.

When both driven wheels run synchronously again ($v_{Dif} = 0$), the single-sided braking force F_{Br} or the corresponding braking torque M_{Br} is reduced again (Fig. 5).

The exact buildup and reduction of M_{Br} depends on the internal implementation of the transversal differential lock controller (the controller functions like a PI-controller).

Low : Drive axle speed controller

If both driven wheels are on a slippery road surface with a low coefficient of friction (e.g. the vehicle is standing on ice) when the vehicle pulls away, the drive axle speed controller of the TCS becomes active.

4 Differential lock effect due to asymmetric brake application

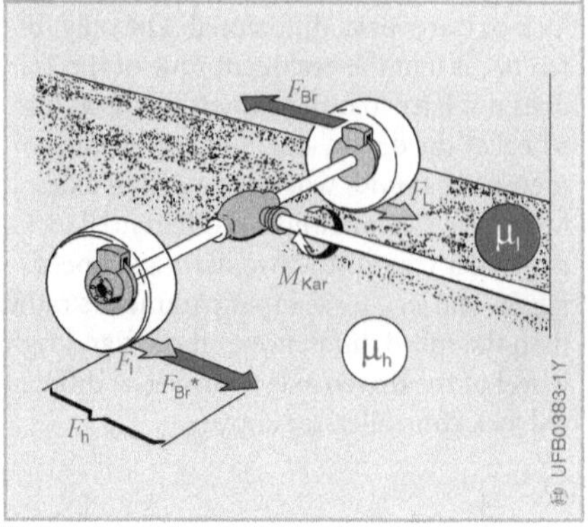

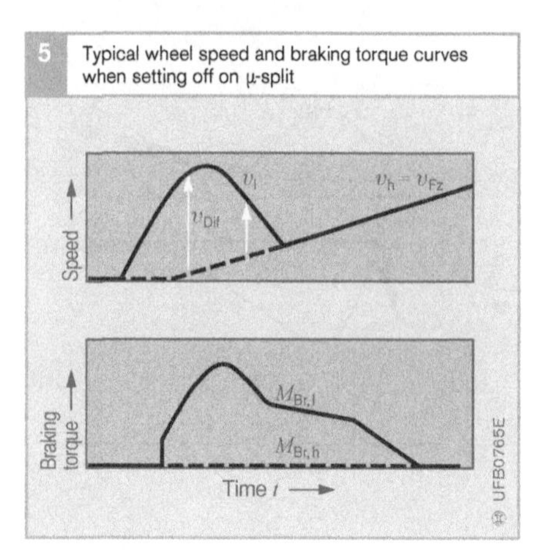

5 Typical wheel speed and braking torque curves when setting off on μ-split

Fig. 4
M_{Kar} Drive torque
F_{Br} Braking force
F_{Br}^* Braking force, based on effective radii
μ_l Low coefficient of friction
μ_h High coefficient of friction
F_l Transferable motive force on μ_l
F_h Transferable motive force on μ_h

Fig. 5
v Wheel speed
M_{Br} Braking torque
l Low-μ wheel
h High-μ wheel
v_{Fz} Vehicle speed
v_{Dif} Differential speed

If the driver increases the driver-specified torque $M_{FahVorga}$, the drive torque M_{Kar} increases almost simultaneously. Both driven wheels then spin at almost the same speed. The differential speed $v_{Dif} = v_L - v_R$ is approximately 0, while the drive axle speed $v_{Kar} = (v_L + v_R)/2 = v_L = v_R$ is considerably greater than a reasonable reference value v_{SoKar} determined by the TCS due to the spinning driven wheels. The drive axle speed controller reacts by reducing the drive torque M_{Kar} to a level below the torque specified by the driver $M_{FahVorga}$ and by initiating brief, symmetric brake application $M_{Br, Sym}$ (Fig. 6). As a result the drive axle speed v_{Kar} is reduced and thus the speed of the spinning wheels. The vehicle begins to accelerate. Since the optimum point of the μ-slip curve (see also "Basic principles of vehicle dynamics") would not be achieved without these TCS interventions, the acceleration would be slower while the wheels were spinning and considerably less lateral stability would be present.

The exact characteristic of M_{Kar} and $M_{Br, Sym}$ depends in turn on the internal implementation of the drive axle speed controller (the controller functions like a PID-controller).

6 Typical wheel speed, engine and braking torque curves when setting off on low-µ

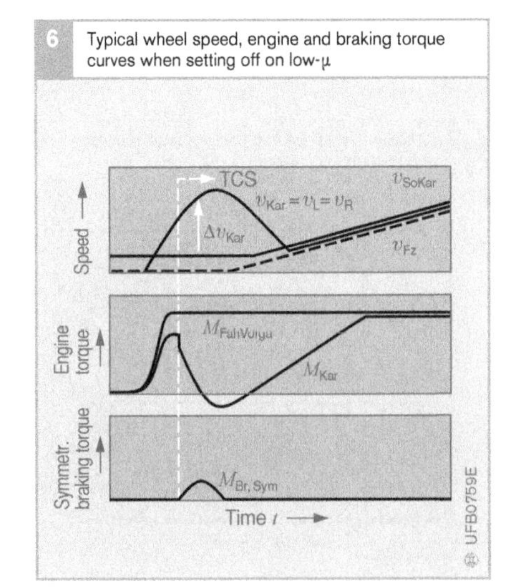

Fig. 6
v Wheel speed
v_{Fz} Vehicle speed
v_{Kar} Drive axle speed
v_{SoKar} Drive axle speed reference value
$M_{Br, Sym}$ Symmetric braking torque
$M_{FahVorga}$ Driver-specified drive torque (via accelerator pedal position)
L Left
R Right

Traction control system (TCS) for four wheel drive vehicles

In recent years four wheel drive vehicles have continued to increase in popularity. Amongst these types of vehicles, Sport Utility Vehicles, or SUVs, are the most popular of all. These are road vehicles with off-road characteristics.

If all four wheels of a vehicle are to be driven, both a second transversal differential and an additional longitudinal differential (also called a central differential) are required (Fig. 7). The first task of the longitudinal differential is to compensate for the differences between the drive axle speed of the front and rear axle $v_{Kar, VA}$ and $v_{Kar, HA}$, respectively. A rigid connection would result in tension between the front and rear axle. The second task is to achieve optimum distribution of the drive torque M_{Kar} to the two axles $M_{Kar, VA}$ and $M_{Kar, HA}$, respectively.

Less expensive SUVs are often equipped with a longitudinal differential with a preset distribution ratio. Unlike with a transversal differential, fixed distribution ratios other than 50:50 are useful – e.g. 60:40 for a design with emphasis on the rear-wheel drive of the vehicle. Brake application by the traction control system (TCS) can be used to mimic the behavior of a longitudinal differential lock.

By applying the brakes to eliminate part of $M_{Kar, VA}$, the distribution ratio $M_{Kar, HA}$ to $M_{Kar, VA}$ can be increased, or can be decreased by applying the brakes to eliminate part of $M_{Kar, HA}$. The principle is the same as previously described for the transversal differential lock or transversal differential. The only difference is that the braking torque of the TCS does not have to be asymmetric, (i.e. at one wheel of the driven axle) but can occur symmetrically at both wheels of a driven axle. Moreover, the longitudinal differential lock controller regards the two drive axle speeds $v_{Kar, VA}$ and $v_{Kar, HA}$ as input parameters rather than the speed difference of the left and right wheel of the driven axle (transversal differential lock controller, see above).

Fig. 8 shows the expansion of the TCS concept from Fig. 3 for an all-wheel drive vehicle. As with a single-axle driven vehicle, the drive axle speed controller uses the drive axle speed v_{Kar} to regulate the drive torque M_{Kar} provided by the engine. As already described, the longitudinal differential lock controller distributes this torque to the front and rear axle ($M_{Kar, VA}$ and $M_{Kar, HA}$, respectively). The transversal differential lock controller uses the differential speed $v_{Dif, XA}$ to regulate the distribution of the drive torque $M_{Kar, XA}$ per axle to the driven wheels. This must now be carried out for both the front and rear axle ("X" = "V" (front) or "X" = "H" (rear)).

Electronic differential locks designed as part of the TCS software have the advantage that they do not require additional hardware. There are therefore very cost-efficient. They are used for road vehicles, which is usually the intended application of SUVs. When used in true off-road cross-country vehicles, electronic differential locks reach their limit in tough off-road conditions, at the latest

8 TCS controller concept for a four-wheel-drive vehicle

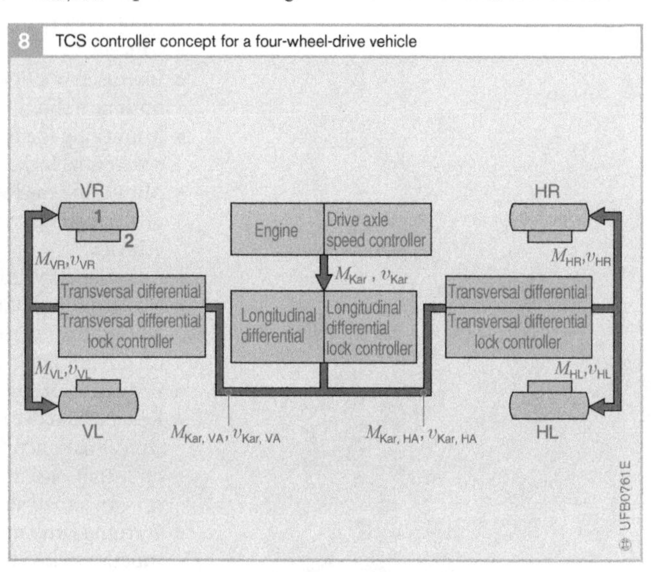

Fig. 8
1 Wheel
2 Wheel brake

v Wheel speed
v_{Kar} Drive axle speed
M_{Kar} Drive axle torque
A Axle
V Front
H Rear
R Right
L Left

7 Drive concept of a four-wheel-drive vehicle with TCS

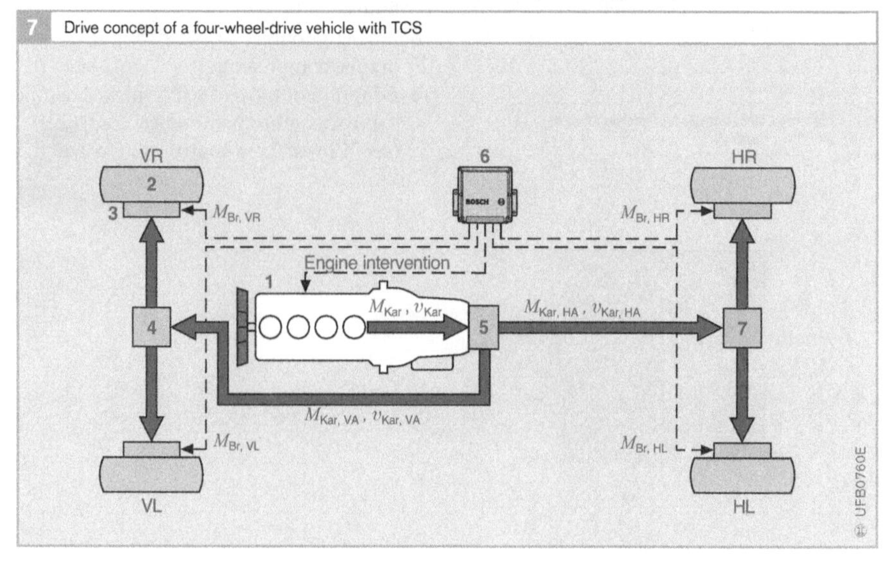

Fig. 7
1 Engine with transmission
2 Wheel
3 Wheel brake
4 Transversal differential
5 Longitudinal differential
6 Control unit with TCS functionality
7 Transversal differential

Engine, transmission, gear ratios of differentials and losses are combined into one unit
v Wheel speed
v_{Kar} Drive axle speed
M_{Kar} Drive axle torque
M_{Br} Braking torque
R Right
L Left
V Front
H Rear
A Axle

when the brakes overheat. Vehicles for these conditions are therefore often fitted with mechanical locks (examples can be seen in Figs. 9 and 10). The lock controllers of the TCS software are then only used as a backup system and they do not intervene during normal operation.

9 Classic solution of a differential lock

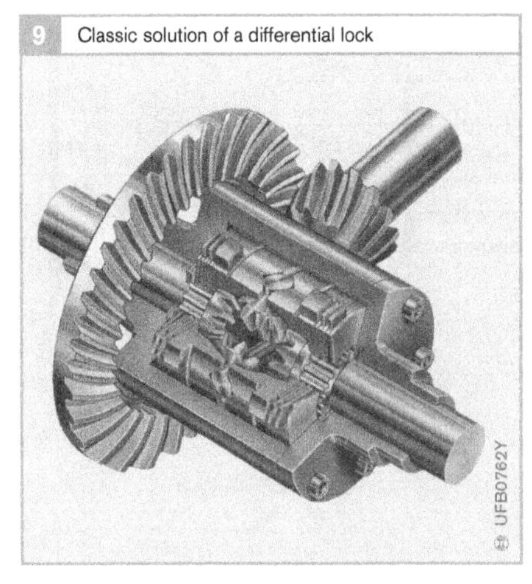

10 Electronically controllable differential lock (Haldex coupling)

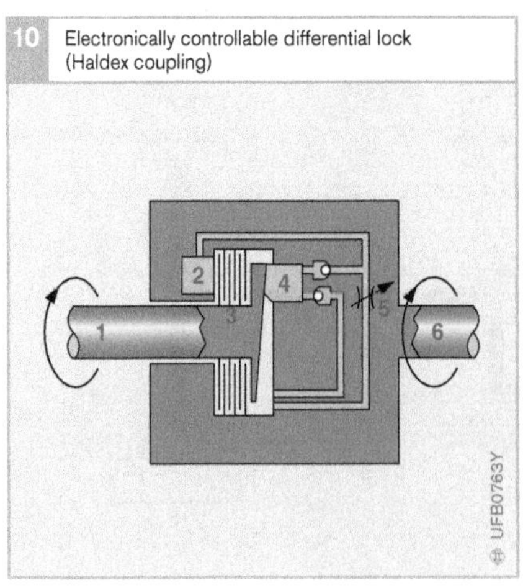

Fig. 10
1 Output shaft
2 Working piston
3 Lamella
4 Axial-piston pump
5 Control valve
6 Input shaft

Summary: Advantages of TCS

Below is a summary of the advantages of using TCS to prevent the driven wheels from spinning when starting off or accelerating on slippery road surfaces under one or both sides of the vehicle, when accelerating when cornering, and when starting off on an incline:

- Unstable vehicle conditions are avoided and therefore the driving safety is enhanced.
- Increased traction due to regulating the optimum slip.
- Mimicking the function of a transversal differential lock.
- Mimicking the function of a longitudinal differential lock with four-wheel-drive vehicles.
- Automatic control of the engine output.
- No "grinding" of the tires when driving around tight corners (unlike with mechanical differential locks).
- Reduction of tire wear.
- Reduction of wear to drive mechanism (transmission, differential, etc.) especially on μ-split or if a wheel suddenly starts to spin on a road surface providing good grip.
- Warning lamp informs the driver during situations close to the physical critical limits.
- Efficient double-use of existing ABS hydraulic components.
- Adoption of tasks of ESP vehicle dynamics control as subordinate wheel controller (see "Closed-loop control system").

Basic principles of automotive control engineering

Many subsystems of a driving safety system (e. g. ESP) influence the driving dynamics of a vehicle by means of a controller i. e., they form a control loop together with the relevant components of the vehicle.

Control loop

A simple standard control loop consists of controllers and a controlled system. The objective is to influence the characteristic of the parameter y_{actual} (controlled variable) of the controlled system using the controller such that the parameter follows a reference characteristic y_{ref} as closely as possible. To do this the controlled variable is measured and passed to the controller. The actual value of the controlled variable is constantly compared with the current reference value by generating the control variance $e = y_{ref} - y_{actual}$.

The main task of the controller is to determine a suitable value for the correcting variable u for every control variance e so that the control variance is decreased, i. e., $y_{actual} = y_{ref}$ at least approximately.

This task may be made more difficult by unknown natural dynamics of the controlled system and other external factors z, which also influence the controlled system.

Example: TCS transversal differential lock controller

The principle of a control loop can be explained using the transversal differential lock controller of the TCS system as an example. The controlled variable $y_{actual} = v_{Dif}$ is the differential speed of the two wheels of a driven axle. The reference value v_{SoDif} is determined by the TCS and adapted to the current driving situation. When driving straight ahead, this value is typically 0. The asymmetric braking torque is used as a correcting variable to influence the controlled variable. The controlled system is the vehicle itself which is affected by external influences such as changing road surfaces.

Standard control loop using TCS transversal differential lock controller as an example

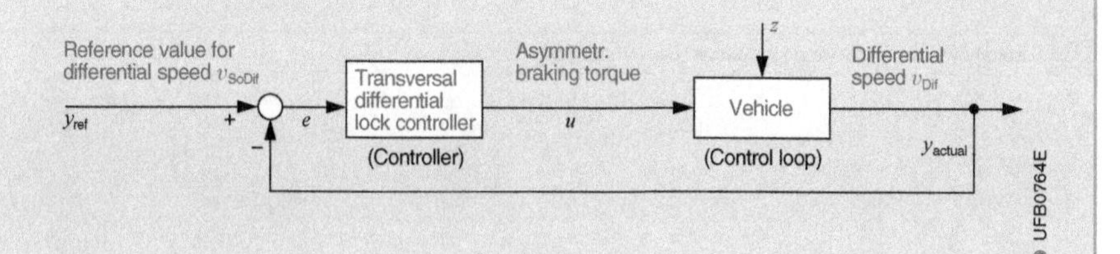

y_{actual} Controlled variable
y_{ref} Reference variable
e Control variance $y_{ref} - y_{actual}$
u Correcting variable
z External disturbance values

Standard controller

Proportional, **I**ntegral and **D**ifferential elements are often used as controllers. The correcting variable u is determined by the current control variance e as shown below:

P-controller	Multiplication	$u(t) = K_P \cdot e(t)$
I-controller	Time integration	$u(t) = K_I \cdot \int e(t)dt$
D-controller	Time derivative	$u(t) = K_D \cdot de(t)/dt$

The degree of counter-reaction of these controllers increases the greater the control variance (P-controller), the longer the control variance lasts (I-controller) or the greater the tendency of the control variance to change (D-controller). Combining these basic controllers gives PI, PD and PID-controllers.

The transversal differential lock controller of the traction control system (TCS) is designed as a PI-controller that contains additional non-linear elements.

Electronic stability program (ESP)

Human error is the cause for a large portion of road accidents. Due to external circumstances, such as an obstacle suddenly appearing on the road or driving at inappropriately high speeds, the vehicle can reach its critical limits and it becomes uncontrollable. The lateral acceleration forces acting on the vehicle reach values that overtax the driver. Electronic systems can make a major contribution towards increasing driving safety.

The Electronic Stability Program (ESP) is a closed-loop system designed to improve vehicle handling and braking response through programmed intervention in the braking system and/or drivetrain. The integrated functionality of the ABS prevents the wheels from locking when the brakes are applied, while TCS inhibits wheel spin during acceleration. In its role as an overall system, ESP applies a unified, synergistic concept to control the vehicle's tendency to "plow" instead of obeying the helm during attempted steering corrections; and at the same time it maintains stability to prevent the vehicle breaking away to the side, provided the vehicle remains within its physical limits.

Requirements

ESP enhances driving safety by providing the following assets:

- Enhanced vehicle stability; the system keeps the vehicle on track and improves directional stability under all operating conditions, including emergency stops, standard braking maneuvers, coasting, acceleration, trailing throttle (overrun), and load shift
- Increased vehicle stability at the limits of traction, such as during sharp steering maneuvers (panic response), to reduce the danger of skidding or breakaway.
- In a variety of different situations, further improvements in the exploitation of traction potential when ABS and TCS come into action, and when engine drag torque control is active, by automatically increasing engine speed to inhibit excessive engine braking. The ultimate effects are shorter braking distances and greater traction along with enhanced stability and higher levels of steering response.

Fig. 1
1 Driver steers, lateral-force buildup.
2 Incipient instability because side-slip angle is too large.
3 Countersteer, driver loses control of vehicle.
4 Vehicle becomes uncontrollable.

M_G Yaw moment
F_R Wheel forces
β Directional deviation from vehicle's longitudinal axis (side-slip angle)

Tasks and method of operation

The electronic stability program is a system that relies on the vehicle's braking system as a tool for "steering" the vehicle. When the stability-control function assumes operation it shifts the priorities that govern the brake system. The basic function of the wheel brakes – to decelerate and/or stop the vehicle – assumes secondary importance as ESP intervenes to keep the vehicle stable and on course, regardless of the conditions.

Specific braking intervention is directed at individual wheels, such as the inner rear wheel to counter understeer, or the outer front wheel during oversteer, as shown in Fig. 2. For optimal implementation of stability objectives, ESP not only initiates braking intervention, but it can also intervene on the engine side to accelerate the driven wheels.

Because this "discriminatory" control concept relies on two *individual intervention* strategies, the system has two options for steering the vehicle: it can brake selected wheels (selective braking) or accelerate the driven wheels. Within the invariable limits imposed by the laws of physics, ESP keeps the vehicle on the road and reduces the risk of accident and overturning. The system enhances road safety by furnishing the driver with effective support.

Below are four examples comparing vehicles with and without ESP during operation "on the limit". Each of the portrayed driving maneuvers reflects actual operating conditions, and is based on simulation programs designed using data from vehicle testing. The results have been confirmed in subsequent road tests.

2 Lateral dynamic response on passenger car with ESP

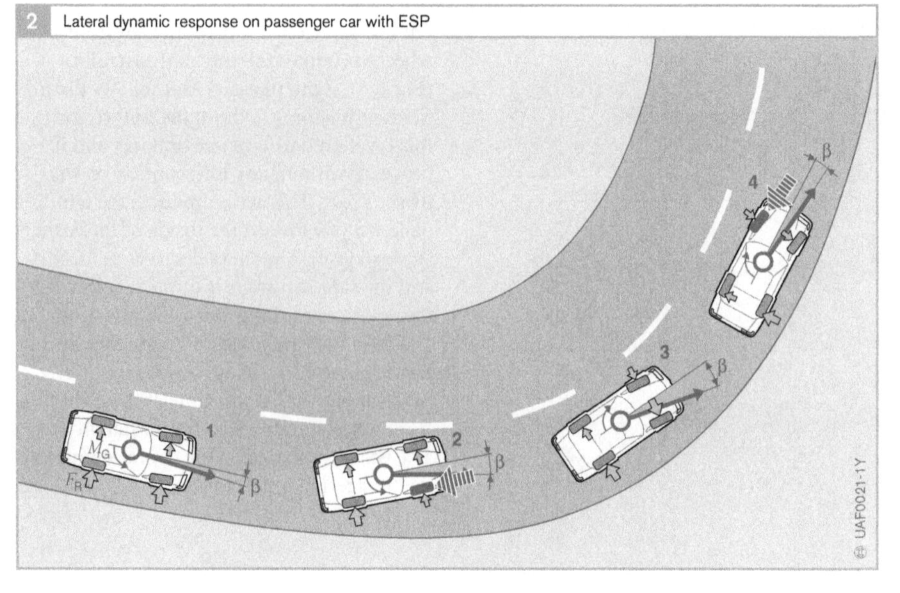

Fig. 2
1 Driver steers, lateral-force buildup.
2 Incipient instability, ESP intervention at right front.
3 Vehicle remains under control.
4 Incipient instability, ESP intervention at left front, complete stabilization.

M_G Yaw moment
F_R Wheel forces
β Directional deviation from vehicle's longitudinal axis (side-slip angle)
⇐ Increased braking force

Maneuvers

Rapid steering and countersteering

This maneuver is similar to lane changes or abrupt steering inputs such as might be expected for instance

- when a vehicle is moving too fast when it enters a series of consecutive S-bends,
- or which have to be initiated when, with oncoming traffic, an obstacle suddenly appears on a country road, or
- which are necessary when an overtaking maneuver on the highway or freeway suddenly has to be aborted.

Figs. 3 and 4 demonstrate the handling response of two vehicles (with and without ESP) negotiating a series of S-bends with rapid steering and countersteering inputs

3 Curves for dynamic response parameters during a right-left cornering sequence

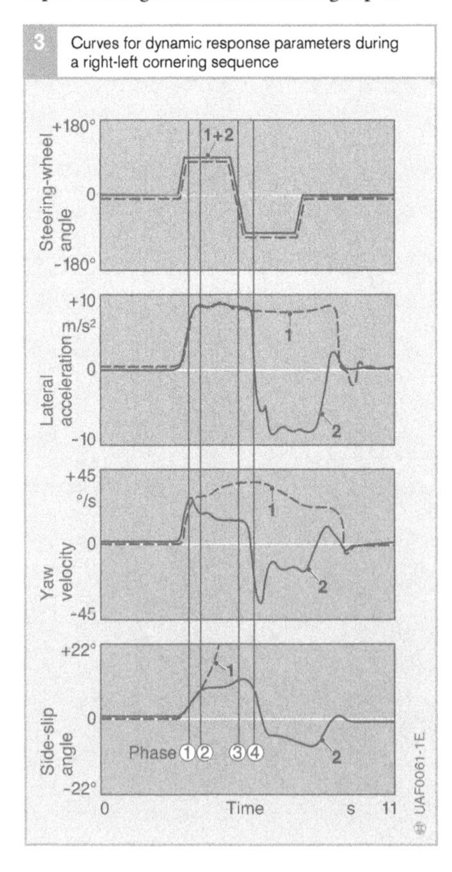

Fig. 3
1 Vehicle without ESP
2 Vehicle with ESP

- on a high-traction road-surface (coefficient of friction $\mu_{HF} = 1$),
- without the driver braking,
- with an initial speed of 144 km/h.

Initially, as they approach the S-bend, the conditions for both vehicles, and their reactions, are identical. Then come the first steering inputs from the drivers (phase 1).

Vehicle without ESP

As can be seen, in the period following the initial, abrupt steering input the vehicle without ESP is already threatening to become unstable (Fig. 4 on left, phase 2). Whereas the steering input has quickly generated substantial lateral forces at the front wheels, there is a delay before the rear wheels start to generate similar forces. The vehicle reacts with a clockwise movement around its vertical axis (inward yaw). The next stage is phase 3 with the second steering input. The vehicle without ESP fails to respond to the driver's attempt to countersteer and goes out of control. The yaw velocity and the side-slip angle rise radically, and the vehicle breaks into a skid (phase 4).

Vehicle with ESP

On this vehicle ESP brakes the left front wheel to counter the threat of instability (Fig. 4 on right, phase 2) that follows the initial steering input. Within the ESP context this is referred to as active braking, and it proceeds without any intervention on the driver's part. This action reduces the tendency to yaw toward the inside of the corner (inward yaw). The yaw velocity is reduced and the side-slip angle is limited. Following the countersteer input, first the yaw moment and then the yaw velocity reverse their directions (phase 3). In phase 4 a second brief brake application – this time at the right front wheel – restores complete stability. The vehicle remains on the course defined by the steering-wheel angle.

4 Vehicle tracking during right-left cornering sequence

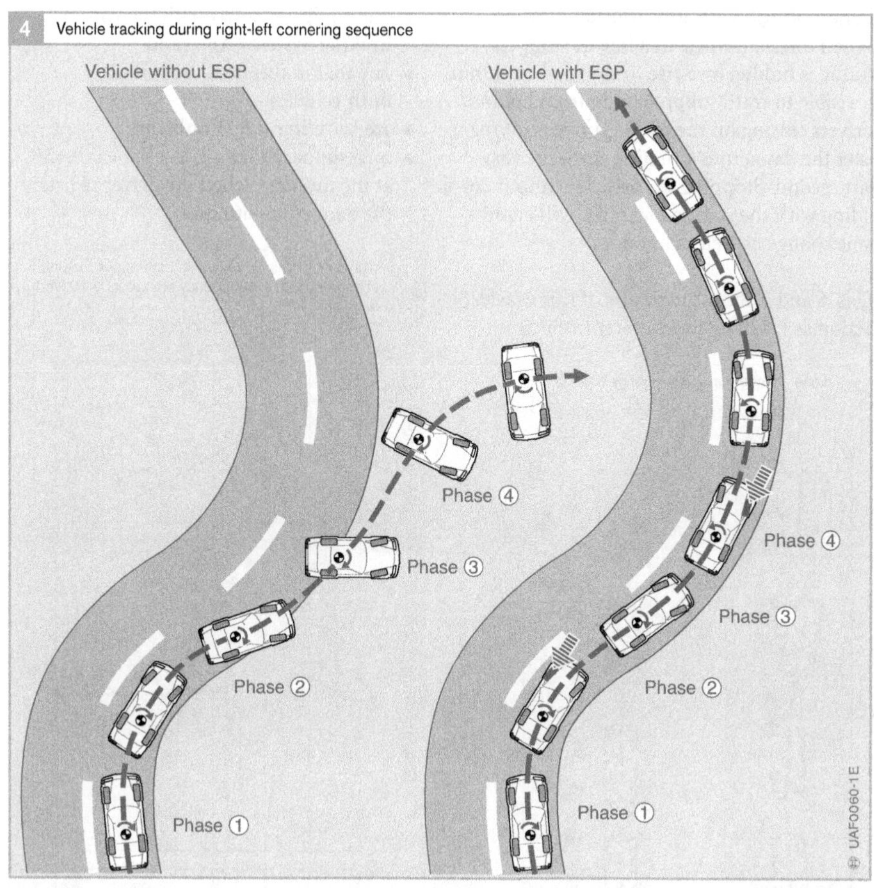

5 Over and understeering behavior when cornering

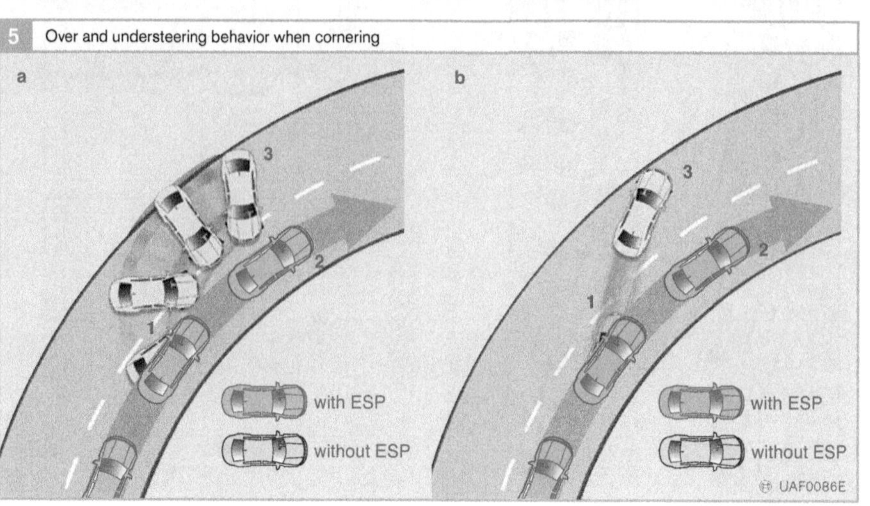

Fig. 4
Increased braking force
① Driver steers, lateral-force buildup.
② Incipient instability
Right: ESP intervention at left front.
③ Countersteer
Left: Driver loses control of vehicle;
Right: Vehicle remains under control.
④ Left: Vehicle becomes uncontrollable,
Right: ESP intervention at right front, complete stabilization.

Fig. 5
a Oversteering behavior.
1 The rear end of the vehicle breaks away.
2 ESP applies the brake at the outer front wheel and this reduces the risk of skidding.
3 The vehicle without ESP breaks into a slide.

b Understeering behavior
1 The front of the vehicle breaks away.
2 ESP applies the brake at the inner rear wheel and this reduces the risk of understeering.
3 The vehicle without ESP is understeered and leaves the road.

Lane change with emergency braking

When the last vehicle in a line of stopped traffic is hidden by a rise in the road, and thus invisible to traffic approaching from behind, drivers closing on the traffic jam cannot register the dangerous situation until the very last second. Stopping the vehicle without colliding with the stationary traffic will entail a lane change as well as braking.

Figs. 6 and 7 show the results of this evasive action as taken by two different vehicles:

- one equipped solely with the Antilock Braking System (ABS) and
- one that is also fitted with ESP.

Both vehicles

- are traveling at 50 km/h and
- on a slippery road surface ($\mu_{HF} = 0.15$) at the moment when the driver registers the dangerous situation.

6 Lane change during emergency braking

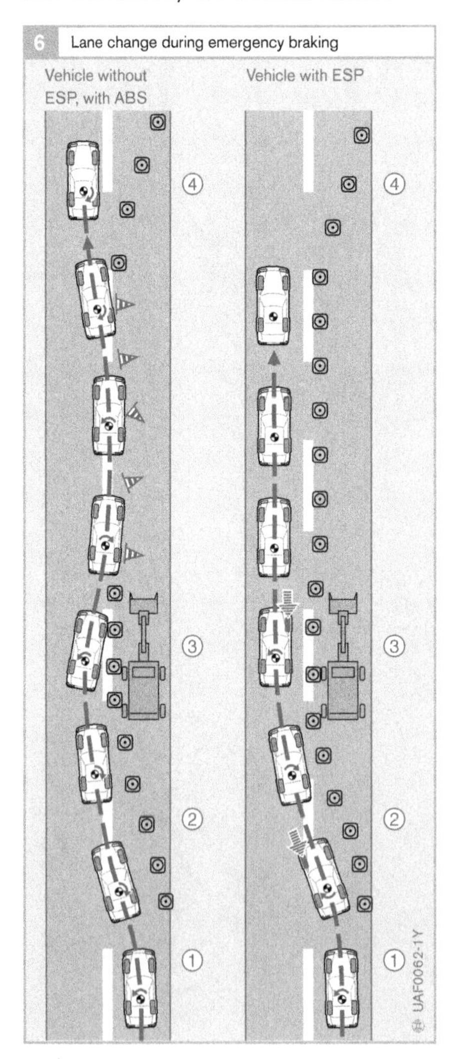

Fig. 6
$v_0 = 50$ km/h
$\mu_{HF} = 0.15$

Increased brake slip

7 Curves for dynamic response parameters for lane change during an emergency stop at $v_0 = 50$ km/h and $\mu_{HF} = 0.15$

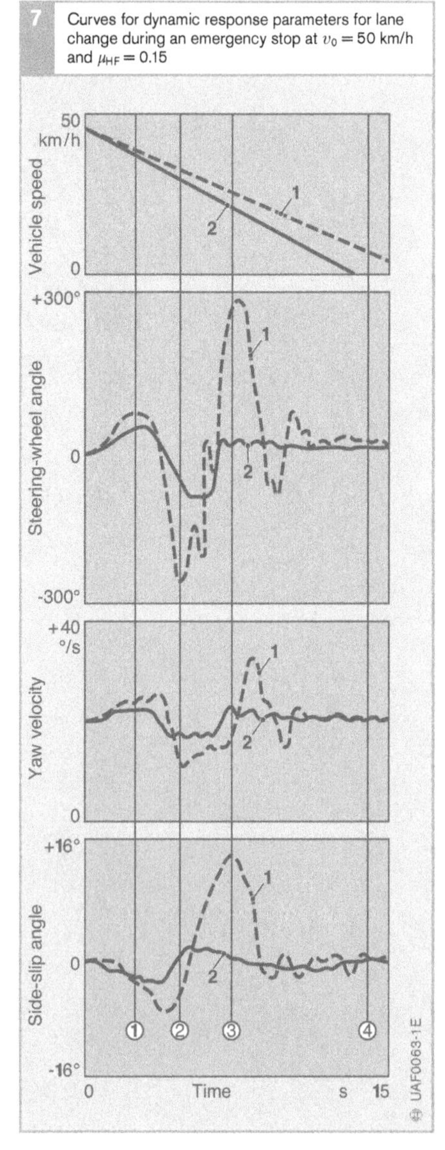

Fig. 7
$v_0 = 50$ km/h
$\mu_{HF} = 0.15$

1 Vehicle without ESP
2 Vehicle with ESP

Vehicle with ABS but without ESP

Immediately after the initial steering input both the side-slip angle and the yaw velocity have increased to the point where driver intervention – in the form of countersteer – has become imperative (Fig. 6, on left). This driver action then generates a side-slip angle in the opposite direction (technically: with the opposite operational sign). This side-slip angle increases rapidly, and the driver must countersteer for a second time. Here the driver is able – but only just – to restabilize the vehicle and bring it to a safe halt.

Vehicle with ESP

Because ESP reduces yaw velocity and side-slip angle to easily controllable levels, this vehicle remains stable at all times. The driver is not confronted with unanticipated instability and can thus continue to devote full attention to keeping the vehicle on course. ESP substantially reduces the complexity of the steering process and lowers the demands placed on the driver. Yet another asset is that the ESP vehicle stops in less distance than the vehicle equipped with ABS alone.

9 Oversteering and understeering when cornering

Fig. 9
1 Vehicle with ESP
2 Oversteered vehicle without ESP
3 Understeered vehicle without ESP

8 Critical obstacle-avoidance maneuver with and without ESP

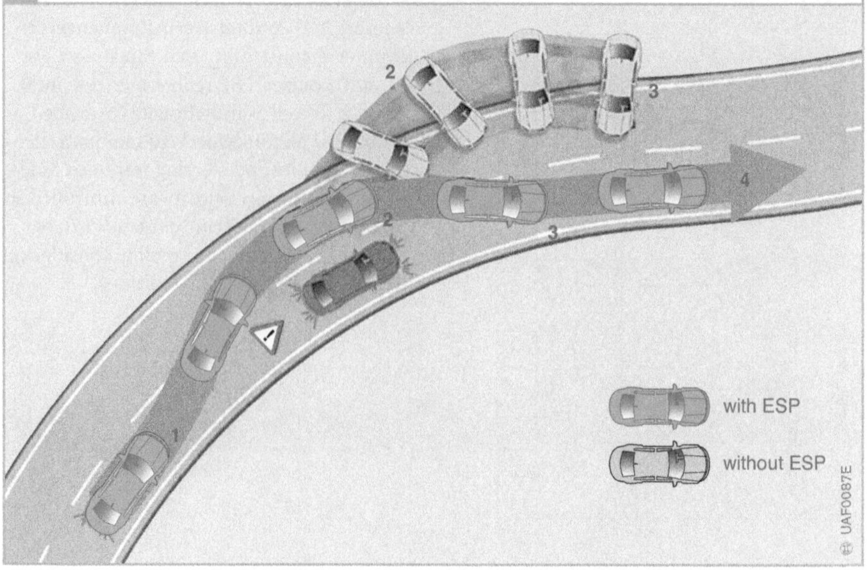

with ESP
without ESP
UAF0087E

Fig. 8
Vehicle without ESP
1 Vehicle approaches an obstacle.
2 Vehicle breaks away and does not follow the driver's steering movements.
3 Vehicle slides uncontrolled off the road.

Vehicle with ESP
1 Vehicle approaches an obstacle.
2 Vehicle almost breaks away → ESP intervention, vehicle follows driver's steering movements.
3 Vehicle almost breaks away again when recentering the steering wheel → ESP intervention.
4 Vehicle is stabilized.

Extended steering and countersteering sequence with progressively greater input angles

A vehicle traversing a series of S-curves (for instance, on a snaking secondary road) is in a situation similar to that encountered on a slalom course. The way the ESP works can be clearly seen during this kind of dynamic maneuver when the steering wheel has to be turned to progressively greater angles to negotiate each turn.

Figs. 10 and 11 illustrate the handling response of two vehicles (one with and one without ESP) under these conditions

- on a snow-covered road ($\mu_{HF} = 0.45$),
- without the driver braking, and
- at a constant velocity of 72 km/h.

10 Curves for dynamic response parameters for rapid steering and countersteering inputs with increasing steering-wheel angles

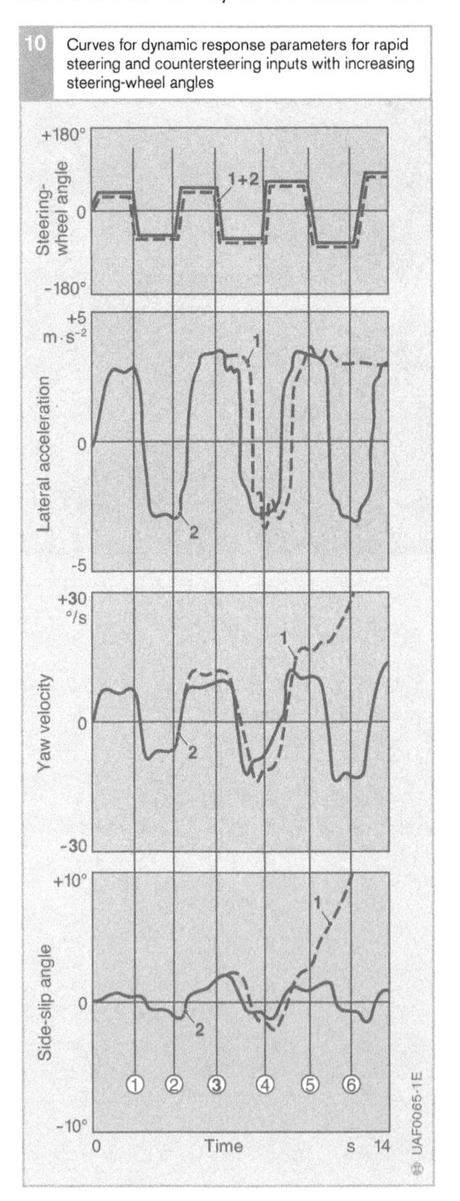

Fig. 10
1 Vehicle without ESP
2 Vehicle with ESP

Vehicle without ESP

Engine output will have to be increased continually in order to maintain a constant road speed. This, in turn, will generate progressively greater slip at the driven wheels. A sequence of steering and countersteering maneuvers with a 40° steering-wheel angle can quickly increase drive slip to such levels that a vehicle without ESP becomes unstable. At some point in this alternating sequence the vehicle suddenly ceases to respond to steering inputs and breaks into a slide. While lateral acceleration remains virtually constant, both side-slip angle and yaw velocity rise radically.

Vehicle with ESP

The Electronic Stability Program (ESP) intervenes at an early stage in this sequence of steering and countersteering maneuvers to counter the instability that threatens right from the outset. ESP relies on engine intervention as well as individually controlled braking of all four wheels to maintain the vehicle's stability and steering response. Side-slip angle and yaw velocity are controlled so that the driver's steering demands can be complied with as far as possible considering the prevailing physical conditions.

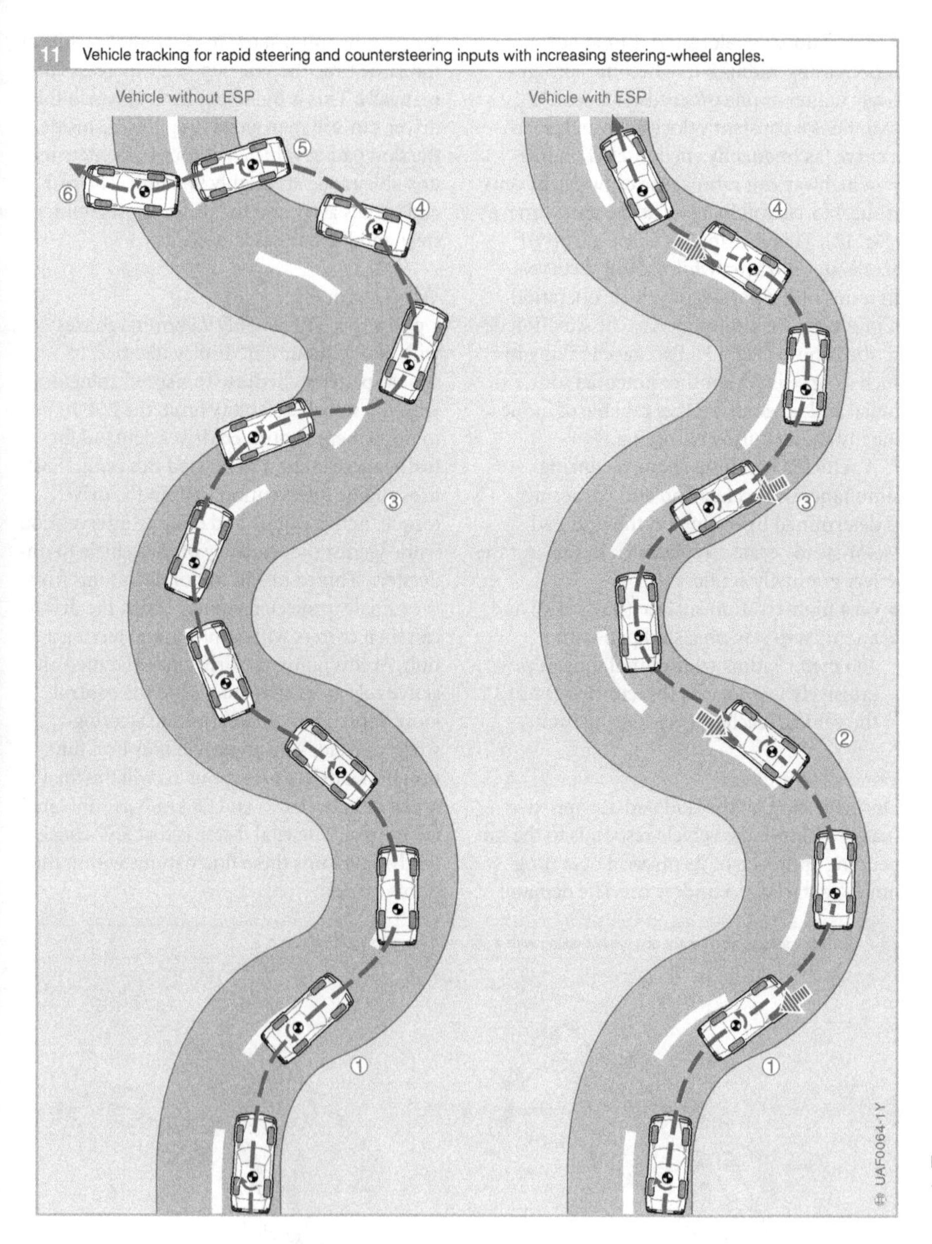

Fig. 11
Increased braking force

Acceleration/deceleration during cornering

A decreasing-radius curve becomes progressively tighter as one proceeds. If a vehicle maintains a constant velocity through such a curve (as frequently encountered on freeway/highway exit ramps) the outward, or centrifugal force, will increase at the same time (Fig. 12). This also applies when the driver accelerates too soon while exiting a curve. In terms of the physics of vehicle operation, it produces the same effects as the situation described above (Fig. 13). Excessive braking in such a curve is yet another potential source of radial and tangential forces capable of inducing instability during cornering.

A vehicle's handling response during simultaneous acceleration and cornering is determined by testing on the skid pad (semi-steady-state circulation). In this test the driver gradually accelerates

- on a high-traction surface ($\mu_{HF} = 1.0$) and
- attempts to stay on a skid pad with a 100 meter radius while circulating at progressively faster rates; this continues until the vehicle reaches its cornering limits.

Vehicle without ESP

During testing on the skid pad at approximately 95 km/h the vehicle responds to the impending approach of its physical operating limits by starting to understeer. The demand for steering input starts to rise rapidly, while at the same time the side-slip angle increases dramatically. This is the upper limit at which the driver can still manage to keep the car inside the skid pad. A vehicle without ESP enters its unstable range at roughly 98 km/h. The rear end breaks away and the driver must countersteer and leave the skid pad.

Vehicle with ESP

Up to a speed of roughly 95 km/h, vehicles with and without ESP display identical response patterns. Because this speed coincides with the vehicle's stability limit, the ESP refuses to implement continued driver demand for further acceleration to beyond this point. ESP uses engine intervention to limit the drive torque. Active engine and braking intervention work against the tendency of the vehicle to understeer. This results in minor deviations from the initially projected course, which the driver can then correct with appropriate steering action. At this point, the driver has assumed an active role as an element within the control loop. Subsequent fluctuations in steering-wheel and side-slip angle will now be a function of the driver's reactions, as will the final speed of between 95 and 98 km/h as stipulated for the test. The vital factor is that ESP consistently maintains these fluctuations within the stable range.

12 Vehicle tracking when cornering while braking with a constant steering-wheel angle

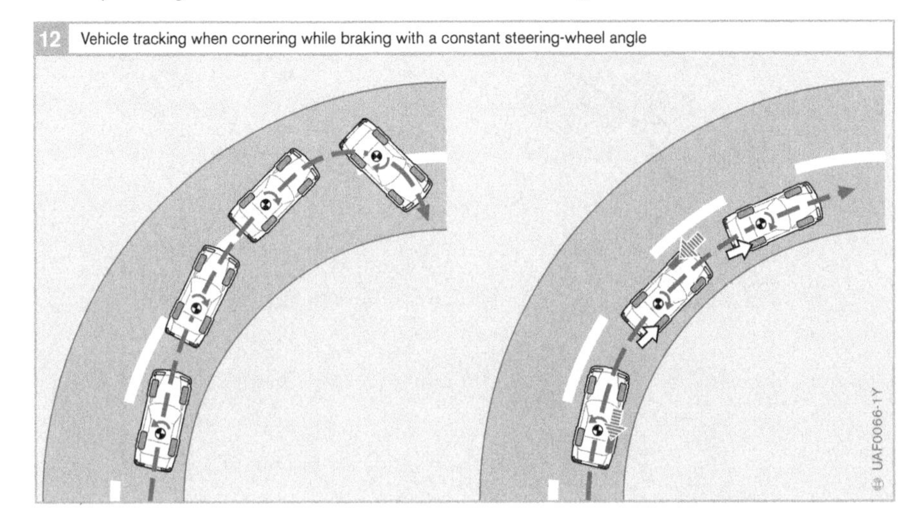

Fig. 12
Increased braking force
Decreased braking force

13 Vehicle tracking when cornering while accelerating

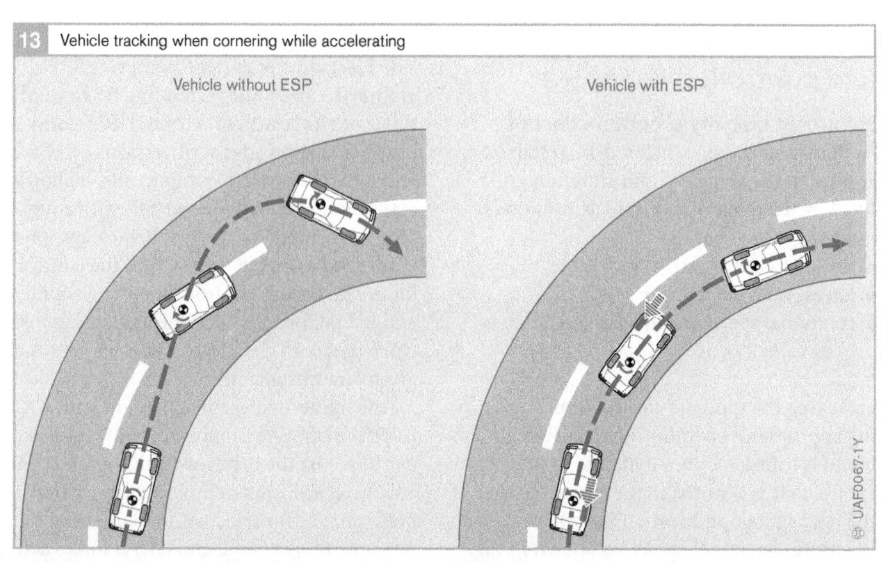

Fig. 13
Increased braking force

14 Comparison of cornering with vehicles with and without ESP

Closed-loop control system and controlled variables

Electronic stability program concept

Application of the ESP closed-loop stability control in the vehicle's limit situation as defined by the dynamics of vehicle motion is intended to prevent the

- linear (longitudinal) velocity, the
- lateral velocity and the
- rotational speed around the vertical axis (yaw velocity),

exceeding the ultimate control limits. Assuming appropriate operator inputs, driver demand is translated into dynamic vehicular response that is adapted to the characteristics of the road in an optimization process designed to ensure maximum safety. As shown in Fig. 1, the first step is to determine how the vehicle should respond to driver demand during operation in the limit range (ideal response), and also how it actually does respond (actual response). Actuators are then applied to minimize the difference between the ideal and the actual response (control deviation) by indirectly influencing the forces acting at the tires.

System and control structure

The Electronic Stability Program (ESP) embraces capabilities extending far beyond those of either ABS or ABS and TCS combined. Based on advanced versions of ABS and ABS/TCS system components, it allows active braking at all four wheels with a high level of dynamic sensitivity. Vehicle response is adopted as an element within the control loop. The system controls braking, propulsive and lateral forces so that the actual response converges with the ideal response under the given circumstances.

An engine-management system with CAN interface can vary engine output torque in order to adjust the driven-wheel slip rates. The advanced ESP system provides highly precise performance for selective adjustment of the dynamic longitudinal and lateral forces acting on each individual wheel.

Fig. 2 shows ESP control in a schematic diagram with

- the sensors that determine the controller input parameters,
- the ESP control unit with its hierarchically-structured controller, featuring a higher-level ESP controller and the subordinate slip controllers,
- the actuators used for ultimate control of braking, drive and lateral forces.

1 Block diagram of electronic stability program (ESP)

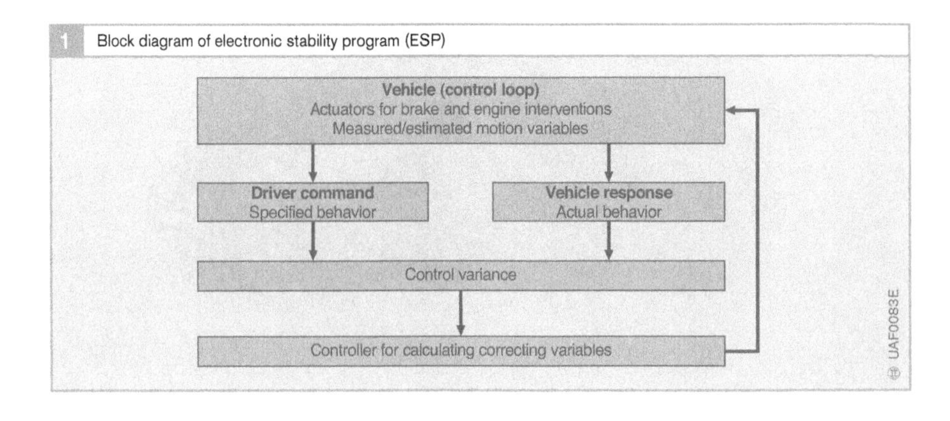

Controller hierarchy

Level 1 ESP controller

Task

The ESP controller is responsible for
- determining the current vehicle status based on the yaw velocity signal and the side-slip angle estimated by the "monitor" and then
- achieving maximum possible convergence between vehicle response in the limit range and its characteristics in the normal operating range (ideal response).

The following components register driver demand and the system processes their signals as the basis for defining ideal response:
- engine-management system (e.g. pressure on accelerator pedal),
- primary-pressure sensor (e.g. activation of brakes), or
- steering-wheel sensor (steering-wheel angle).

At this point the specified response is defined as driver demand. The coefficient of friction and the vehicle speed are also included in the processing calculations as supplementary parameters. The "monitor" estimates these factors based on signals transmitted by the sensors for
- wheel speed,
- lateral acceleration,
- braking pressures, and
- yaw velocity.

The desired vehicle response is brought about by generating a yaw moment acting on the vehicle. In order to generate the desired yaw moment, the controller intervenes in the tire-slip rates to indirectly influence the longitudinal and lateral forces. The system influences the tire slip by varying the specifications for slip rate, which must then be executed by the subordinated ABS and TCS controllers.

The intervention process is designed to maintain the handling characteristics that the manufacturer intended the vehicle to have and to serve as the basis for ensuring consistently reliable control.

2 ESP control loop in vehicle

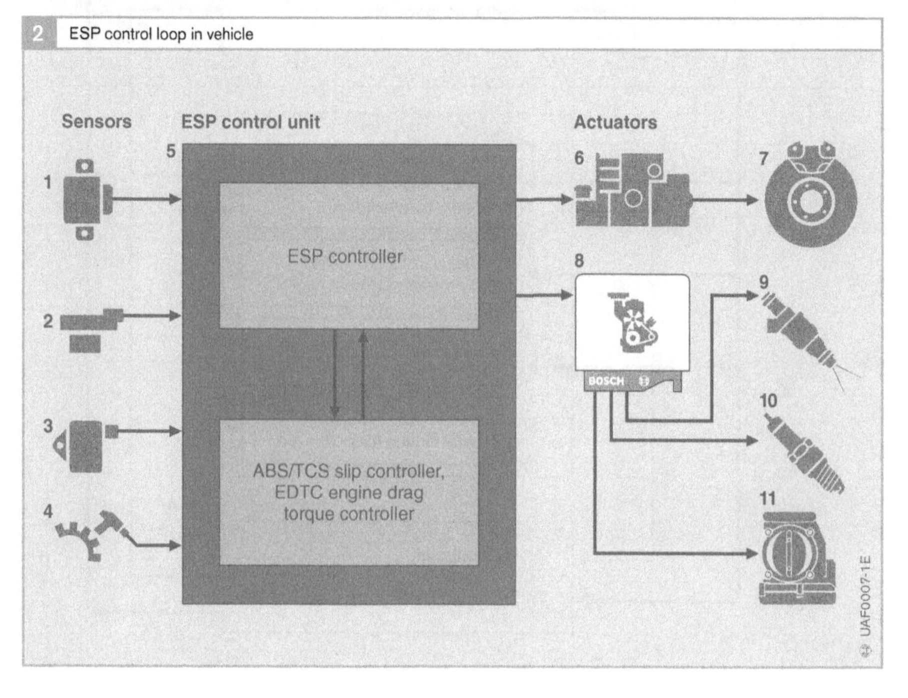

Fig. 2
1 Yaw-rate sensor with lateral-acceleration sensor
2 Steering-wheel-angle sensor
3 Primary-pressure sensor
4 Wheel-speed sensors
5 ESP control unit
6 Hydraulic modulator
7 Wheel brakes
8 Engine management ECU
9 Fuel injection

only for gasoline-engines:
10 Ignition-timing intervention
11 Throttle-valve intervention (ETC)

The ESP controller generates the specified yaw moment by relaying corresponding slip-modulation commands to the selected wheels.

The subordinate-level ABS and TCS controllers trigger the actuators governing the brake hydraulic system and the engine-management system using the data generated in the ESP controller.

Design
Fig. 3 is a simplified block diagram showing the design structure of the ESP controller. It portrays the signal paths for input and output parameters. Based on the

- yaw velocity (measured parameter),
- steering-wheel angle (measured parameter),
- lateral acceleration (measured parameter),
- vehicle's linear velocity (estimated parameter), and
- longitudinal tire forces and slip rates (estimated parameters)

the monitor determines the following:
- lateral forces acting on the wheel,
- slip angle,
- side-slip angle, and
- vehicle lateral speed.

3 Simplified block diagram showing ESP controller with input and output variables

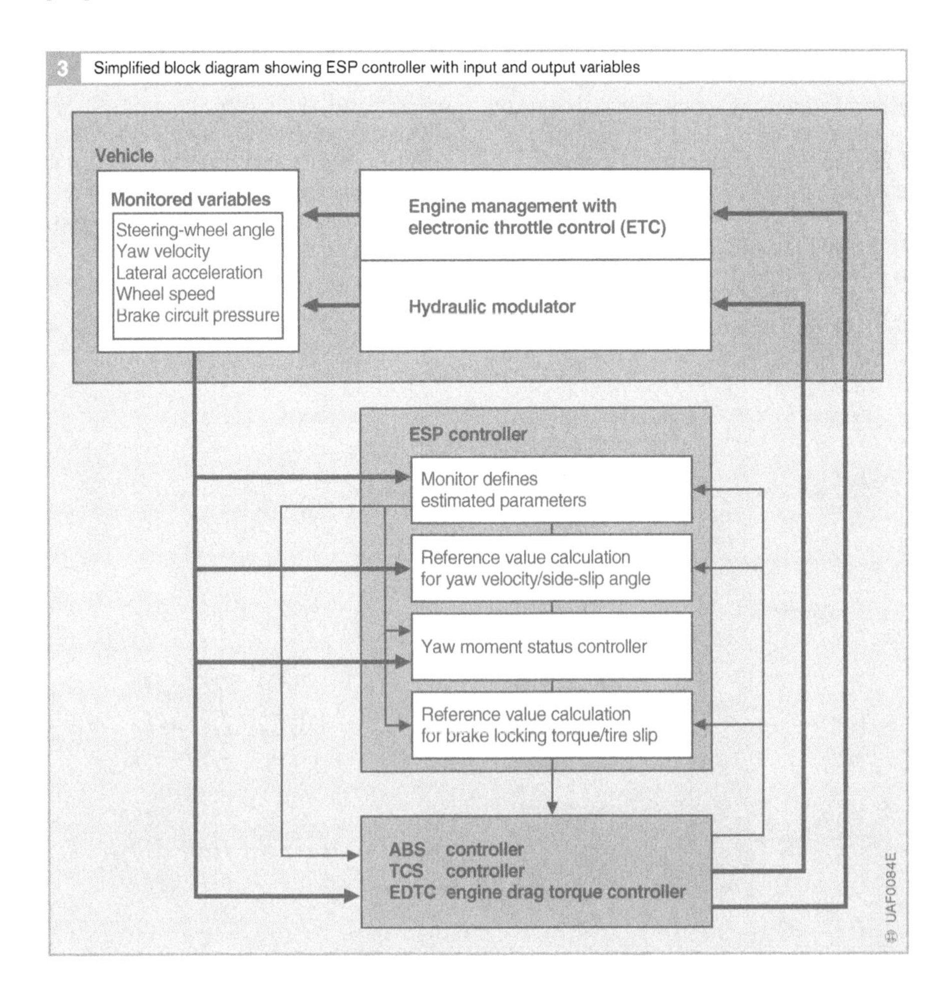

The specifications for side-slip angle and yaw velocity are determined on the basis of the following parameters, which may be either directly or indirectly defined by driver input:

- steering-wheel angle,
- estimated vehicle speed,
- coefficient of friction, which is determined on the basis of the longitudinal acceleration (estimated parameter) and lateral acceleration (measured parameter), and
- accelerator-pedal travel (engine torque) or brake-circuit pressure (force on brake pedal).

These processes also take into account the special characteristics related to vehicle dynamics, as well as unusual situations, such as a crowned road or μ-split surface (e.g. high traction on left of lane with right side slippery).

Method of Operation

The ESP controller governs the two status parameters "yaw velocity" and "side-slip angle" while calculating the yaw moment required to make the actual and desired-state parameters converge. As the side-slip angle increases, so does its significance for the controller.

The control program is based upon data concerning the maximum potential lateral acceleration and other data selected to reflect the vehicle's dynamic response patterns. These are determined for each vehicle in *steady-state skid-pad testing*. In subsequent steady-state vehicle operation, as well as during braking and acceleration, this data – defining how the steering angle and vehicle speed relate to the yaw velocity – serves as the basis for defining the desired vehicle motion. The required data (nominal yaw velocity) is stored in the program in the form of a single-track model.

The nominal yaw velocity must be limited in line with the friction coefficient so as to keep the vehicle on the predefined physically feasible track.

For example, if a vehicle breaks into oversteer while coasting into a right-handed

Single-track model

Ranges of lateral acceleration

Passenger cars can achieve lateral accelerations up to 10 m/s². Lateral acceleration in the small signal range (0...0.5 m/s²) can be caused by road conditions such as ruts or by crosswinds.

The linear range extends from 0.5...4 m/s². Typical lateral maneuvers include changing lane or load-change reactions when cornering. The response of the vehicle in these situations can be described using the linear single-track model.

In the transition range (4...6 m/s²) some vehicles still respond linearly while others do not.

The critical range above 6 m/s² is only reached in extreme situations such as situations that almost result in an accident. In this case the vehicle response is highly non-linear.

Assumptions with the single-track model

The linear single-track model can be used to obtain important information about the lateral behavior of a vehicle. In the single-track model the lateral properties of an axle and its wheels are summarized into one effective wheel. In the simplest version, the properties taken into account are modeled using linear equations; thus the model is referred to as a linear single-track model.

The most important model assumptions are:

- Kinematics and elastokinematics of the axle are only modeled linearly.
- The lateral force buildup of the tire is linear and the aligning torque is ignored.
- The center of gravity is assumed to be at road level. The only rotational degree of freedom of the vehicle is therefore the yaw motion. Rolling, pitching and lifting (translational movements in z-axis) are not taken into account.

curve, and the specified yaw velocity is exceeded (the vehicle evinces a tendency to rotate too quickly around its vertical axis), ESP responds by braking the left front wheel to generate a defined brake slip which shifts the yaw moment toward greater counter-clockwise rotation thus suppressing the vehicle's tendency to break away.

If a vehicle breaks into understeer while coasting into a right-handed curve, and the yaw velocity is below the specified yaw rate (the vehicle evinces a tendency to rotate too slowly around its vertical axis), ESP responds by braking the right rear wheel to generate a defined brake slip which shifts the yaw moment toward greater clockwise rotation thus suppressing the vehicle's tendency to push over the front axle.

ESP controller functions during ABS and TCS operation

The entire spectrum of monitored and estimated data is relayed to the subordinate controllers for continuous processing. This guarantees maximum exploitation of the traction available between tire and road surface for the basic ABS and TCS functions under all operating conditions.

During active ABS operation (with wheels tending to lock) the ESP controller provides the subordinate ABS controller with the following data:

- lateral vehicle velocity,
- yaw velocity,
- steering-wheel angle, and
- wheel speeds as the foundation for providing the desired ABS slip.

When TCS is active (wheels threatening to break into uncontrolled spin when moving off or during rolling acceleration) the ESP controller transmits the following offset values to the subordinate TCS controller:

- change in the specified value for the drive-slip,
- change in the slip tolerance range, and
- change in a value to influence the torque reduction.

Level 2 ABS controller

Task

The hierarchically subordinated ABS controller goes into operation whenever the desired slip rate is exceeded during braking, and it becomes necessary for ABS to intervene. During both ABS and "active" braking, the closed-loop control of wheel-slip rates as applied for various dynamic-intervention functions must be as precise as possible. The system needs precise data on slip as a precondition for dialing in the specified slip rates. It must be pointed out that the system does not measure the vehicle's longitudinal speed directly. Instead, this parameter is derived from the rotation rates of the wheels.

Design and method of operation

By briefly "underbraking" one wheel, the ABS controller performs an indirect measurement of vehicle speed. It interrupts the slip control to lower the current braking torque by a defined increment. The torque is then maintained at this level for a given period. Assuming that the wheel has stabilized and is turning freely with no slip at the end of this period, it can serve as a suitable source for determining (no-slip) wheel speed.

The calculated velocity at the center of gravity can be used to determine the effective (free-rolling) wheel speeds at all four wheels. These data, in turn, form the foundation for calculating the actual slip rates at the remaining three – controlled – wheels.

Level 2 engine drag-torque controller (EDTC)

Task

Following downshifts and when the accelerator is suddenly released, inertia in the engine's moving parts always exerts a degree of braking force at the drive wheels. Once this force and the corresponding reactive torque rise beyond a certain level, the tires will lose their ability to transfer the resulting loads to the road. Engine drag-torque control intervenes under these conditions (by "gently" accelerating the engine).

Design and method of operation
Factors such as variations in the character of the road surface can lead to conditions under which the engine-braking torque is suddenly too high. The result is a tendency for the wheels to lock. One available countermeasure is judicious throttle application. Here, the ECU transmits signals to trigger the corresponding actuators in the engine-management system (with ETC function) for an increase in drive torque. Intervention at the engine-management level is employed to regulate the driven wheel within the prescribed limits.

Level 2 TCS controller

Task
The hierarchically subordinated TCS controller is triggered in case of excessive slip during (for example) standing-start and during rolling acceleration when it becomes necessary for TCS to intervene. TCS is intended to prevent the driven wheels from breaking into a free spin. It functions by limiting engine torque to a level corresponding to the drive torque that the wheels can transfer to the road surface.

Intervention at the driven wheels is carried out by means of brake application or by including the engine management system in the process. In diesel-engine vehicles, the electronic diesel control system (EDC) reduces the engine torque by modifying the quantity of fuel injected. With gasoline-engines, the engine torque can be reduced by varying the throttle valve aperture (ETC), modifying the ignition timing, or suppressing individual injection pulses.

Active braking at non-driven wheels is governed directly by the ABS controller. Unlike with ABS, TCS receives values from the ESP controller for changing the specified slip and permissible slip difference of the driven axle(s). These changes take effect in the form of an offset applied to the basic values defined in the TCS.

Design
The specified propshaft speed and wheel-speed differential data are calculated from the specified slip values and speeds for the "coasting" wheels. The propshaft speed and wheel-speed differential controlled variables are derived from the driven wheel speeds.

Method of Operation
The TCS module calculates the desired braking torque for both driven wheels and the desired value for engine torque reduction to be implemented by the engine-management system.

Because the propshaft speed is affected by inertial forces originating from the drivetrain as a whole (engine, transmission, driven wheels, and the propshaft itself), a relatively large time constant is employed to describe its corresponding leisurely rate of dynamic response. In contrast, the time constant for the wheel-speed differential is relatively small, reflecting the fact that the wheels' own inertial forces are virtually the sole determining factor for their dynamic response. Another relevant factor is that the wheel-speed differential – unlike propshaft speed – is not affected by the engine.

The torques prescribed for the propshaft and wheel-speed differential are taken as the basis for defining the actuators' positioning forces. The system achieves the specified difference in braking torque between the left and right-side driven wheels by transmitting the appropriate control signals to valves in the hydraulic modulator.

Propshaft torques are regulated to the desired level using symmetrical braking as well as engine intervention. With a gasoline-engine, adjustments undertaken through the throttle valve are relatively slow to take effect (lag and the engine's transition response). Retarding the ignition timing and, as a further option, selective suppression of injection pulses are employed for rapid engine-based intervention, while symmetrical braking can be applied for brief transitional support of engine torque reduction.

Micromechanical yaw-rate sensors

Applications

In vehicles with Electronic Stability Program (ESP), the rotation of the vehicle about its vertical axis is registered by micromechanical yaw-rate (or yaw-speed) sensors (also known as gyrometers) and applied for vehicle-dynamics control. This takes place during normal cornering, but also when the vehicle breaks away or goes into a skid.

These sensors are reasonably priced as well as being very compact. They are in the process of forcing out the conventional high-precision mechanical sensors.

1 Structure of the MM1 yaw-rate sensor

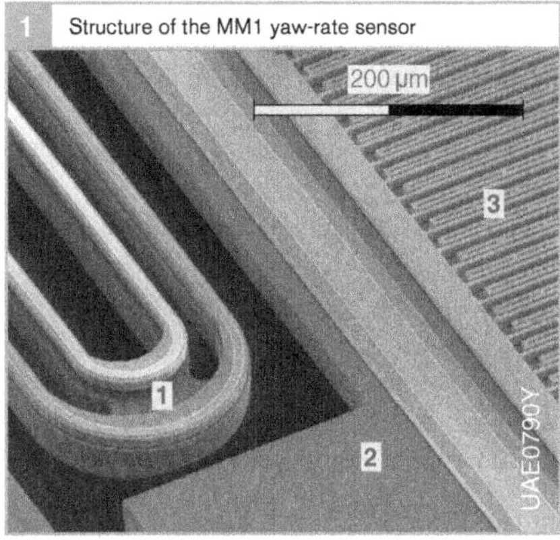

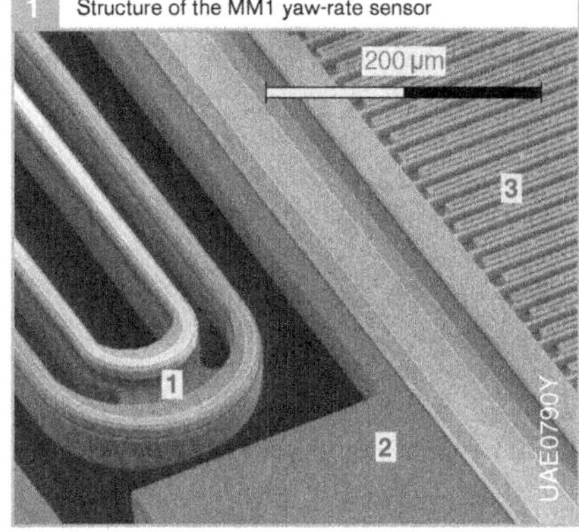

Fig. 1
1 Retaining/guide spring
2 Part of the oscillating element
3 Coriolis acceleration sensor

2 MM1 micromechanical yaw-rate sensor

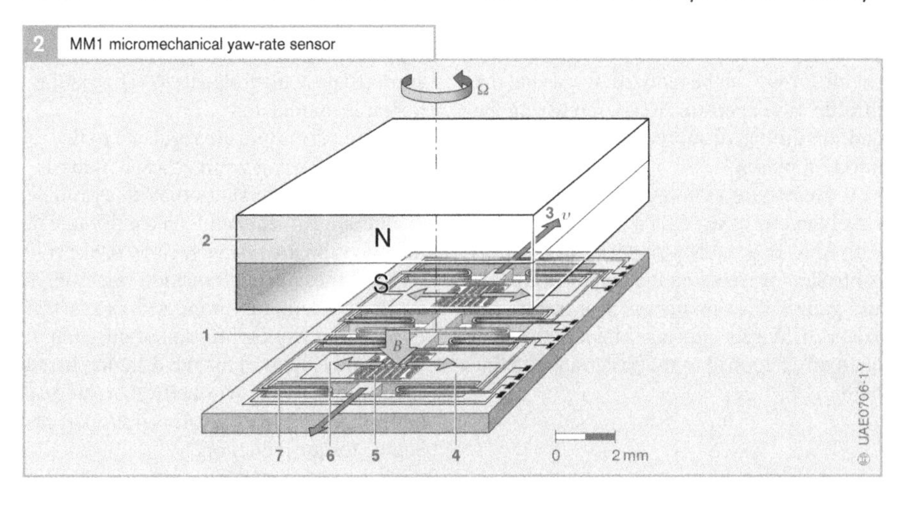

Fig. 2
1 Frequency-determining coupling spring
2 Permanent magnet
3 Direction of oscillation
4 Oscillating element
5 Coriolis acceleration sensor
6 Direction of Coriolis acceleration
7 Retaining/guide spring
Ω Yaw rate
v Oscillating velocity
B Permanent-magnet field

Design and construction

MM1 micromechanical yaw-rate sensor

A mixed form of technology is applied in order to achieve the high accuracies needed for vehicle-dynamics systems. That is, two somewhat thicker oscillating elements (mass plates) which have been machined from a wafer using bulk micromechanics oscillate in counter-phase to their resonant frequency which is defined by their mass and their coupling springs (>2 kHz). On each of these oscillating elements, there is a miniature, surface-type micromechanical capacitive acceleration sensor. When the sensor chip rotates about its vertical axis at yaw rate Ω, these register the Coriolis acceleration in the wafer plane vertical to the direction of oscillation (Figs. 1 and 2). These accelerations are proportional to the product of yaw rate and and the oscillatory velocity which is maintained electronically at a constant value.

To drive the sensor, all that is required is a simple, current-carrying printed conductor on each oscillating element. In the permanent-magnet field B vertical to the chip surface, this oscillating element is subjected to an electrodynamic (Lorentz) force. Using a further, simple printed conductor (which saves on chip surface), the same magnetic field is used to directly measure the oscillation velocity by inductive means. The different physical construction of drive system and sensor system serves to avoid undesirable coupling between the two sections. In order to suppress unwanted external acceleration effects, the opposing sensor signals are subtracted from each other. The external acceleration effects can be measured by applying summation. The high-precision micromechanical construction helps to suppress the effects of high oscillatory acceleration which is several factors of 10 higher than the low-level Coriolis acceleration (cross sensitivity far below 40 dB). Here, the drive and measurement systems are rigorously decoupled from each other.

MM2 micromechanical yaw-rate sensor

Whereas this silicon yaw-rate sensor is produced completely using surface-micromechanic techniques, and the magnetic drive and control system have been superseded by an electrostatic system, absolute decoupling of the power/drive system and measuring system is impossible. Comb-like structures (Figs. 3 and 4) electrostatically force a centrally mounted rotary oscillator to oscillate. The amplitude of these oscillations is held constant by means of a similar capacitive pick-off. Coriolis forces result at the same time in an out-of-plane tilting movement, the amplitude of which is proportional to the yaw rate Ω, and which is detected capacitively by the electrodes underneath the oscillator.

To avoid excessive damping of this movement, the sensor must be operated in a vacuum. Although the chip's small size and the somewhat simpler production process result in considerable cost reductions, this miniaturisation is at the expense of reductions in the measuring effect, which in any case is not very pronounced, and therefore of the achievable precision. It also places more severe demands on the electronics. The system's high flexural stability, and mounting in the axis of gravity, serve to mechanically suppress the effects of unwanted acceleration from the side.

4 MM2 yaw-rate sensor: Structure

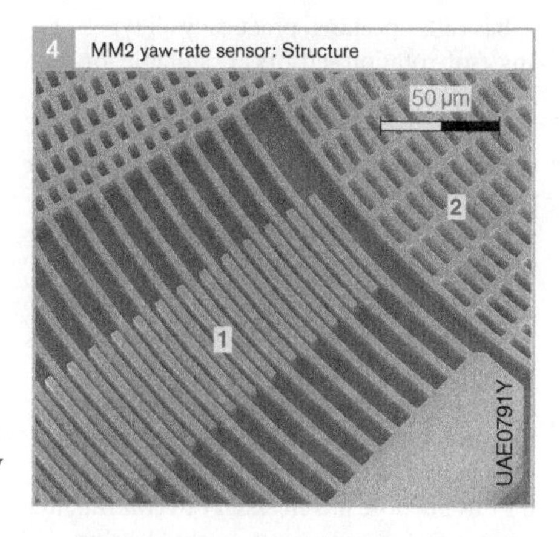

Fig. 4
1 Comb-like structure
2 Rotary oscillator

3 MM2 surface-micromechanical yaw-rate sensor

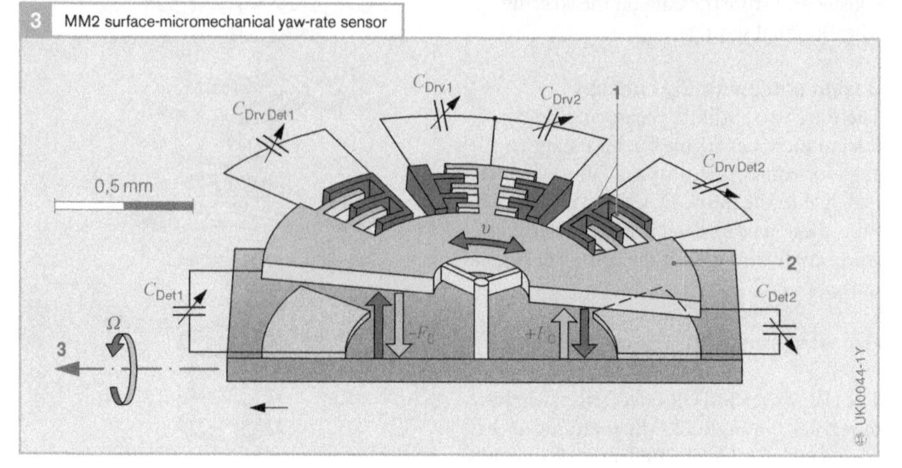

Fig. 3
1 Comb-like structure
2 Rotary oscillator
3 Measuring axis
C_{Drv} Drive electrodes
C_{Det} Capactive pick-off
F_C Coriolis force
v Oscillatory velocity
$\Omega = \Delta C_{Det}$, measured yaw rate

Steering-wheel-angle sensors

Application

The Electronic Stability Program (ESP) applies the brakes selectively to the individual wheels in order to keep the vehicle on the desired track selected by the driver. Here, the steering-wheel angle and the applied braking pressure are compared with the vehicle's actual rotary motion (around its vertical axis) and its road speed. If necessary, the brakes are applied at individual wheels. These measures serve to keep the float angle (deviation between the vehicle axis and the actual vehicle movement) down to a minimum and, until the physical limits are reached, prevent the vehicle breaking away.

Basically speaking, practically all types of angle-of-rotation sensors are suitable for registering the steering-wheel angle. Safety considerations, though, dictate that only those types are used which can be easily checked for plausibility, or which in the ideal case automatically check themselves. Potentiometer principles are used, as well as optical code-registration and magnetic principles. Whereas a passenger-car steering wheel turns through ±720° (a total of 4 complete turns), conventional angle-of-rotation sensors can only measure maximum 360°. This means that with the majority of the sensors actually used for this purpose it is necessary to continually register and store the data on the steering wheel's actual setting.

Design and operating concept

There are two absolute-measuring (in contrast to incremental-measuring) magnetic angle-of-rotation sensors available which are matched to the Bosch ECUs. At any instant in time, these sensors can output the steering-wheel angle throughout the complete angular range.

Hall-effect steering-wheel-angle sensor (LWS1)

The LWS1 uses 14 Hall-effect vane switches to register the angle and the rotations of the steering wheel. The Hall-effect vane switch is similar in operation to a light barrier. A Hall-effect element measures the magnetic field of an adjacent magnet. A magnetic code disc rotates with the steering shaft and strongly reduces the magnet's field or screens it off completely. In this manner, with nine Hall ICs it is possible to obtain the steering wheel's angular position in digital form. The remaining five Hall-effect sensors register the particular steering-wheel revolution which is transformed to the final 360° range by 4:1 step-down gearing.

The first item from the top in the exploded view of the LWS 1 steering-wheel-angle sensor (Fig. 1) shows the nine permanent magnets. These are screened individually by the magnetically-soft code disc beneath them when this rotates along with the steering shaft, and depending upon steering-wheel movement. The PCB immediately below the code disc contains Hall-effect switches (IC), and a microprocessor in

1 Exploded view of the digital LWS1 steering-wheel-angle sensor

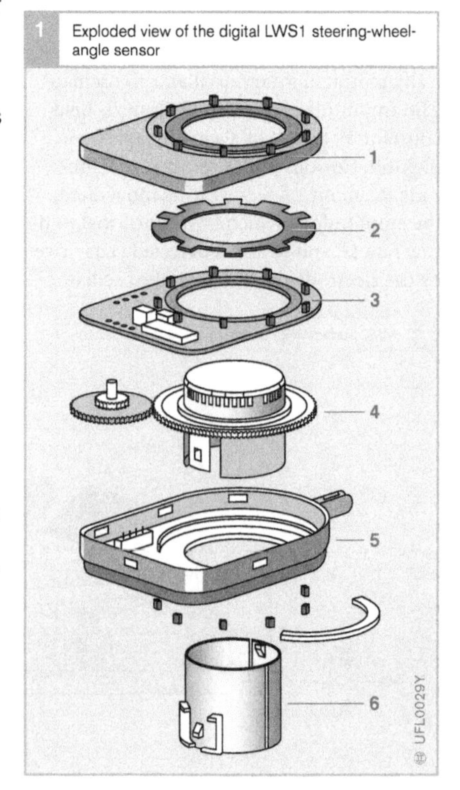

Fig. 1
1 Housing cover with nine equidistantly spaced permanent magnets
2 Code disc (magnetically soft material)
3 PCB with 9 Hall-effect switches and microprocessor
4 Step-down gearing
5 Remaining 5 Hall-effect vane switches
6 Fastening sleeve for steering column

which plausibility tests are performed and information on angular position decoded and conditioned ready for the CAN-Bus. The bottom half of the assembly contains the step-down gearing and the remaining five Hall-effect vane switches.

The LWS1 was superseded by the LWS3 due to the large number of sensor elements required, together with the necessity for the magnets to be aligned with the Hall-IC.

Magnetoresistive steering-wheel-angle sensor LWS3

The LWS 3 also depends upon AMR (anisotropic magnetoresistive sensors) for its operation. The AMR's electrical resistance changes according to the direction of an external magnetic field. In the LWS3, the information on angle across a range of four complete rotations is provided by measuring the angles of two gearwheels which are rotated by a third gearwheel on the steering-column shaft. The first two gearwheels differ by one tooth which means that a definite pair of angular variables is associated with every possible steering-wheel position.

By applying a mathematical algorithm (a computing process which follows a defined step-by-step procedure) referred to here as a modified vernier principle, it is possible to use the above AMR method for calculating the steering-wheel angle in a microcomputer. Here, even the measuring inaccuracy of the two AMR sensors can be compensated for. In addition, a self-check can also be implemented so that a highly plausible measured value can be sent to the ECU.

Fig. 2 shows the schematic representation of the LWS3 steering-wheel-angle sensor. The two gearwheels, with magnets inserted, can be seen. The sensors are located above them togther with the evaluation electronics. With this design too, price pressure forces the development engineers to look for innovative sensing concepts. In this respect, investigation is proceeding on whether, since it only measures up to 360°, a single AMR angle-of-rotation sensor (LWS4) on the end of the steering shaft would be accurate enough for ESP (Fig. 4).

2 AMR steering-wheel-angle sensor LWS3 (principle)

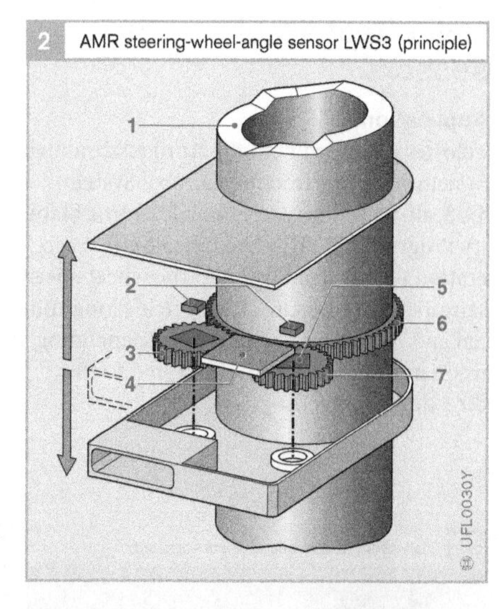

Fig. 2
1 Steering-column shaft
2 AMR sensor elements
3 Gearwheel with m teeth
4 Evaluation electronics
5 Magnets
6 Gearwheel with n > m teeth
7 Gearwheel with m + 1 teeth

3 AMR steering-wheel-angle sensor LWS3

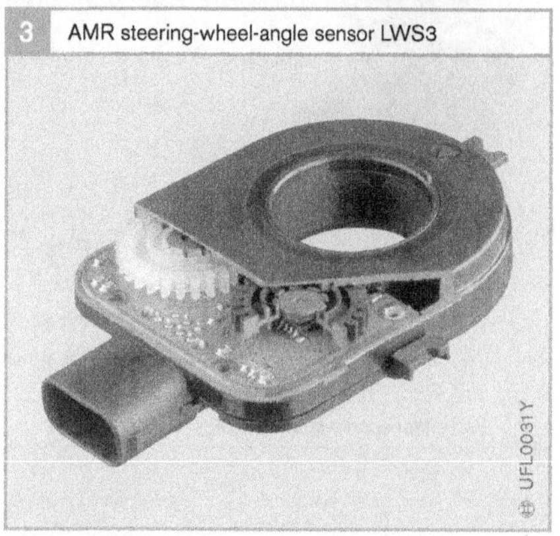

4 AMR steering-wheel-angle sensor LWS4 for attachment to the end of the steering-column shaft

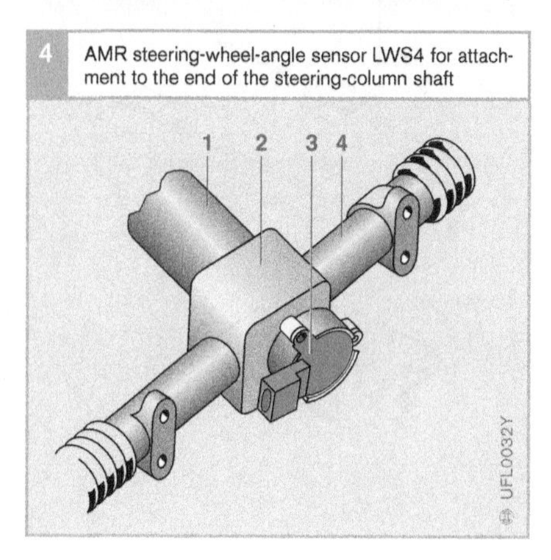

Fig. 4
1 Steering column
2 Steering box
3 Steering-wheel-angle sensor
4 Steering rack

Hall-effect acceleration sensors

Applications

Vehicles equipped with the Antilock Braking System ABS, the Traction Control System TCS, all-wheel drive, and/or Electronic Stability Program ESP, also have a Hall-effect acceleration sensor in addition to the wheel-speed sensors. This measures the vehicle's longitudinal and transverse accelerations (depending upon installation position referred to the direction of travel).

1 Hall-effect acceleration sensor (opened)

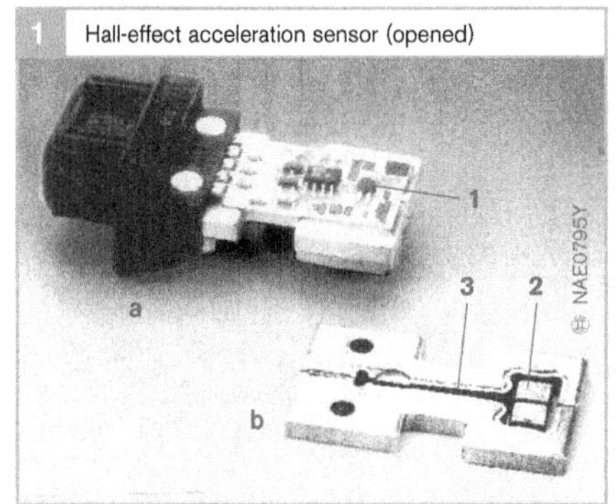

Fig. 1
a Electronic circuitry
b Spring-mass system
1 Hall-effect sensor
2 Permanent magnet
3 Spring

Design and construction

A resiliently mounted spring-mass system is used in the Hall-effect acceleration sensors (Figs. 1 and 2).

It comprises an edgewise-mounted strip spring (3) tightly clamped at one end. Attached to its other end is a permanent magnet (2) which acts as the seismic mass. The actual Hall-effect sensor (1) is located above the permanent magnet together with the evaluation electronics. There is a small copper damping plate (4) underneath the magnet.

Operating concept

When the sensor is subjected to acceleration which is lateral to the spring, the spring-mass system changes its neutral position accordingly. Its deflection is a measure for the acceleration. The magnetic flux F from the moving magnet generates a Hall voltage U_H in the Hall-effect sensor. The output voltage U_A from the evaluation circuit is derived from this Hall voltage and climbs linearly along with acceleration (Fig. 3, measuring range approx. 1 g).

This sensor is designed for a narrow bandwidth of several Hz and is electrodynamically damped.

2 Hall-effect acceleration sensor

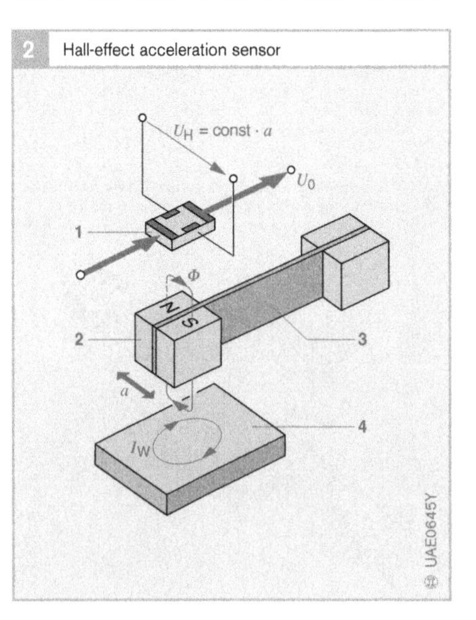

Fig. 2
1 Hall-effect sensor
2 Permanent magnet
3 Spring
4 Damping plate
I_W Eddy currents (damping)
U_H Hall voltage
U_0 Supply voltage
Φ Magnetic flux
a Applied (transverse) acceleration

3 Hall-effect acceleration sensor (example of curve)

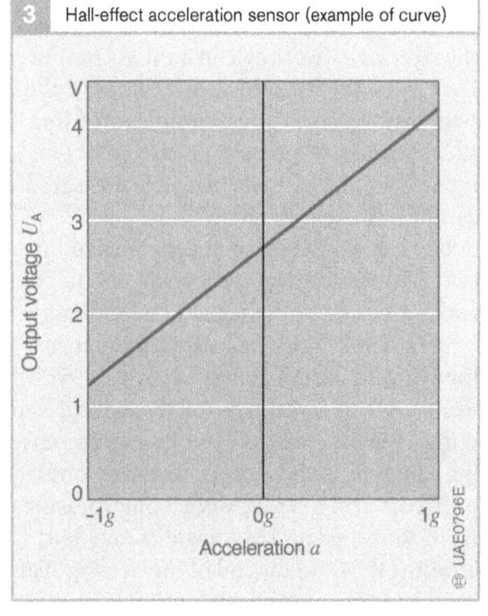

Miniaturization

Thanks to micromechanics it has become possible to locate sensor functions in the smallest possible space. Typically, the mechanical dimensions are in the micrometer range. Silicon, with its characteristics has proved to be a highly suitable material for the production of the very small, and often very intricate mechanical structures. With its elasticity and electrical properties, silicon is practically ideal for the production of sensors. Using processes derived from the field of semiconductor engineering, mechanical and electronic functions can be integrated with each other on a single chip or using other methods.

Bosch was the first to introduce a product with a micromechanical measuring element for automotive applications.

This was an intake-pressure sensor for measuring load, and went into series production in 1994. Micromechanical acceleration and yaw-rate sensors are more recent developments in the field of miniaturisation, and are used in driving-safety systems for occupant protection and vehicle dynamics control (Electronic Stability Program ESP). The illustrations below show quite clearly just how small such components really are.

Micromechanical acceleration sensor

Electric circuit

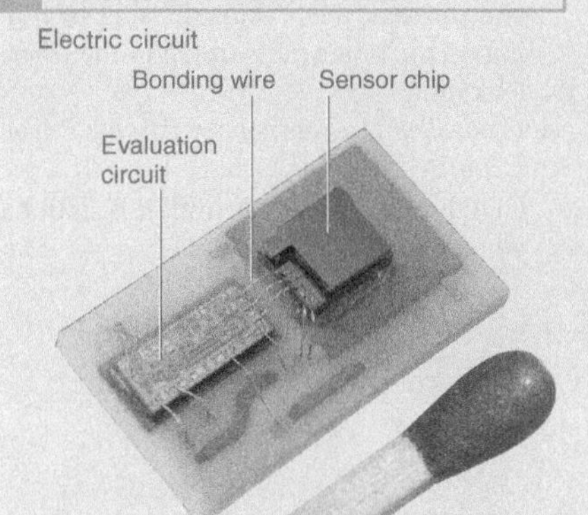

Comb-like structure compared to an insect's head

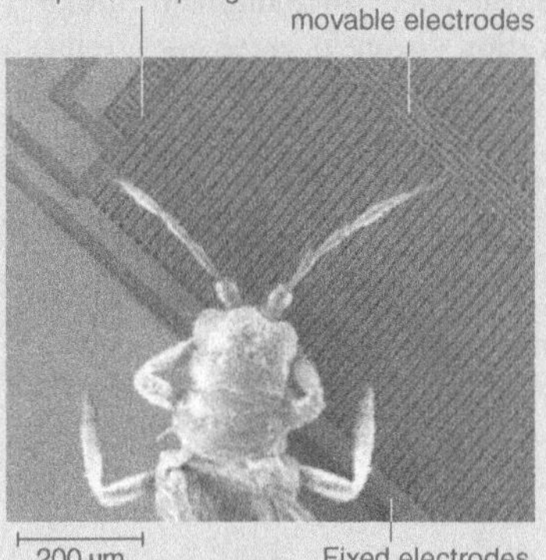

Micromechanical yaw-rate sensor

DRS-MM1 vehicle-dynamics control (ESP)

DRS-MM2 roll-over sensing, navigation

Automatic brake functions

The possibilities of today's electronic brake systems go far beyond the tasks for which they were originally designed. Originally the antilock braking system (ABS) was only used to prevent the wheels of a vehicle from locking up and therefore to ensure the steerability of the vehicle even during emergency braking. Today, the brake system also controls the distribution of the braking-force. The electronic stability program (ESP), with its ability to build up brake pressure independently of the position of the brake pedal, offers a whole series of possibilities for active brake intervention. The ESP is intended to assist the driver by applying the brakes automatically and to therefore provide the driver with a higher level of comfort and convenience. Some functions, however, enhance the vehicle safety since automatic brake application during an emergency results in shorter braking distances.

Overview

The main function of the electronic brake system is the Electronic Braking-force Distribution (EBD) function which replaces the mechanical components for braking-force distribution between the front and rear axles. This function not only cuts costs, but also makes the electronic distribution of the braking force extremely flexible.

Additional functions are gradually being integrated into electronic brake systems. The following additional functions are currently available:

- *Hydraulic Brake Assist (HBA):* HBA detects emergency braking situations and shortens the braking distance by building up the brake pressure up to the wheel-lock limit.
- *Controlled Deceleration for Parking Brake (CDP):* CDP brakes the vehicle until it is stationary when requested by the driver.

1 Block diagram

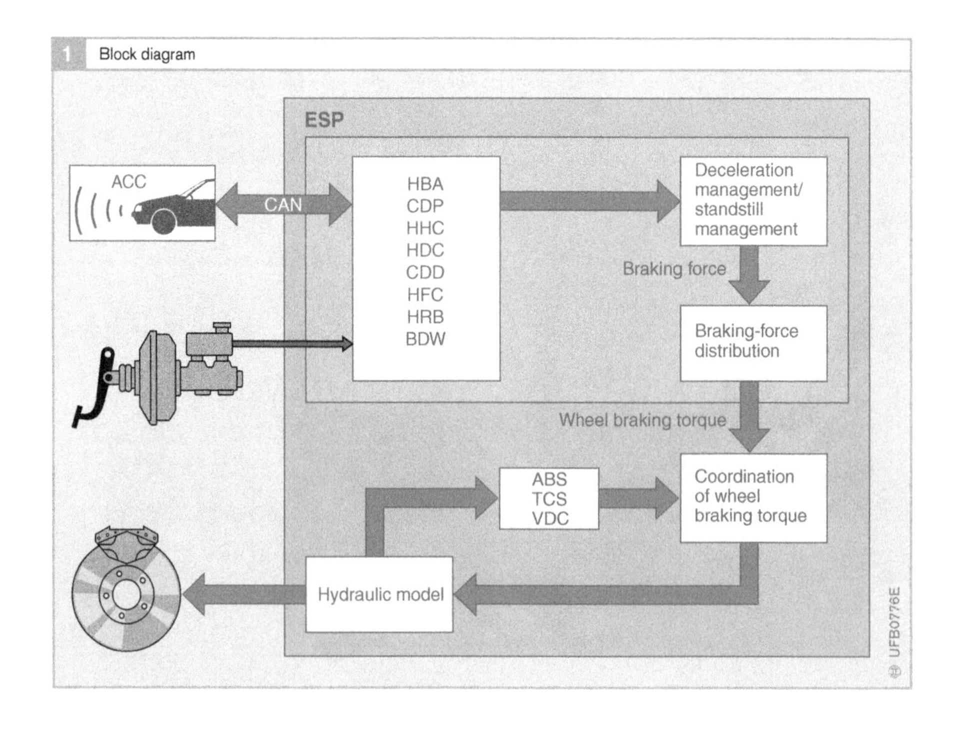

- *Hill Hold Control (HHC):*
  HHC intervenes in the brake system when pulling away on a hill and prevents the vehicle from rolling backwards.
- *Hill Descent Control (HDC):*
  HDC assists the driver when driving downhill on steep terrain by automatically applying the brakes.
- *Controlled Deceleration for Driver Assistance Systems (CDD):*
  CDD brakes the vehicle if required in combination with automatic vehicle-to-vehicle ranging.
- *Hydraulic Fading Compensation (HFC):*
  HFC intervenes if the maximum possible vehicle deceleration is not achieved even though the driver is forcefully pressing the brake pedal, e.g. due to high brake disk temperatures.
- *Hydraulic Rear Wheel Boost (HRB):*
  HRB also increases the brake pressure in the rear wheels up to the wheel-lock limit during ABS brake application.
- *Brake Disk Wiping (BDW):*
  BDW removes splash water from the brake disks by briefly applying the brakes. This brake application is not noticed by the driver.

These functions work together with the Electronic Stability Program (ESP). Some of these functions may also be available with the Antilock Braking System (ABS) or the Traction Control System (TCS).

Most of the additional functions operate with the sensor technology of the existing electronic brake systems. Some functions, however, require additional sensors.

2 System layout of additional functions

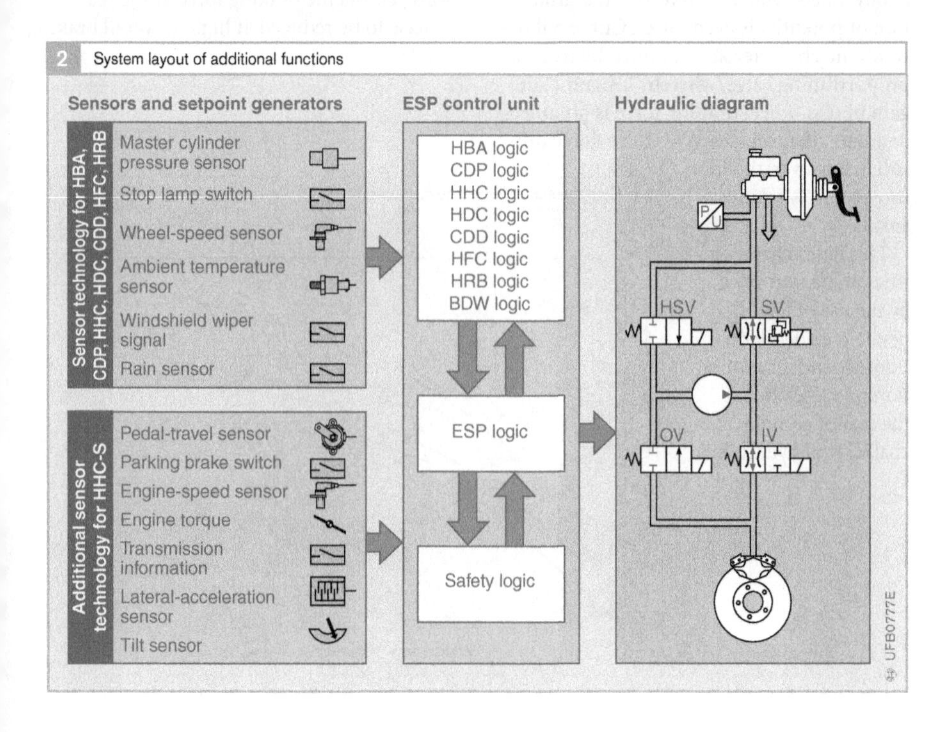

Standard function

Electronic Braking-force Distribution (EBD)

Requirements

Legal requirements demand that the braking systems of road vehicles are designed in such a way that they ensure a deceleration up to of $0.83\,g$[1] and provide stable driving behavior during all types of maneuvers so that the vehicle does not exhibit unstable handling characteristics (i. e. a tendency to skid).

[1] Gravitational acceleration $g = 9.81\ \text{m/s}^2$

Conventional Braking-force Distribution

On vehicles without ABS, this is achieved by a fixed braking-force distribution between the front and rear brakes or by the use of proportioning valves for the rear brakes (Fig. 1).

Curve 2 shows the pattern for a vehicle with a fixed braking-force distribution which, within the range $0...0.83\,g$ is below the ideal braking-force distribution curve (1l) for an unladen vehicle and fails to utilize the potential for higher rear-wheel braking forces. With a fully laden vehicle, (curve 1b), the utilization of potential is even lower. Curve 3 illustrates the characteristic obtained by using a proportioning valve, whereby a significant gain in rear-wheel braking force is obtained in an unladen vehicle. With the vehicle fully laden, however, the improvement is relatively small.

This latter characteristic can be improved by the use of load-dependent or deceleration-dependent proportioning valves, but at the cost of complex mechanics and hydraulics.

Electronic distribution

Electronic Braking-force Distribution (EBD) allows the distribution between front and rear brakes to be adjusted according to conditions. Handling response is continuously monitored and a greater proportion of the overall braking force can be applied to the rear brakes when conditions allow because the proportioning valve is dispensed with or else larger dimensioned rear wheel brakes are fitted. This releases additional braking potential for the front wheels which can be utilized on vehicles with a high forward weight bias in particular.

Design

The vehicle is designed in such a way that without a proportioning valve, the fixed braking-force distribution curve intersects the ideal braking-force distribution curve (curve 1) at a point P (Fig. 2) at a lower overall braking force, e.g. $0.5\,g$. The use of the existing ABS system's hydraulics, sensors and electronics, but with modified valves and software, allows the braking force at the rear wheels to be reduced at higher overall braking levels.

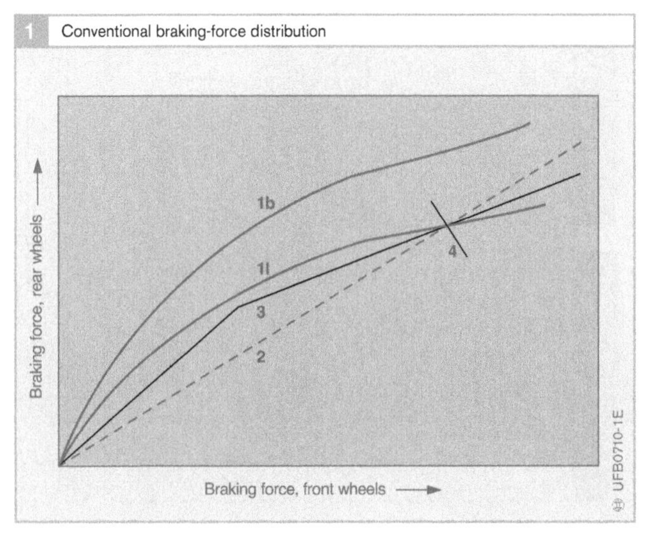

Fig. 1
1 Ideal braking-force distribution of a vehicle:
1l Unladen vehicle
1b Fully laden vehicle
2 Fixed braking-force distribution
3 Braking-force distribution with proportioning valve
4 Straight line for retardation of $0.83\,g$ (g: gravitational acceleration)

Method of operation
The ECU continuously calculates the slip difference between the front and rear wheels in all driving situations. If the ratio of front to rear wheel slip exceeds a defined stable-handling threshold when braking, the ABS pressure inlet valve for the appropriate rear wheel is closed. This prevents further increase of brake pressure at that wheel.

If the driver then further increases the force applied to the brake pedal, and therefore the brake pressure, the degree of slip at the front wheels also increases. The difference between front and rear wheel slip diminishes and the pressure inlet valve is opened again so that the pressure at the rear wheel rises once more. This process may then be repeated a number of times depending on the brake pedal force and the maneuver being performed. The electronic braking-force distribution curve then takes on a staircase appearance (curve 3) which approximates to the ideal braking-force distribution curve.

For the Electronic Braking-force Distribution (EBD) function, only the rear-brake valves of the ABS system are activated, the return pump motor in the hydraulic modulator unit remains de-energized.

Advantages
The characteristics of the EBD system outlined above provide the following advantages:

- Optimized vehicle handling stability under all payload conditions, in all cornering situations, on uphill or downhill gradients, and in any drivetrain status (clutch engaged/disengaged, automatic transmission),
- No need for conventional proportioning valves or limiting valves,
- Reduced thermal stresses on the front brakes,
- Even wear between front and rear brake pads,
- Better vehicle deceleration with the same pedal force,
- Constant braking-force distribution patterns over the entire life of the vehicle,
- Only minor modifications to existing ABS components are required.

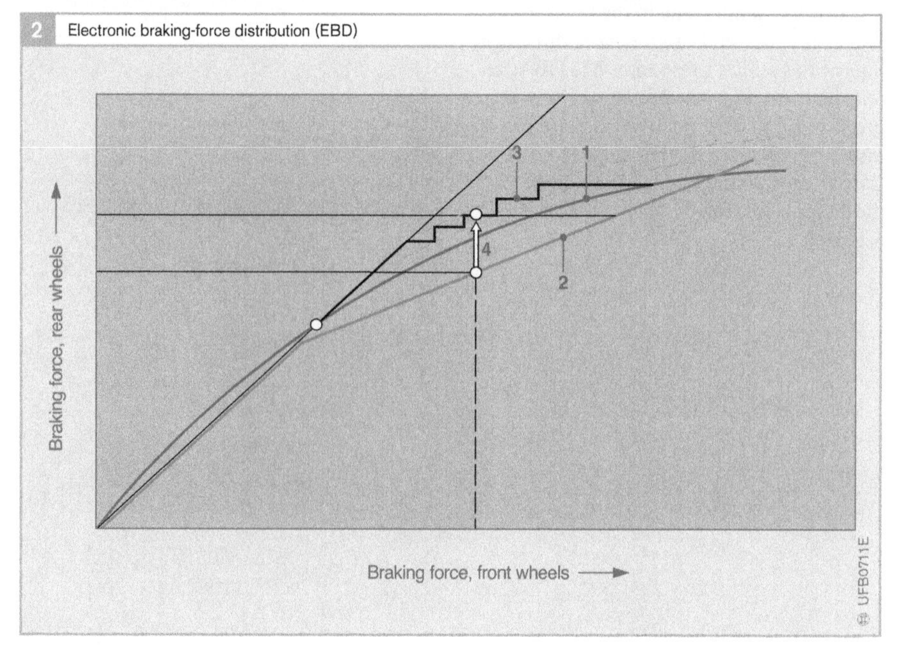

Fig. 2
1 Ideal braking-force distribution
2 Design braking-force distribution
3 Electronic braking-force distribution
4 Gain in rear-wheel braking force

Additional functions

Hydraulic Brake Assist (HBA)

The main task of the hydraulic brake assist is to detect an emergency braking situation and as a result to automatically increase the vehicle deceleration. The vehicle deceleration is only limited by the intervention of ABS control and is therefore close to the optimum level possible within physical limits. A normal driver is therefore able to achieve the short braking distances that could previously only be achieved by specially trained drivers. If the driver reduces the desired level of braking, the vehicle deceleration is reduced in accordance with the force applied to the brake pedal. The driver can therefore precisely modulate the vehicle deceleration when the emergency braking situation has passed.

The driver's desired level of braking is determined by the force or pressure he/she applies to the brake pedal. The pedal pressure is derived from the measured master cylinder pressure taking into account the current hydraulic control.

The driver can intervene in the brake application at any time and can therefore directly influence the vehicle's response. The HBA can only increase the brake pressure. The primary pressure applied by the driver is therefore the minimum pressure used by the system. If a system error occurs, the HBA is shutoff and the driver is informed that an error has occurred.

Controlled Deceleration for Parking Brake CDP

The Electromechanical Parking Brake (EPB) is an automated parking brake system. It replaces the conventional hand brake lever or foot-operated parking brake lever with an electric motor. The disadvantage of the parking brake is that it only affects the rear axle and its braking force is limited in the event of emergency braking. The CDP function increases braking deceleration and simultaneously enables an ESP system controller to be used to ensure vehicle stability.

The CDP function is an additional function to actively increase the brake pressure in vehicles fitted with an hydraulic braking system and ESP system. On the driver's request, the CDP function automatically decelerates the vehicle until it is stationary. Once the vehicle has come to a standstill, the ESP hydraulics system briefly adopts all static parking brake processes.

1 Comparison of braking with and without brake assist function

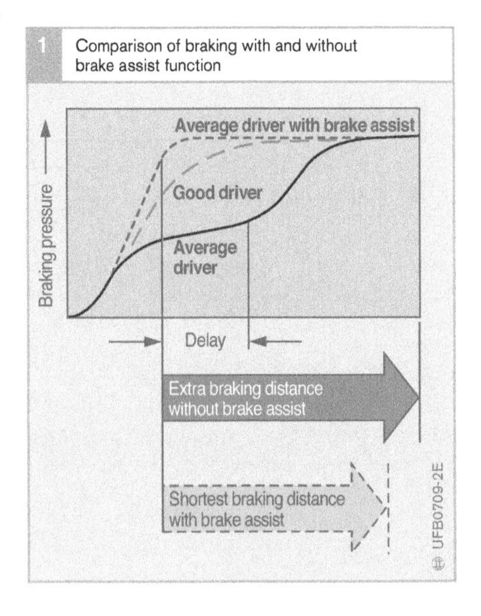

Hill Hold Control HHC

The Hill **Hold Control** (HHC) starting-off assistant is a comfort and convenience function that prevents the vehicle from rolling backwards when pulling away on hills and inclines. The gradient of the hill or slope is measured by a tilt sensor (longitudinal acceleration sensor). To operate, the starting-off assistant requires the brake pressure available when the vehicle is stationary that was built up when the driver pressed the foot brake.

The brake pressure specified by the driver when stopping the vehicle is maintained in the brake system when the system detects that the vehicle is stationary, even if the driver releases the brake pedal. The brake pressure is reduced after a pressure holding time of up to a maximum of two seconds. During this time the driver can press the accelerator pedal and pull away. The brake pressure is reduced when the system detects the driver's intention to pull away.

The system detects that the driver wishes to pull away if the engine torque is sufficient for the vehicle to move in the desired direction. This state can be triggered by the driver pressing the accelerator pedal and/or clutch as well as by the transmission outputting an engine torque (e. g. automatic/continuously variable transmission).

The brake pressure is not maintained in the brake system if there is already sufficient engine torque when the vehicle is stationary (e.g. due to the accelerative force of the automatic transmission).

If the driver presses the accelerator pedal during the vehicle-specific holding time, the holding time is extended until sufficient engine torque is available for the vehicle to pull away.

If the driver does not press the accelerator or the brake pedal, the function is deactivated at the latest after two seconds. The vehicle then starts rolling.

The HHC function is also designed as an additional function for the ESP system and it uses parts of this system. The function is activated automatically.

2 Function description of HHC

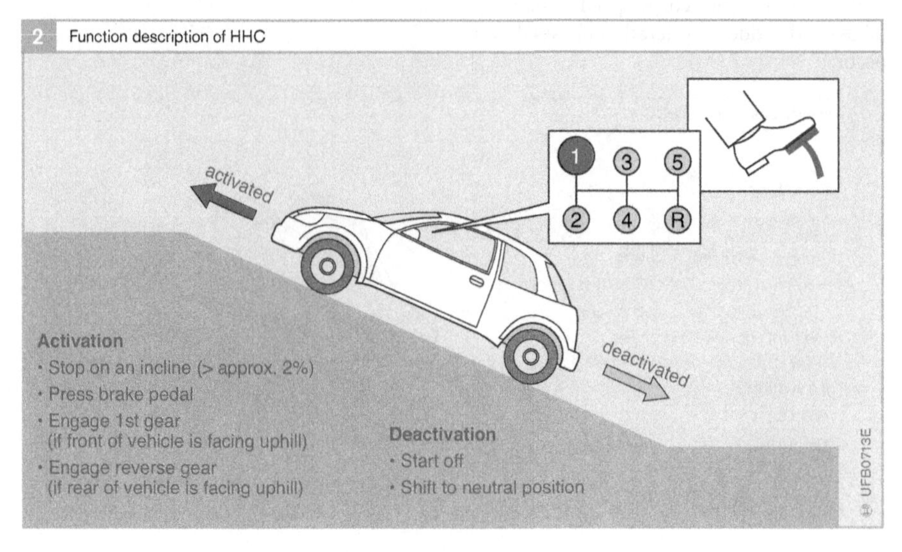

Hill Descent Control HDC

The Hill Descent Control (HDC) is a comfort and convenience function that assists the driver when driving down hill (on gradients of up to 50%) by automatically applying the brakes. Once this function is activated a predetermined, low desired-speed is introduced and regulated without any necessary intervention from the driver.

The driver must activate and deactivate the HDC function by pressing the HDC push-button.

If required the driver can vary the predetermined speed by pressing the brake and accelerator pedal or using the control buttons of a speed control system.

If the brake slip of the wheels is too high during HDC control, the ABS system intervenes automatically. If the wheels are on different road surfaces, the braking torque of the slipping wheels is automatically distributed to the wheels with a higher coefficient of friction.

If engine braking torque is available it is used automatically. Compared to the exclusive use of engine braking torque in order to maintain the vehicle speed, the HDC function has the advantages that if the wheels leave the surface of the road (loss of engine braking torque) the vehicle speed is maintained and sudden acceleration phases do not occur.

Another advantage of the HDC function is the variable distribution of the braking force which is coupled to the automatic driving direction recognition function. When reversing the rear axle is braked more forcefully to ensure optimum steerability even if the front axle is relieved of the load.

The level ground detection function integrated in the HDC only permits brake application by the HDC when driving downhill. If the vehicle is on the flat or driving uphill, the HDC switches to a standby mode and is reactivated automatically as soon as the system detects that the vehicle is traveling downhill.

To prevent the driver misusing this function, the HDC also switches to its standby mode if the accelerator pedal is pressed passed a certain threshold or if the vehicle exceeds a maximum control speed. The HDC function is deactivated if the vehicle accelerates beyond a preset cutoff-speed.

The status of the HDC function is indicated by the HDC indicator lamp. The brake lamp indicates when the HDC function applies the brakes.

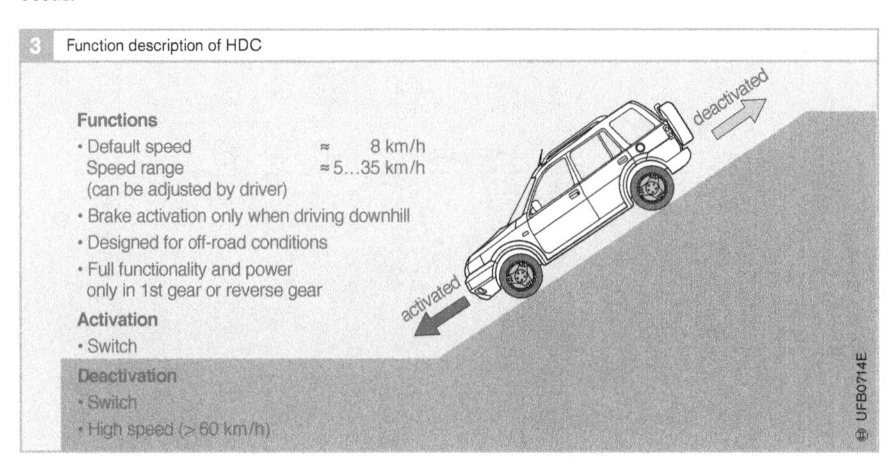

3 Function description of HDC

Controlled Deceleration for Driver Assistance Systems CDD

The basic CDD function is an additional function for active brake application with the Adaptive Cruise Control (ACC), i. e. for automatic vehicle-to-vehicle ranging. The brakes are applied automatically without the driver pressing the brake pedal as soon as the distance to the vehicle in front falls below a predetermined distance. CDD is based on a hydraulic braking system and an ESP system.

The CDD function receives a request to decelerate the vehicle by a desired amount (input). CDD then sends the actual amount of deceleration (output) which is achieved by controlling the pressure using hydraulics. The requested amount of deceleration is specified by the cruise control system which is connected upstream.

4 Function description of CDD

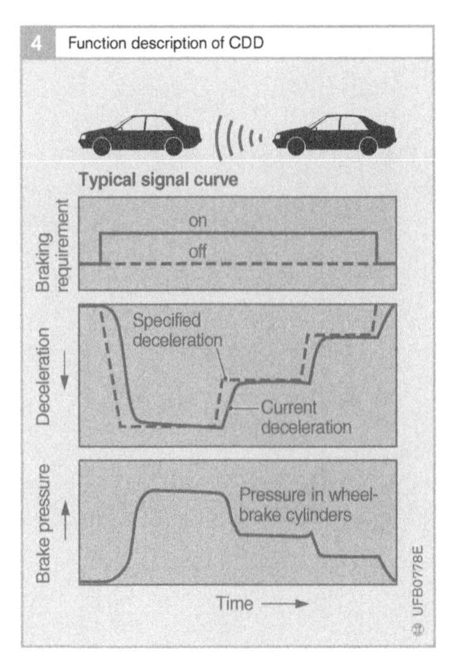

Hydraulic Fading Compensation HFC

The Hydraulic Fading Compensation (HFC) function offers the driver additional brake servo assistance. The function is activated if the maximum possible vehicle deceleration is not achieved even if the driver forcefully presses the brake pedal to the point that would normally cause the lockup pressure to be reached (primary pressure over approx. 80 bar). This is the case, for example at high brake disk temperatures or if the brake pads have a considerably reduced coefficient of friction.

When the HFC is activated, the wheel pressures are increased until all wheels have reached the lockup pressure level and ABS control is initiated. The brake application is therefore at the physical optimum. The pressure in the wheel-brake cylinders can then exceed the pressure in the master cylinder, also during ABS control.

If the driver reduces the desired level of braking to a value below a particular threshold value, the vehicle deceleration is reduced in accordance with the force applied to the brake pedal. The driver can therefore precisely modulate the vehicle deceleration when the braking situation has passed. The HFC cuts off if the primary pressure or vehicle speed falls below the respective cutoff threshold.

5 Function description of HFC

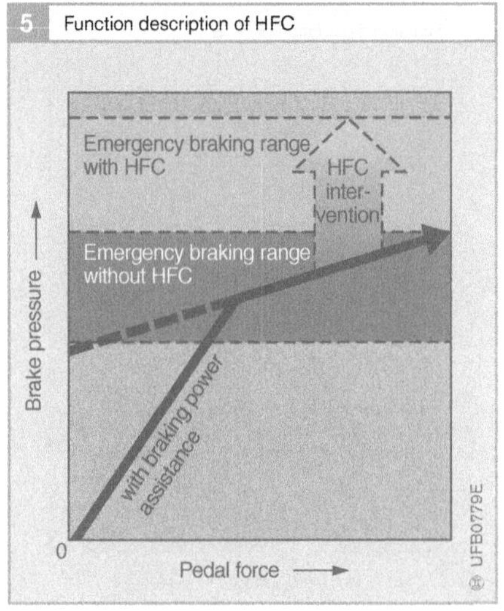

Hydraulic Rear Wheel Boost HRB

The Hydraulic Rear wheel Boost (HRB) provides the driver with additional brake servo assistance for the rear wheels if the front wheels are controlled by the ABS system. This function was introduced because many drivers do not increase the pedal force at the start of ABS control even though the situation would require this. When the HRB function is active, the wheel pressures are increased at the rear wheels until they also reach the lockup pressure and ABS control is initiated. The brake application is therefore at the physical optimum. The pressure in the rear axle wheel-brake cylinders can then exceed the pressure in the master cylinder, also during ABS control.

The HRB cuts off when the wheels at the front axle are no longer under ABS control or when the primary pressure falls below the cutoff threshold.

Brake Disk Wiping BDW

The Brake Disk Wiping (BDW) function detects rain or a wet road by evaluating windshield wiper or rain sensor signals and then actively increases the brake pressure in the service brake. The brake pressure buildup is used to remove splash water from the disk brake to ensure minimum brake response times when driving in wet conditions. The pressure level when dry-braking is adjusted so that the vehicle deceleration cannot be perceived by the driver.

Dry-braking is repeated at a defined interval for as long as the system detects rain or a wet road. If required just the disks at the front axle can be wiped.

The BDW function interrupts the wiping procedure as soon as the driver applies the brakes.

6 Function description of HRB

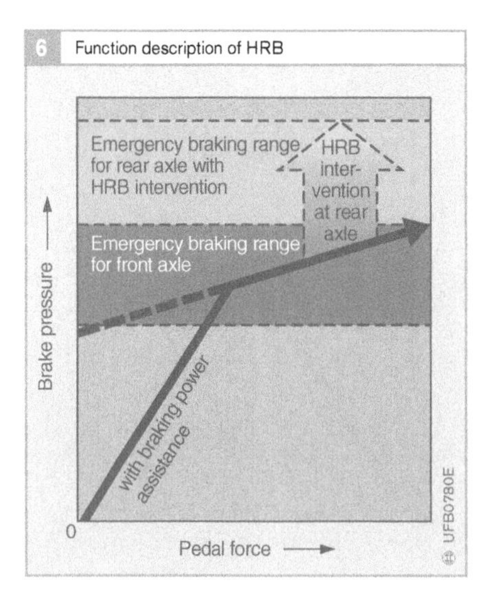

Directional stability

Satisfactory handling is defined according to whether a vehicle maintains a path that accurately reflects the steering angle while at the same time remaining stable. To meet this stability criterion, the vehicle must remain consistently secure and sure-footed, without "plowing" or breaking away.

Dynamic lateral response is a critical factor in the overall equation. The response pattern is defined based on the vehicle's lateral motion (characterized by the side-slip angle) and its tendency to rotate around its vertical axis (yaw velocity) (Fig. 1).

Fig. 2 illustrates the dynamic lateral response of a vehicle being driven with a fixed steering angle (skid-pad circulation). Position 1 represents the instant when steering input is applied. Curve 2 is the vehicle's subsequent course on a high-grip road surface; this track is an accurate reflection of the steering angle. This is the case when the coefficient of friction is sufficient to transfer the lateral acceleration forces to the road surface. If the coefficient of friction is lower e.g. due to a slippery road surface, the side-slip angle becomes excessive (curve 3). Although with a controlled yaw velocity the vehicle will rotate just as far around its vertical axis as in curve 2, the larger side-slip angle is now a potential source of instability. For this reason the electronic stability program controls the yaw velocity and limits the side-slip angle β (curve 4).

2 Dynamic lateral response

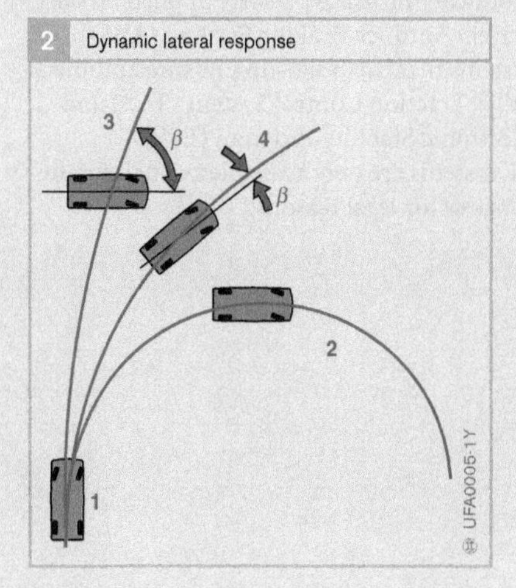

Fig. 2
1 Steering input, fixed steering-wheel angle
2 Vehicle path on high-grip surface
3 Vehicle path on low-grip surface with yaw velocity control
4 Vehicle path on low-grip surface with additional control of the side-slip angle β (ESP)

1 Vehicle travel directions

Vehicle vertical axis
Yaw velocity
Vehicle longitudinal axis
Linear acceleration
Vehicle horizontal axis
Lateral acceleration
Rolling
Steering movement
UFA0013-1E

Hydraulic modulator

The hydraulic modulator forms the hydraulic connection between the master cylinder and the wheel-brake cylinders and is therefore the central component of electronic brake systems. It converts the control commands of the electronic control unit and uses solenoid valves to control the pressures in the wheel brakes.

A distinction is made between systems that modulate the brake pressure applied by the driver (Antilock Braking System, (ABS)) and systems that can build-up pressure automatically (Traction Control System (TCS) and Electronic Stability Program (ESP)).
All systems are only available as dual-circuit versions for legal reasons.

Development history

The transition from 3/3 to 2/2 solenoid valve was a milestone in the development of ABS. With 3/3 valves, which were used in generation 2, the control functions for building up, maintaining and reducing pressure could be carried out using just one valve. The valves had three hydraulic connections to carry out these functions. The disadvantages of this valve design were extremely expensive electrical control and great mechanical complexity. Control with the 2/2 valves of the current generations offers a less expensive solution. The following section describes how these valves work.

Generation 8, which was introduced onto the market in 2001, is designed as a fully modular system. The hydraulic system can therefore be tailored to meet the requirements of the respective vehicle manufacturer such as with regard to value added functions, comfort and convenience, vehicle segment (up to lightweight commercial vehicles), etc. Generation 8 is immersion-proof which means that the hydraulic modulator can withstand brief immersion in water.

1 ESP 8 hydraulic modulator

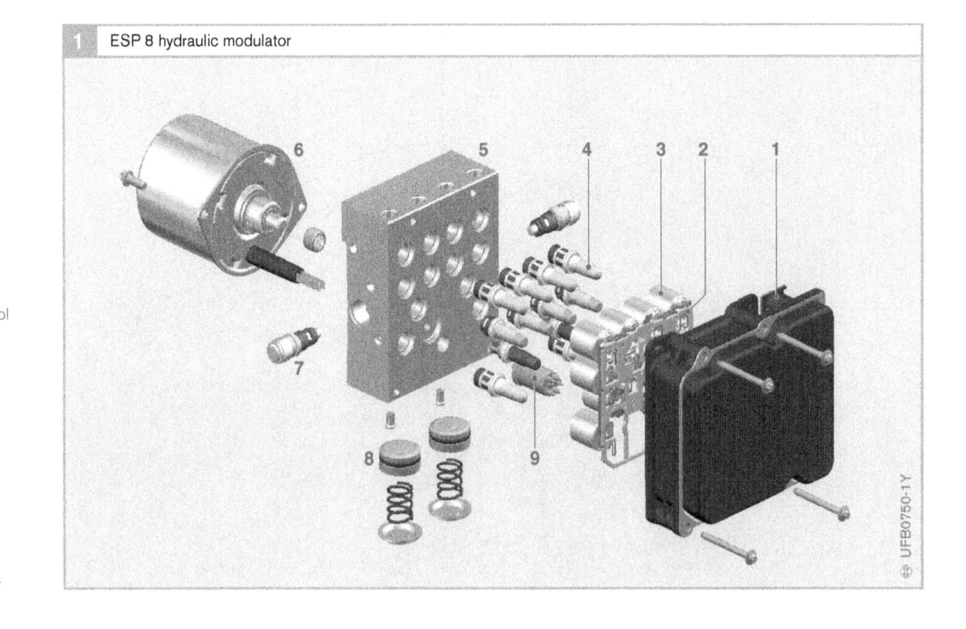

Fig. 1
1 Electronic control unit
2 Coil grid
3 Coils/solenoid group
4 Solenoid valves
5 Hydraulic block
6 DC motor
7 Plunger pump
8 Low-pressure reservoir
9 Pressure sensor

Design

Mechanical system

A hydraulic modulator for ABS/TCS/ESP consists of an aluminum block into which the hydraulic layout is drilled (Fig. 2). This block also houses the necessary hydraulic function elements (Fig. 1) presented below.

ABS hydraulic modulator

With a 3-channel ABS system, this block features one inlet valve and one outlet valve for each front wheel and one inlet valve and one outlet valve for the rear axle – a total of six valves. This system can only be used in vehicles featuring a dual brake-circuit configuration. The two wheels at the rear axle are not controlled individually, rather both wheels are controlled according to the select-low principle. This means that the wheel with the greater slip determines the controllable pressure of the axle.

With a 4-channel ABS system (for dual or X-braking-force distribution), one inlet and one outlet valve is used for each wheel resulting in a total of eight valves. This system allows each wheel to be controlled individually.

Moreover, the system features one pump element (return pump) and one low-pressure reservoir for each brake circuit. Both pump elements are operated by a shared DC motor.

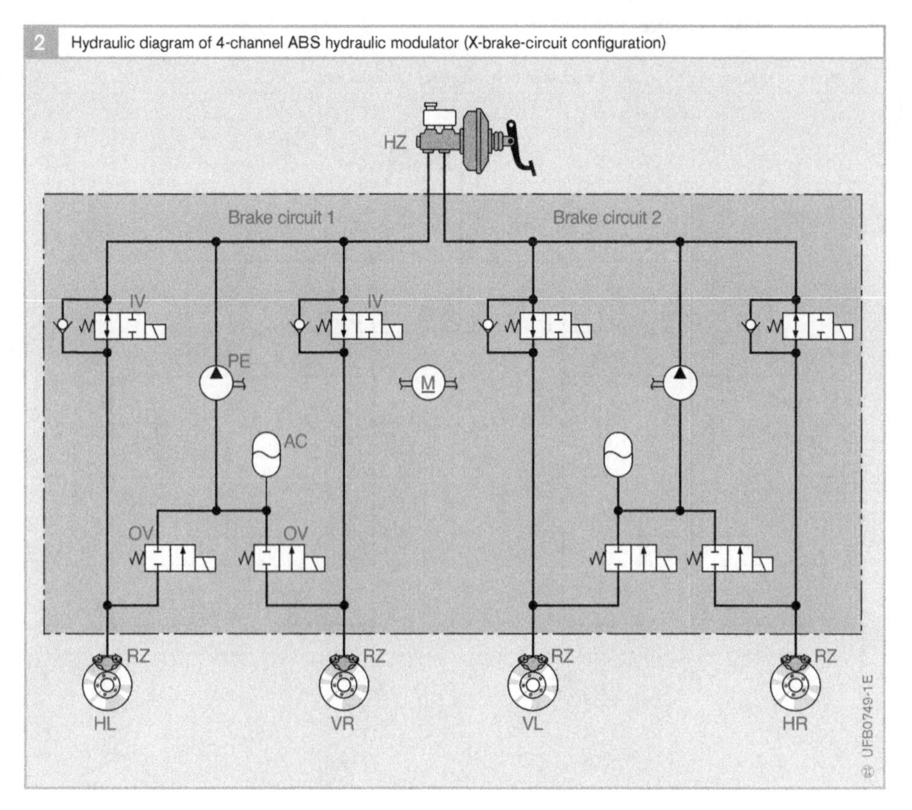

Fig. 2
HZ Master cylinder
RZ Wheel-brake cylinder
IV Inlet valve
OV Outlet valve
PE Return pump
M Pump motor
AC Low-pressure reservoir
V Front
H Rear
R Right
L Left

TCS hydraulic modulator

Unlike an ABS unit, a TCS system with II-brake-circuit configuration also has a switchover valve and an inlet valve at the rear axle (driven wheels) resulting in a total of 10 valves.

In TCS systems with X-brake-circuit configuration an additional switchover valve and inlet valve are required for each circuit (a total of 12 valves).

ESP hydraulic modulator

ESP systems require 12 valves irrespective of the brake-circuit configuration (Fig. 3). In these systems, the two inlet valves, as used in the TCS hydraulic modulator, are replaced by two high-pressure switching valves. The difference between the two types of valves is that the high-pressure switching valve can switch against higher differential pressures (> 0.1 MPa). With ESP it may be necessary to increase the brake pressure specified by the driver in order to stabilize the vehicle. During such a maneuver (partially-active maneuver), the suction path of the pump must be opened despite the high primary pressure.

An integrated pressure sensor is used exclusively in ESP systems. The sensor detects the brake pressure in the master cylinder i.e. the driver command. This is also required for a partially-active ESP control maneuver since it is important to know the primary pressure applied by the driver pressing the brake pedal.

Since TCS/ESP systems are required to generate pressure automatically, the return pump is replaced by a self-priming pump in both these systems. An additional non-return valve with a specific closing pressure is required to prevent the pump from drawing unwanted media from the wheels.

3 Hydraulic diagram of ESP hydraulic modulator (X-brake-circuit configuration)

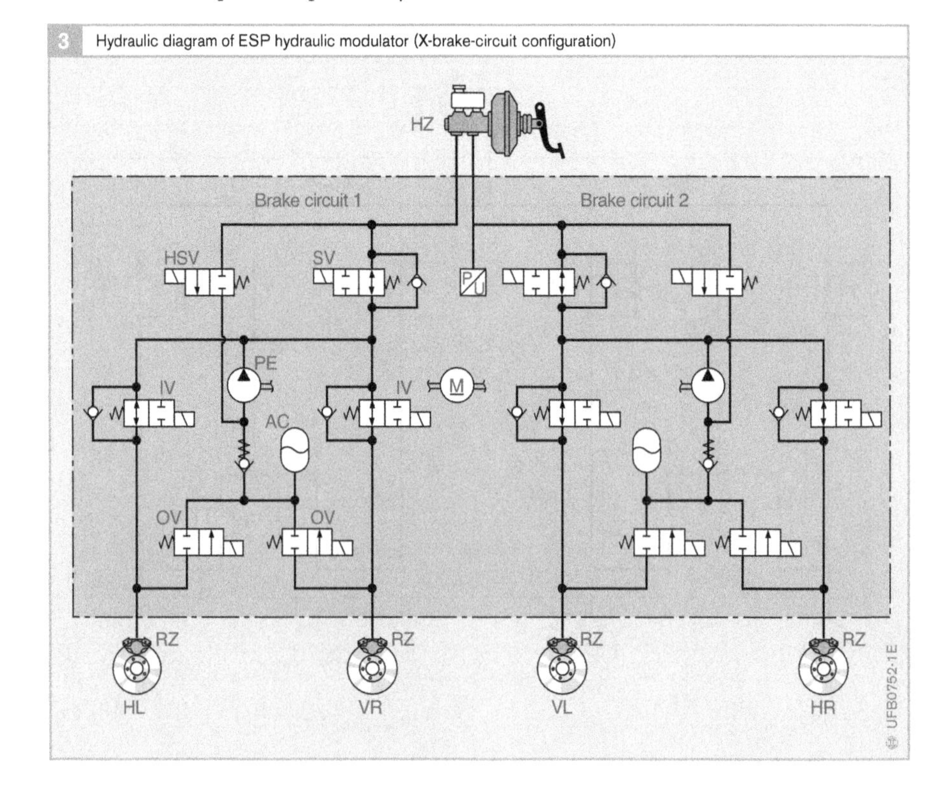

Fig. 3
HZ Master cylinder
RZ Wheel-brake cylinder
IV Inlet valve
OV Outlet valve
SV Switchover valve
HSV High-pressure switching valve
PE Return pump
M Pump motor
AC Low-pressure reservoir
V Front
H Rear
R Right
L Left

ABS versions

Evolution of the ABS system

Technological advances in the areas of

- solenoid-valve design and manufacturing,
- assembly and component integration,
- electronic circuitry (discrete components replaced by hybrid and integrated circuits with microcontrollers),
- testing methods and equipment (separate testing of electronic and hydraulic systems before combination in the hydraulic modulator), and
- sensor and relay technology

have enabled the weight and dimensions of ABS systems to be more than halved since the first-generation ABS2 in 1978. As a result, modern systems can now be accommodated even in vehicles with the tightest space restrictions. Those advances have also lowered the cost of ABS systems to the extent that it has now become standard equipment on all types of vehicle.

1 Evolution of ABS configurations

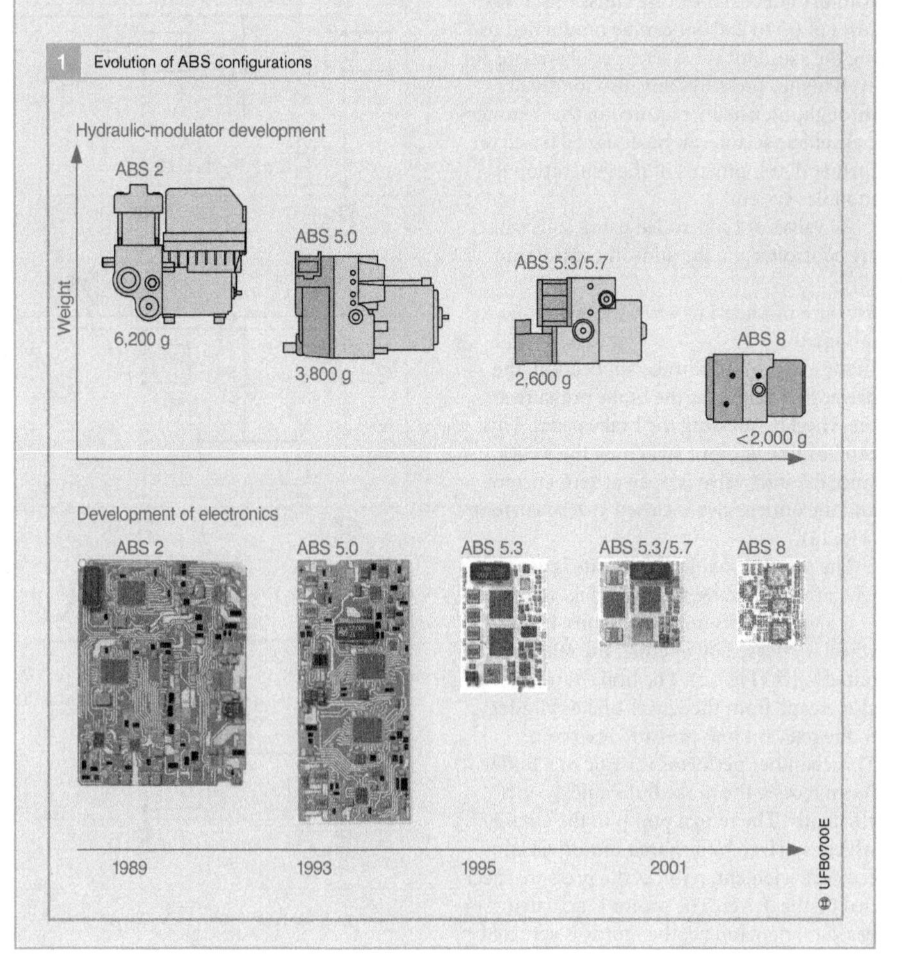

Fig. 1
Historical development of ABS showing technological advances: Decreasing weight accompanied by increasing processing power.

Pressure modulation

Modulation with ABS control

The pressure of an ABS/TCS/ESP system is modulated using solenoid valves. The outlet valves and the inlet and high-pressure switching valves (for TCS and ESP systems) are switching valves that are closed at zero current and can adopt two positions – closed or open.

In contrast, the inlet valves and switchover valves are both open at zero current and are used as control valves for the first time in generation 8. This is advantageous not only for instantaneous braking power and braking comfort but also for noise emissions. Pressures of up to 200 bar can be modulated using the standard valve set. Special systems for even higher pressures and also for greater throughput, usually required in the commercial vehicle sector, can be designed based on further developments of the generation 8 modular system.

All valves are controlled using coils which are controlled via the add-on control unit.

Pressure modulation with ABS hydraulic modulator

In the event of ABS brake application, the driver first generates the brake pressure at the wheel by pressing the brake pedal. This can be done without switching the valves since the inlet valve is open at zero current and the outlet valve is closed at zero current (Fig. 1a).

The *pressure maintenance* state is generated when the inlet valve is closed (Fig. 1b).

If a wheel locks up, the pressure from this wheel is released by opening the relevant outlet valve (Fig. 1c). The brake fluid can also escape from the wheel-brake cylinder to the relevant low-pressure reservoir. This chamber performs the role of a buffer. It can receive the brake fluid quickly and efficiently. The return pump in the circuit, which is driven by a shared motor via an eccentric element, reduces the pressure specified by the driver. The motor is actuated based on demand i.e. the motor is actuated according to the rotational speed. The ABS system can of course control several wheels if they lock up simultaneously.

1 Pressure modulation in ABS hydraulic modulator

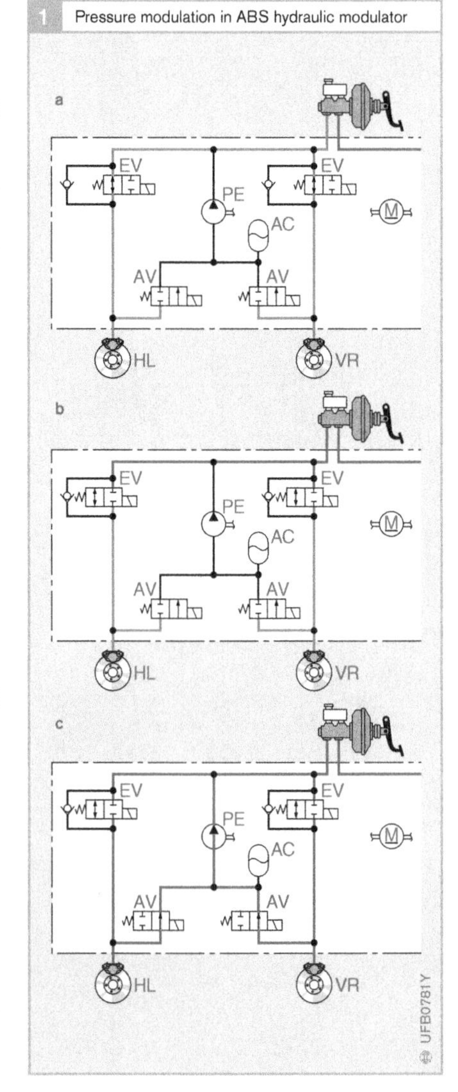

Fig. 1
a Pressure build-up
b Pressure maintenance
c Pressure reduction

IV Inlet valve
OV Outlet valve
PE Return pump
M Pump motor
AC Low-pressure reservoir
V Front
H Rear
R Right
L Left

Pressure modulation with ESP hydraulic modulators

Pressure is modulated with ESP control using ESP hydraulics in the same way as described for ABS. Unlike with ABS, however, the wheel-brake cylinder and master cylinder are also connected via a switchover valve that is open at zero current and a high-pressure switching valve that is closed at zero current. These two valves are required for active/partially-active brake intervention (Fig. 2).

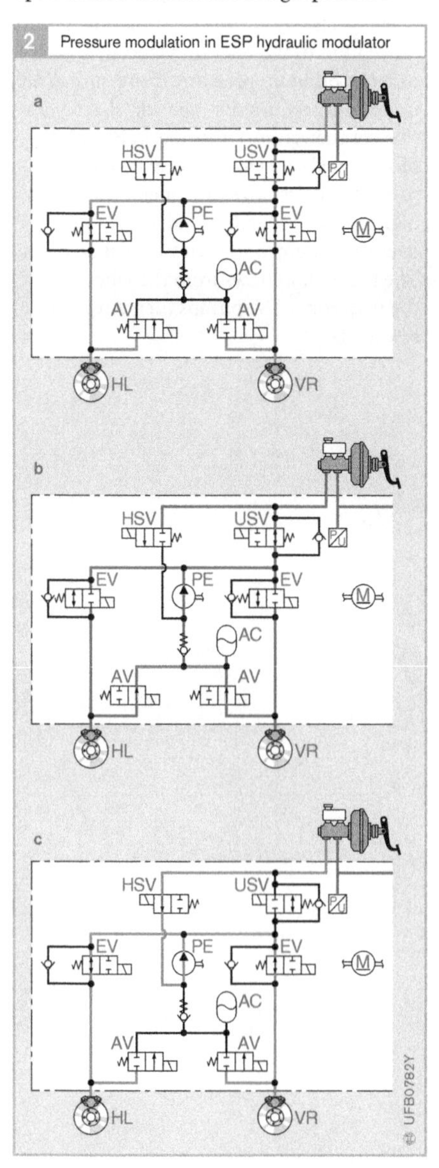

2 Pressure modulation in ESP hydraulic modulator

Pressure generation with ESP

The pressure generation chain consists of two self-priming pumps and a motor. Plunger pumps are used, as for ABS, but these pumps can generate pressure without requiring the primary pressure applied when the driver presses the brake pedal. These pumps are driven by a DC motor based on demand. The motor drives an eccentric bearing located on the shaft of the motor.

TCS/ESP pumps can build up pressure independently of the driver or increase the braking pressure already generated by the driver. These systems are therefore able to initiate brake application. To do this, the switchover valve is closed and the inlet valve or high-pressure switching valve is opened. Fluid can then be drawn from the brake fluid reservoir via the master cylinder and pressure can be built up in the wheel-brake cylinders (Fig. 2c). This is required not only for TCS/ESP functions, but also for many additional convenience functions (value added functions such as the brake assist (HBA)).

Demand-based control of the pump motor reduces noise emission during pressure generation and regulation. The pumps can be equipped with damping elements to satisfy vehicle manufacturer's strict low noise emission requirements.

Fig. 3

- a Pressure build-up when braking
- b Pressure reduction with ABS control
- c Pressure build-up via self-priming pump due to TCS or ESP intervention

- IV Inlet valve
- OV Outlet valve
- SV Switchover valve
- HSV High-pressure switching valve
- PE Return pump
- M Pump motor
- AC Low-pressure reservoir
- V Front
- H Rear
- R Right
- L Left

With ESP, there are basically three different applications:

- Passive – as previously described for ABS control.
- Partially-active – when the pressure specified by the driver is insufficient to stabilize the vehicle.
- Fully-active – when pressure is generated to stabilize the vehicle without the driver pressing the brake pedal.

Both pressure generation cases above, in addition to ESP control, are used for a whole series of additional functions such as adaptive cruise control and the brake assist.

Partially-active control

For partially-active control, the high-pressure switching valve must be able to open the suction path of the pump against high pressures. This is required since the driver has already generated a high pressure, but this pressure is insufficient to stabilize the vehicle.

The high-pressure switching valve is designed with two stages so that the valve can open against the high pressure. The first stage of the valve is opened via the magnetic force of the energized coil; the second stage via the hydraulic area difference.

If the ESP controller detects an unstable vehicle state, the switchover valves (open at zero current) are closed and the high-pressure switching valve (closed at zero current) is opened. The two pumps then generate additional pressure in order to stabilize the vehicle.

Once the vehicle is stabilized, the outlet valve is opened and the excess pressure in the controlled wheel escapes to the reservoir. As soon as the driver releases the brake pedal, the fluid is pumped from the reservoir back to the brake fluid reservoir.

Fully-active control

If the ESP controller detects an unstable vehicle state, the switchover valves are closed. This prevents the pumps from being hydraulically short-circuited via the switchover/ high-pressure switching valve which would prevent pressure generation. The high-pressure switching valves are simultaneously opened. The self-priming pump now pumps brake fluid to the relevant wheel or wheels in order to build up pressure. If pressure generation is only required in one wheel, (for yaw rate compensation), the inlet valves of the other wheels are closed. To reduce the pressure, the outlet valves are opened and the high-pressure switching valves and switchover valves return to their original position. The brake fluid flows from the wheels to the reservoirs. The pumps then empty the reservoirs.

Development of hydraulic modulators

Control of hydraulic modulators
An electronic control unit processes the information received from the sensors and generates the control signals for the hydraulic modulator. The hydraulic modulator incorporates a series of solenoid valves that can open or close the hydraulic circuits between the master cylinder and the wheel-brake cylinders.

Hydraulic modulators with 3/3 solenoid valves
In 1978 version ABS2S was the first antilock braking system to go into series production. In this ABS system the electronic control unit switches the 3/3 solenoid valves of the hydraulic modulator to three different valve positions. There is a solenoid valve for each wheel-brake cylinder (Fig. 1a).

- The first (zero-current) position connects the master cylinder and the wheel-brake cylinder; the wheel brake pressure can rise.
- The second position (excitation at half the maximum current) separates the wheel brake from the master cylinder and return line so that the wheel brake pressure remains constant.
- The third position (excitation at the maximum current) separates the master cylinder and simultaneously connects the wheel brake and return line so that the wheel brake pressure decreases.

By applying these settings in the appropriate sequences, the brake pressure can thus be increased and reduced either continuously or incrementally (and therefore gradually).

Hydraulic modulators with 2/2 solenoid valves
While version ABS2S operates with 3/3 solenoid valves, the successor systems ABS5 and ABS8 feature 2/2 solenoid valves with two hydraulic connections and two valve positions. The inlet valve between the master cylinder and the wheel-brake cylinder controls pressure build-up, while the outlet valve between the wheel-brake cylinder and the return pump controls pressure release. There is a solenoid valve pair for each wheel-brake cylinder (Fig. 1b).

- In the "pressure build-up" position, the inlet valve connects the master cylinder to the wheel-brake cylinder so that the brake pressure built up in the master cylinder can be applied to the wheel-brake cylinder when the brakes are applied.
- In the "pressure maintenance" position, the inlet valve blocks the connection between the master cylinder and wheel-brake cylinder during rapid wheel deceleration (risk of locking up) and thus prevents the brake pressure from rising any higher. The outlet valve is also closed.
- If the wheel deceleration continues to increase, the inlet valve continues to block in the "decrease pressure" position. In addition the return pump pumps out the brake fluid via the open outlet valve so that the brake pressure in the wheel-brake cylinder drops.

1 Comparison of ABS systems

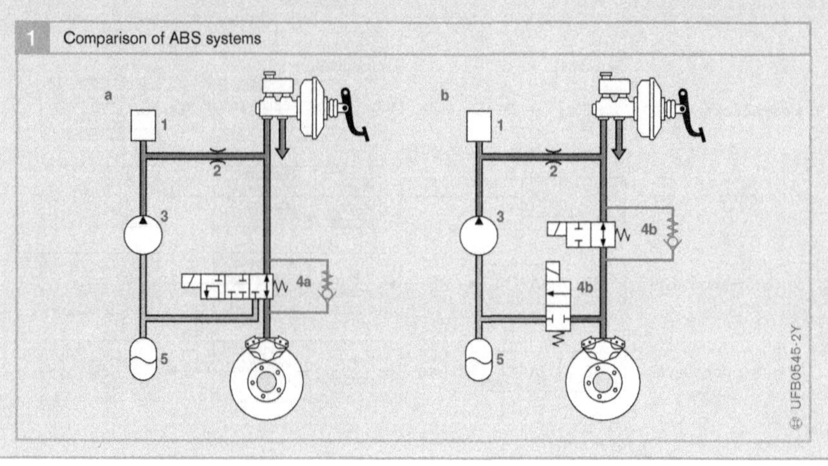

Fig. 1
a ABS2
b ABS5

1 Damping chamber
2 Constrictor
3 Return pump
4a 3/3 solenoid valve
4b 2/2 solenoid valves
5 Accumulator chamber

Sensors for Brake Control

Sensors register operating states (e. g. engine speed) and setpoint/desired values (e. g. accelerator-pedal position). They convert physical quantities (e. g. pressure) or chemical quantities (e. g. exhaust-gas concentration) into electric signals.

Automotive applications

Sensors and actuators represent the interfaces between the ECU's, as the processing units, and the vehicle with its complex drive, braking, chassis, and bodywork functions (for instance, the engine management, the electronic stability program (ESP), and the air conditioner). As a rule, a matching circuit in the sensor converts the signals so that they can be processed by the ECU.

The field of mechatronics, where mechanical, electronic and data-processing components operate closely together, is also becoming increasingly important for sensors. Sensors are integrated in modules (e.g. in the crankshaft CSWS (Composite Seal with Sensor) module complete with rpm sensor).

Sensors are becoming smaller and smaller. At the same time they are also required to become faster and more precise since their output signals directly affect not only the engine's power output, torque, and emissions, but also vehicle handling and safety. These stipulations can be complied with thanks to mechatronics.

Depending upon the level of integration, signal conditioning, analog/digital conversion, and self-calibration functions can all be integrated in the sensor (Fig. 1), and in future a small microcomputer for further signal processing will be added. The advantages are as follows:

- less computing power is required by the electronic control unit,
- one uniform, flexible and bus-compatible interface for all sensors,
- direct multi-purpose use of a sensor via the data bus,
- smaller effects can be measured, and
- easy sensor calibration.

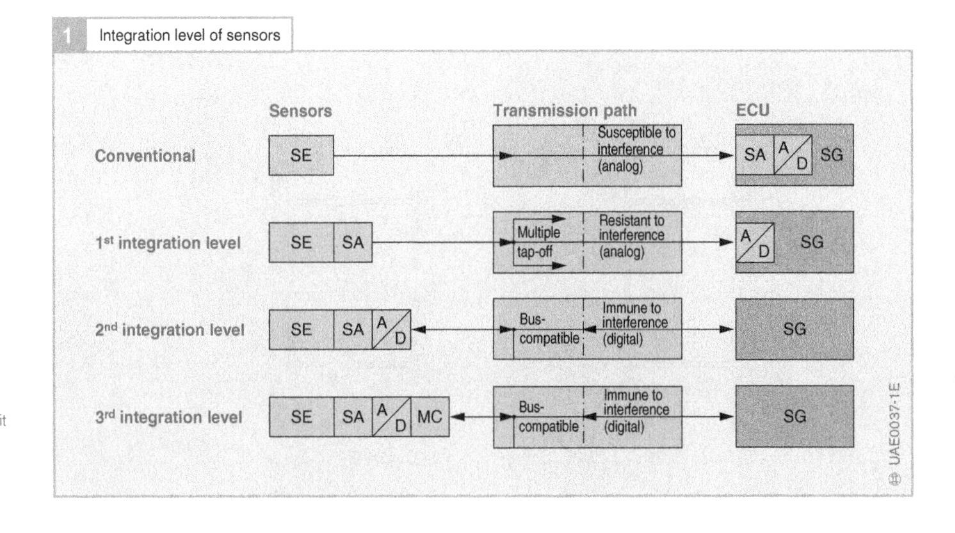

Fig. 1
SE Sensor(s)
SA Analog signal conditioning
A/D Analog-digital converter
SG Digital control unit
MC Microcomputer (evaluation electronics)

Wheel-speed sensors

Application

Wheel-speed sensors are used to measure the rotational speed of the vehicle wheels (wheel speed). The speed signals are transmitted via cables to the ABS, TCS or ESP control unit of the vehicle which controls the braking force individually at each wheel. This control loop prevents the wheels from locking up (with ABS) or from spinning (with TCS or ESP) so that the vehicle's stability and steerability are maintained.

Navigation systems also use the wheel speed signals to calculate the distance traveled (e. g. in tunnels or if satellite signals are unavailable).

Design and method of operation

The signals for the wheel-speed sensor are generated by a steel pulse generator that is fixed to the wheel hub (for passive sensors) or by a multipole magnetic pulse generator (for active sensors). This pulse generator has the same rotational speed as the wheel and moves past the sensitive area of the sensor head without touching it. The sensor "reads" without direct contact via an air gap of up to 2 mm (Fig. 2).

The air gap (with strict tolerances) ensures interference-free signal acquisition. Possible interference caused for instance by oscillation patterns in the vicinity of the brakes, vibrations, temperature, moisture, installation conditions at the wheel, etc. is therefore eliminated.

Since 1998 active wheel-speed sensors have been used almost exclusively with new developments instead of passive (inductive) wheel-speed sensors.

Passive (inductive) wheel-speed sensors

A passive (inductive) speed sensor consists of a permanent magnet (Fig. 2, 1) with a soft-magnetic pole pin (3) connected to it, which is inserted into a coil (2) with several thousand windings. This setup generates a constant magnetic field.

The pole pin is installed directly above the pulse wheel (4), a gear wheel attached to the wheel hub. As the pulse wheel turns, the continuously alternating sequence of teeth and gaps induces corresponding fluctuations in the constant magnetic field. This changes the magnetic flux through the pole pin and therefore also through the coil winding. These fluctuations induce an alternating current in the coil suitable for monitoring at the ends of its winding.

The frequency and amplitude of this alternating current are proportional to wheel speed (Fig. 3) and when the wheel is not rotating, the induced voltage is zero.

Tooth shape, air gap, rate of voltage rise, and the ECU input sensitivity define the smallest still measurable vehicle speed and thus, for ABS applications, the minimum response sensitivity and switching speed.

1 Passive (inductive) wheel-speed sensors

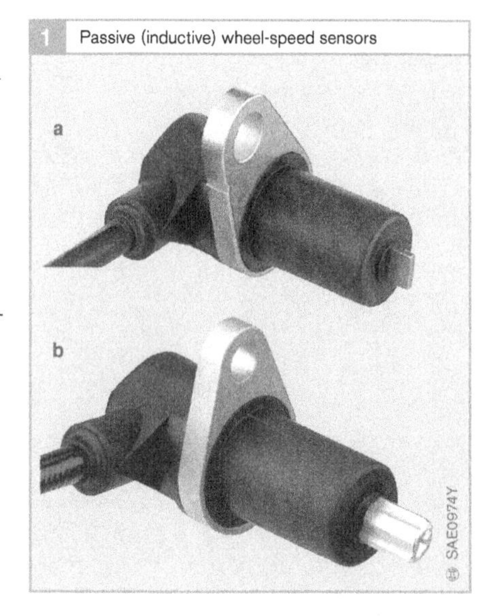

Fig. 1
a Chisel-type pole pin (flat pole pin)
b Rhombus-type pole pin (lozenge-shaped pole pin)

2 Figure illustrating the principle of the passive wheel-speed sensor

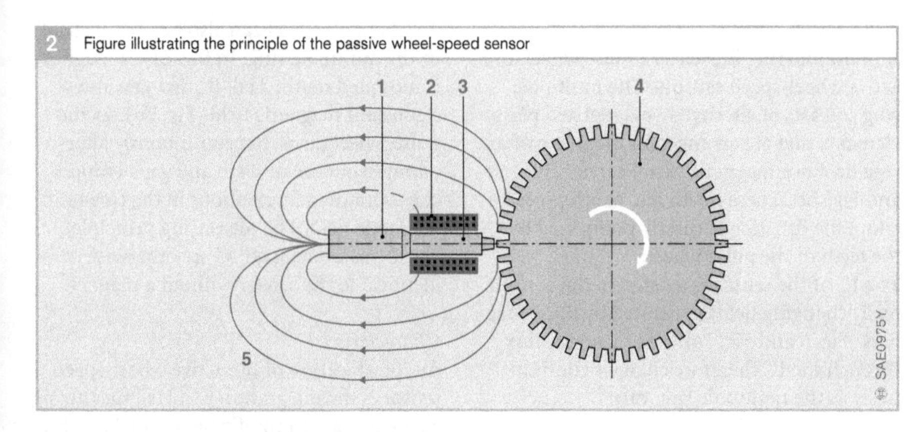

Fig. 2
1 Permanent magnet
2 Solenoid coil
3 Pole pin
4 Steel pulse wheel
5 Magnetic field lines

Various pole-pin configurations and installation options are available to adapt the system to the different installation conditions encountered with various wheels. The most common variants are the chisel-type pole pin (Fig. 1a, also called a flat pole pin) and the rhombus-type pole pin (Fig. 1b, also called a lozenge-shaped pole pin). Both pole-pin designs necessitate precise alignment to the pulse wheel during installation.

Active wheel-speed sensors

Sensor elements

Active wheel-speed sensors are used almost exclusively in today's modern brake systems (Fig. 4). These sensors usually consist of a hermetic, plastic-cast silicon IC that sits in the sensor head.

In addition to magnetoresistive ICs (the electrical resistance changes as the magnetic field changes) Bosch now uses Hall sensor elements almost exclusively. These sensors react to the smallest changes in the magnetic field and therefore allow greater air gaps compared to passive wheel-speed sensors.

3 Signal output voltage of passive wheel-speed sensor

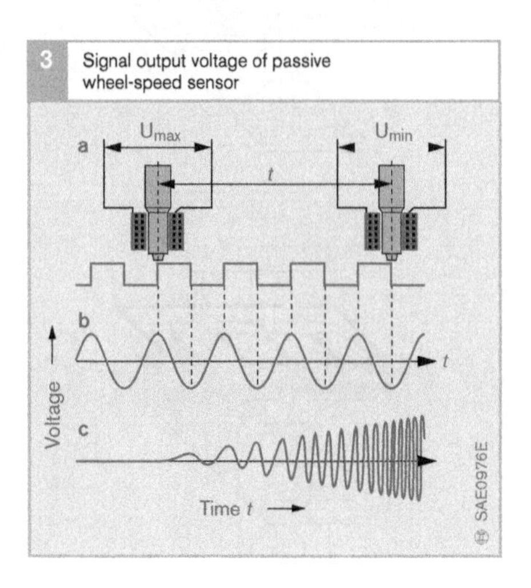

4 Active wheel-speed sensor

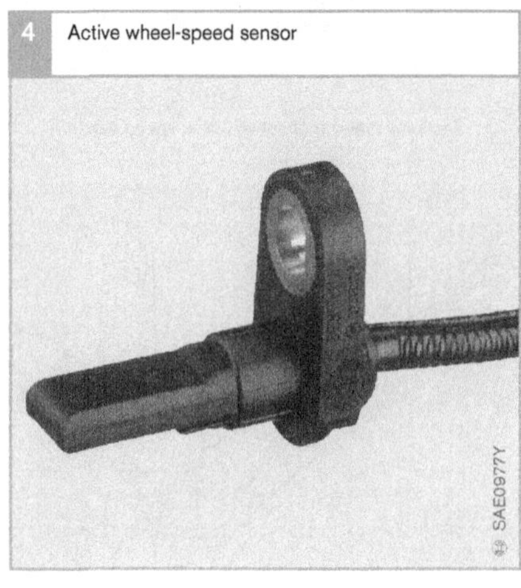

Fig. 3
a Passive wheel-speed sensor with pulse wheel
b Sensor signal at constant wheel speed
c Sensor signal at increasing wheel speed

Pulse wheels

A multipole ring is used as a pulse wheel for active wheel-speed sensors. The multipole ring consists of alternately magnetized plastic elements that are arranged in the shape of a ring on a nonmagnetic metal carrier (Fig. 6 and Fig. 7a). These north and south poles adopt the function formerly performed by the teeth of the pulse wheel.

The IC of the sensor is located in the continuously changing fields generated by these magnets (Fig. 6 and Fig. 7a). The magnetic flux through the IC therefore changes continuously as the multipole ring turns.

A steel pulse wheel can also be used instead of the multipole ring. In this case a magnet is mounted on the Hall IC that generates a constant magnetic field (Fig. 7b). As the pulse wheel turns, the continuously alternating sequence of teeth and gaps induces corresponding fluctuations in the constant magnetic field. The measuring principle, signal processing and IC are otherwise identical to the sensor without a magnet.

Characteristics

A typical feature of the active wheel-speed sensor is the integration of a Hall measuring element, signal amplifier and signal conditioning in an IC (Fig. 8). The wheel-speed data is transferred as an impressed current in the form of square-wave pulses (Fig. 9). The frequency of the pulses is proportional to the wheel speed and the speed can be detected until the wheel is practically stationary (0.1 km/h).

The supply voltage is between 4.5 and 20 volts. The square-wave output signal level is 7 mA (low) and 14 mA (high).

5 Explosion diagram with multipole pulse generator

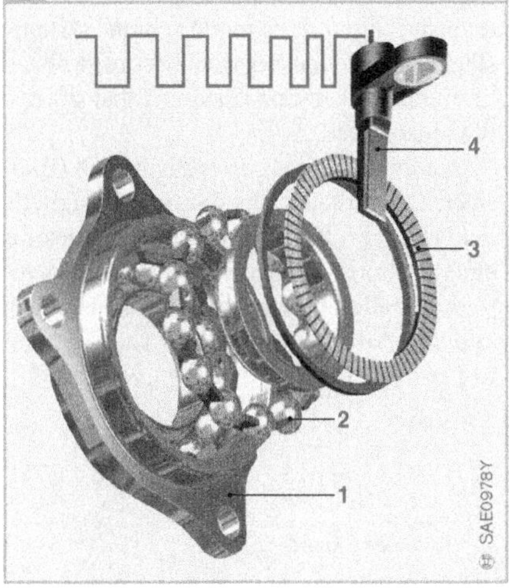

Fig. 5
1 Wheel hub
2 Roller bearing
3 Multipole ring
4 Wheel-speed sensor

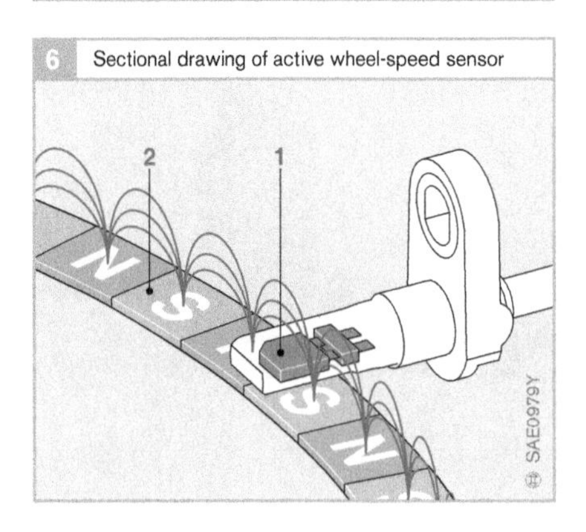

6 Sectional drawing of active wheel-speed sensor

Fig. 6
1 Sensor element
2 Multipole ring with alternating north and south magnetization

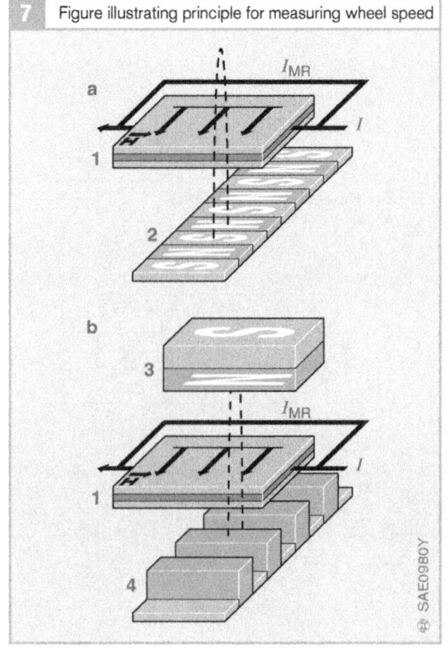

7 Figure illustrating principle for measuring wheel speed

Fig. 7
a Hall IC with multipole pulse generator
b Hall IC with steel pulse generator and magnet in sensor

1 Sensor element
2 Multipole ring
3 Magnet
4 Steel pulse wheel

This type of data-transmission using digital signals is less sensitive to interference than the signals from passive inductive sensors. The sensor is connected to the ECU by a two-conductor wire.

8 Block diagram of Hall IC

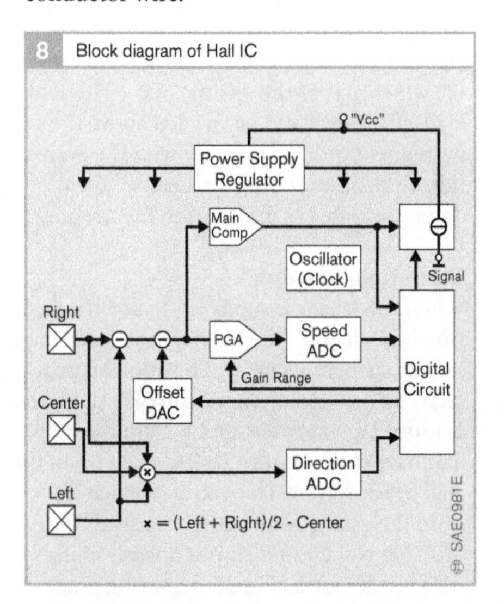

9 Signal conversion in Hall IC

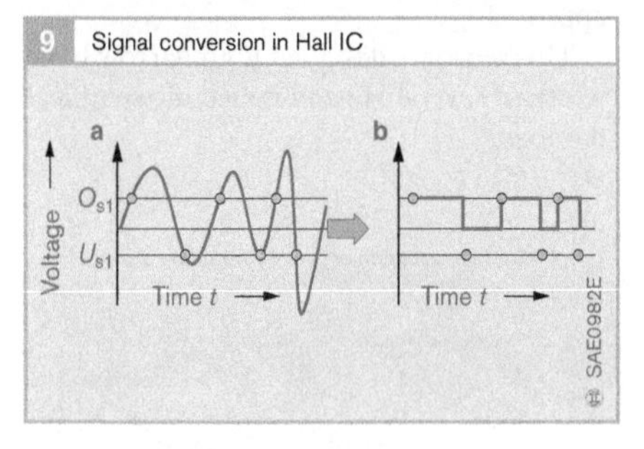

10 Wheel bearing with wheel-speed sensor

Compact dimensions combine with low weight to make the active wheel-speed sensor suitable for installation on and even within the vehicle's wheel-bearing assemblies (Fig. 10). Various standard sensor head shapes are suitable for this.

Digital signal conditioning makes it possible to transfer coded additional information using a pulse-width-modulated output signal (Fig. 11):

- Direction of wheel rotation recognition: This is especially significant for the "hill hold control" feature, which relies on selective braking to prevent the vehicle from rolling backwards when starting off on a hill. The direction of rotation recognition is also used in vehicle navigation systems.
- Standstill recognition: This information can also be evaluated by the "hill hold control" function. The information is also used for self-diagnosis.
- Signal quality of the sensor: Information about the signal quality of the sensor can be relayed in the signal. If a fault occurs the driver can be advised that service is required.

11 Coded information transfer with pulse-width-modulated signals

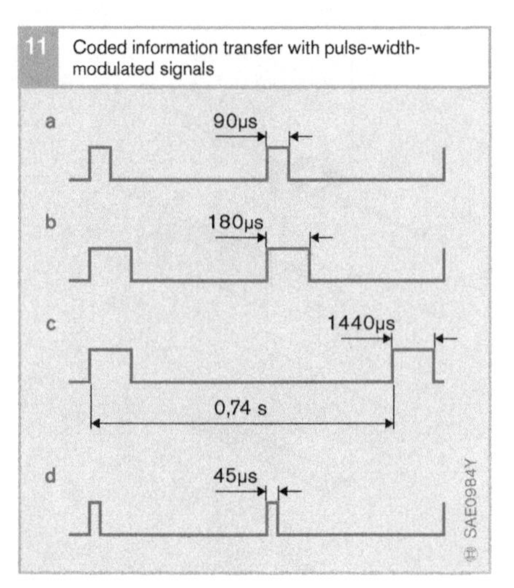

Fig. 9
a Raw signal
b Output signal

O_{S1} Upper switching threshold
U_{S1} Lower switching threshold

Fig. 10
1 Wheel-speed sensor

Fig. 11
a Speed signal when reversing
b Speed signal when driving forwards
c Signal when vehicle is stationary
d Signal quality of sensor, self-diagnosis

Hall-effect acceleration sensors

Applications

Vehicles equipped with the Antilock Braking System ABS, the Traction Control System TCS, all-wheel drive, and/or Electronic Stability Program ESP, also have a Hall-effect acceleration sensor in addition to the wheel-speed sensors. This measures the vehicle's longitudinal and transverse accelerations (depending upon installation position referred to the direction of travel).

1 Hall-effect acceleration sensor (opened)

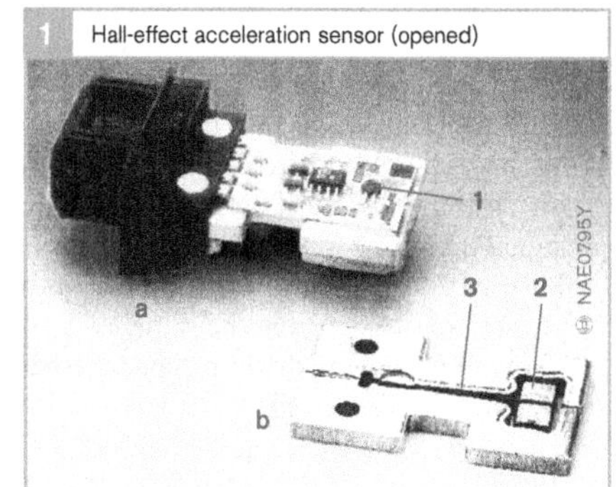

Fig. 1
a Electronic circuitry
b Spring-mass system
1 Hall-effect sensor
2 Permanent magnet
3 Spring

Design and construction

A resiliently mounted spring-mass system is used in the Hall-effect acceleration sensors (Figs. 1 and 2).

It comprises an edgewise-mounted strip spring (3) tightly clamped at one end. Attached to its other end is a permanent magnet (2) which acts as the seismic mass. The actual Hall-effect sensor (1) is located above the permanent magnet together with the evaluation electronics. There is a small copper damping plate (4) underneath the magnet.

Operating concept

When the sensor is subjected to acceleration which is lateral to the spring, the spring-mass system changes its neutral position accordingly. Its deflection is a measure for the acceleration. The magnetic flux F from the moving magnet generates a Hall voltage U_H in the Hall-effect sensor. The output voltage U_A from the evaluation circuit is derived from this Hall voltage and climbs linearly along with acceleration (Fig. 3, measuring range approx. 1 g).

This sensor is designed for a narrow bandwidth of several Hz and is electrodynamically damped.

2 Hall-effect acceleration sensor

3 Hall-effect acceleration sensor (example of curve)

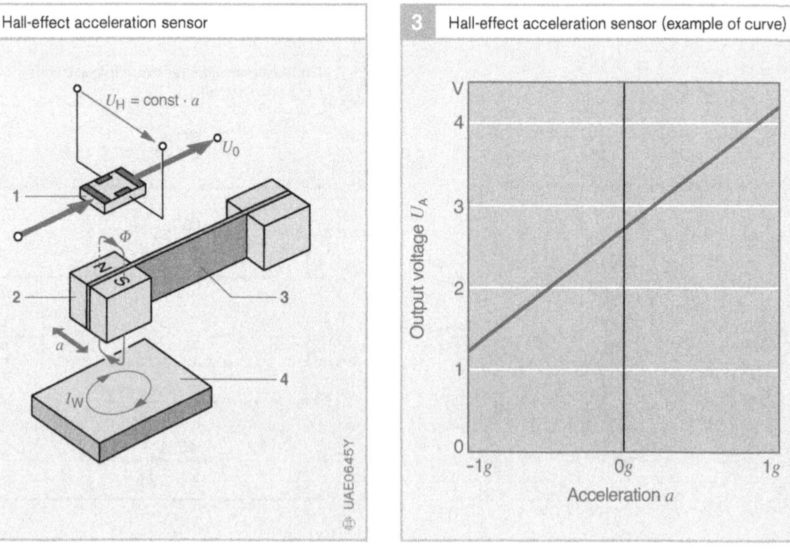

Fig. 2
1 Hall-effect sensor
2 Permanent magnet
3 Spring
4 Damping plate
I_W Eddy currents (damping)
U_H Hall voltage
U_0 Supply voltage
Φ Magnetic flux
a Applied (transverse) acceleration

Micromechanical yaw-rate sensors

Applications
In vehicles with Electronic Stability Program (ESP), the rotation of the vehicle about its vertical axis is registered by micromechanical yaw-rate (or yaw-speed) sensors (also known as gyrometers) and applied for vehicle-dynamics control. This takes place during normal cornering, but also when the vehicle breaks away or goes into a skid.

These sensors are reasonably priced as well as being very compact. They are in the process of forcing out the conventional high-precision mechanical sensors.

1 Structure of the MM1 yaw-rate sensor

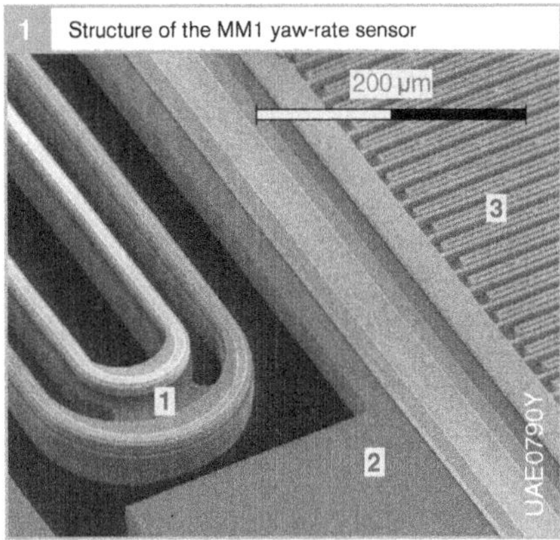
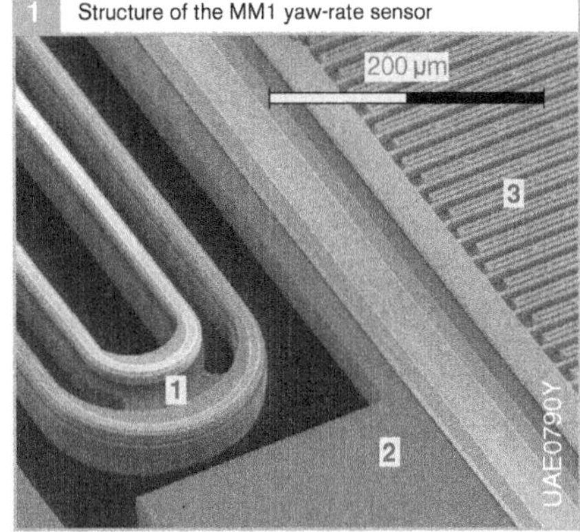

Fig. 1
1 Retaining/guide spring
2 Part of the oscillating element
3 Coriolis acceleration sensor

Design and construction

MM1 micromechanical yaw-rate sensor
A mixed form of technology is applied in order to achieve the high accuracies needed for vehicle-dynamics systems. That is, two somewhat thicker oscillating elements (mass plates) which have been machined from a wafer using bulk micromechanics oscillate in counter-phase to their resonant frequency which is defined by their mass and their coupling springs (>2 kHz). On each of these oscillating elements, there is a miniature, surface-type micromechanical capacitive acceleration sensor. When the sensor chip rotates about its vertical axis at yaw rate Ω, these register the Coriolis acceleration in the wafer plane vertical to the direction of oscillation (Figs. 1 and 2). These accelerations are proportional to the product of yaw rate and and the oscillatory velocity which is maintained electronically at a constant value.

To drive the sensor, all that is required is a simple, current-carrying printed conductor on each oscillating element. In the permanent-magnet field B vertical to the chip surface, this oscillating element is subjected to an electrodynamic (Lorentz) force. Using a further, simple printed conductor (which saves on chip surface), the same magnetic field is used to directly measure the oscillation velocity by inductive means. The different physical construction of drive system and sensor sys-

2 MM1 micromechanical yaw-rate sensor

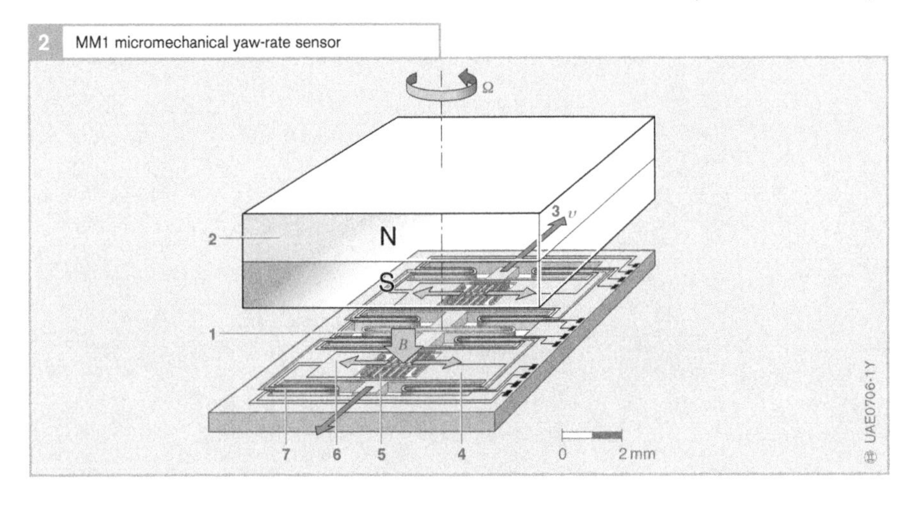

Fig. 2
1 Frequency-determining coupling spring
2 Permanent magnet
3 Direction of oscillation
4 Oscillating element
5 Coriolis acceleration sensor
6 Direction of Coriolis acceleration
7 Retaining/guide spring
Ω Yaw rate
υ Oscillating velocity
B Permanent-magnet field

tem serves to avoid undesirable coupling between the two sections. In order to suppress unwanted external acceleration effects, the opposing sensor signals are subtracted from each other. The external acceleration effects can be measured by applying summation. The high-precision micromechanical construction helps to suppress the effects of high oscillatory acceleration which is several factors of 10 higher than the low-level Coriolis acceleration (cross sensitivity far below 40 dB). Here, the drive and measurement systems are rigorously decoupled from each other.

MM2 micromechanical yaw-rate sensor

Whereas this silicon yaw-rate sensor is produced completely using surface-micromechanic techniques, and the magnetic drive and control system have been superseded by an electrostatic system, absolute decoupling of the power/drive system and measuring system is impossible. Comb-like structures (Figs. 3 and 4) electrostatically force a centrally mounted rotary oscillator to oscillate. The amplitude of these oscillations is held constant by means of a similar capacitive pick-off. Coriolis forces result at the same time in an out-of-plane tilting movement, the amplitude of which is proportional to the yaw rate Ω, and which is detected capacitively by the electrodes underneath the oscillator. To avoid excessive damping of this movement, the sensor must be operated in a vacuum. Although the chip's small size and the somewhat simpler production process result in considerable cost reductions, this miniaturisation is at the expense of reductions in the measuring effect, which in any case is not very pronounced, and therefore of the achievable precision. It also places more severe demands on the electronics. The system's high flexural stability, and mounting in the axis of gravity, serve to mechanically suppress the effects of unwanted acceleration from the side.

4 MM2 yaw-rate sensor: Structure

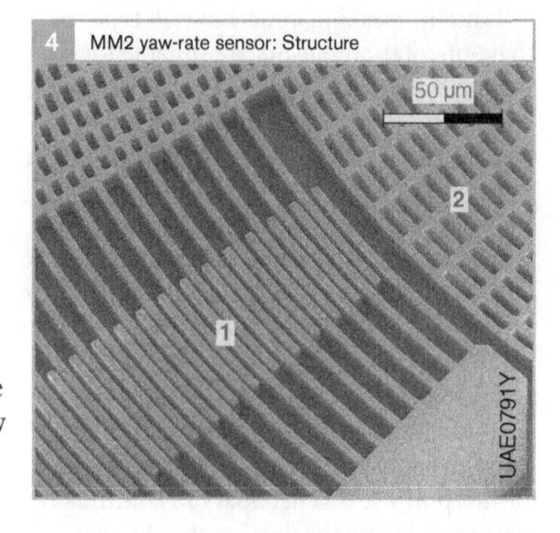

Fig. 4
1 Comb-like structure
2 Rotary oscillator

3 MM2 surface-micromechanical yaw-rate sensor

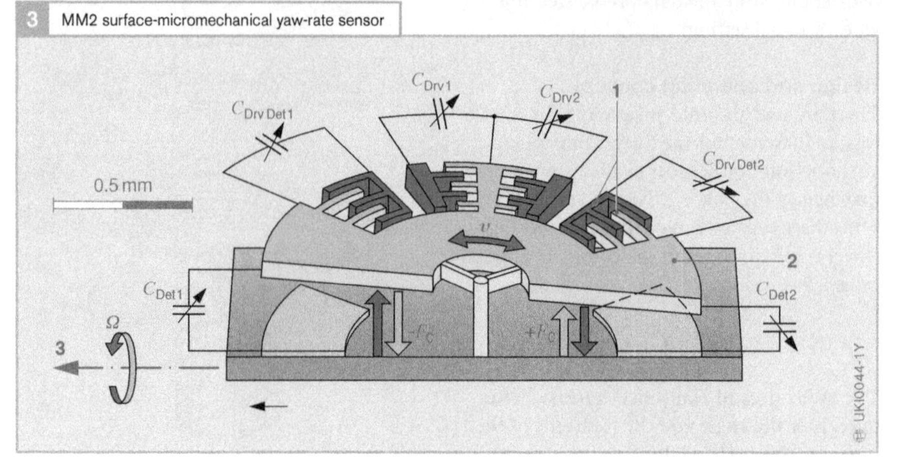

Fig. 3
1 Comb-like structure
2 Rotary oscillator
3 Measuring axis
C_{Drv} Drive electrodes
C_{Det} Capactive pick-off
F_C Coriolis force
v Oscillatory velocity
$\Omega = \Delta C_{Det}$, measured yaw rate

Steering-wheel-angle sensors

Application

The Electronic Stability Program (ESP) applies the brakes selectively to the individual wheels in order to keep the vehicle on the desired track selected by the driver. Here, the steering-wheel angle and the applied braking pressure are compared with the vehicle's actual rotary motion (around its vertical axis) and its road speed. If necessary, the brakes are applied at individual wheels. These measures serve to keep the float angle (deviation between the vehicle axis and the actual vehicle movement) down to a minimum and, until the physical limits are reached, prevent the vehicle breaking away.

Basically speaking, practically all types of angle-of-rotation sensors are suitable for registering the steering-wheel angle. Safety considerations, though, dictate that only those types are used which can be easily checked for plausibility, or which in the ideal case automatically check themselves. Potentiometer principles are used, as well as optical code-registration and magnetic principles. Whereas a passenger-car steering wheel turns through ±720° (a total of 4 complete turns), conventional angle-of-rotation sensors can only measure maximum 360°. This means that with the majority of the sensors actually used for this purpose it is necessary to continually register and store the data on the steering wheel's actual setting.

Design and operating concept

There are two absolute-measuring (in contrast to incremental-measuring) magnetic angle-of-rotation sensors available which are matched to the Bosch ECUs. At any instant in time, these sensors can output the steering-wheel angle throughout the complete angular range.

Hall-effect steering-wheel-angle sensor (LWS1)

The LWS1 uses 14 Hall-effect vane switches to register the angle and the rotations of the steering wheel. The Hall-effect vane switch is similar in operation to a light barrier. A Hall-effect element measures the magnetic field of an adjacent magnet. A magnetic code disc rotates with the steering shaft and strongly reduces the magnet's field or screens it off completely. In this manner, with nine Hall ICs it is possible to obtain the steering wheel's angular position in digital form. The remaining five Hall-effect sensors register the particular steering-wheel revolution which is transformed to the final 360° range by 4:1 step-down gearing.

The first item from the top in the exploded view of the LWS 1 steering-wheel-angle sensor (Fig. 1) shows the nine permanent magnets. These are screened individually by the magnetically-soft code disc beneath them when this rotates along with the steering shaft, and depending upon steering-wheel movement. The pcb immediately below the code disc contains Hall-effect switches (IC), and a microprocessor in

1 Exploded view of the digital LWS1 steering-wheel-angle sensor

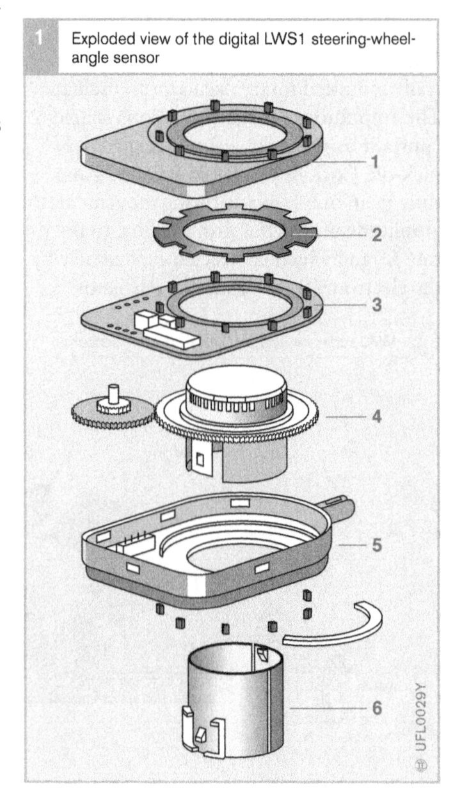

Fig. 1
1 Housing cover with nine equidistantly spaced permanent magnets
2 Code disc (magnetically soft material)
3 PCB with 9 Hall-effect switches and microprocessor
4 Step-down gearing
5 Remaining 5 Hall-effect vane switches
6 Fastening sleeve for steering column

which plausibility tests are performed and information on angular position decoded and conditioned ready for the CAN-Bus. The bottom half of the assembly contains the step-down gearing and the remaining five Hall-effect vane switches.

The LWS1 was superseded by the LWS3 due to the large number of sensor elements required, together with the necessity for the magnets to be aligned with the Hall-IC.

Magnetoresistive steering-wheel-angle sensor LWS3

The LWS 3 also depends upon AMR (anisotropic magnetoresistive sensors) for its operation. The AMR's electrical resistance changes according to the direction of an external magnetic field. In the LWS3, the information on angle across a range of four complete rotations is provided by measuring the angles of two gearwheels which are rotated by a third gearwheel on the steering-column shaft. The first two gearwheels differ by one tooth which means that a definite pair of angular variables is associated with every possible steering-wheel position.

By applying a mathematical algorithm (a computing process which follows a defined step-by-step procedure) referred to here as a modified vernier principle, it is possible to use the above AMR method for calculating the steering-wheel angle in a microcomputer. Here, even the measuring inaccuracy of the two AMR sensors can be compensated for. In addition, a self-check can also be implemented so that a highly plausible measured value can be sent to the ECU.

Fig. 2 shows the schematic representation of the LWS3 steering-wheel-angle sensor. The two gearwheels, with magnets inserted, can be seen. The sensors are located above them togther with the evaluation electronics. With this design too, price pressure forces the development engineers to look for innovative sensing concepts. In this respect, investigation is proceeding on whether, since it only measures up to 360°, a single AMR angle-of-rotation sensor (LWS4) on the end of the steering shaft would be accurate enough for ESP (Fig. 4).

2 AMR steering-wheel-angle sensor LWS3 (principle)

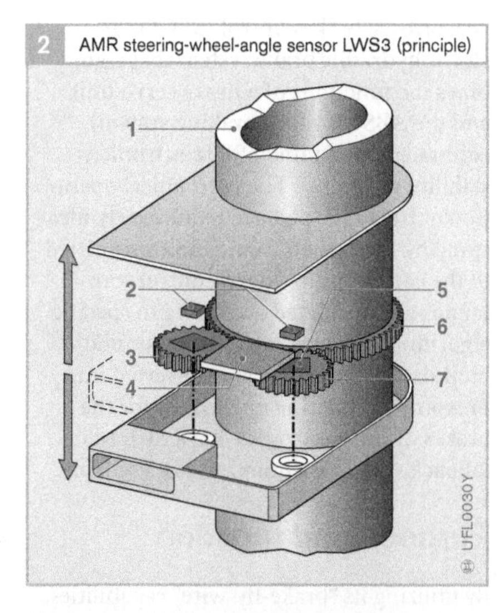

Fig. 2
1 Steering-column shaft
2 AMR sensor elements
3 Gearwheel with m teeth
4 Evaluation electronics
5 Magnets
6 Gearwheel with n > m teeth
7 Gearwheel with m + 1 teeth

3 AMR steering-wheel-angle sensor LWS3

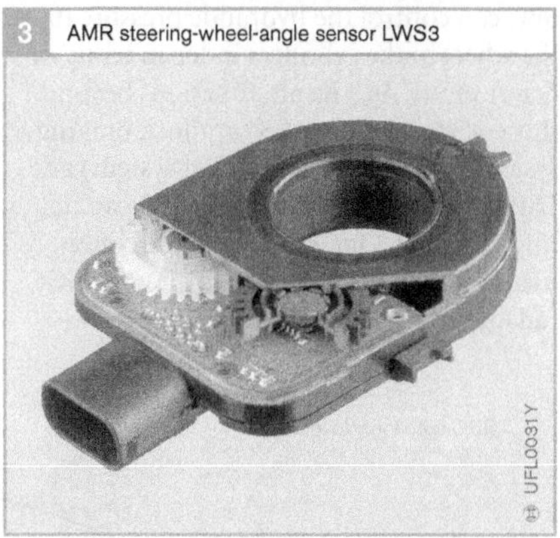

4 AMR steering-wheel-angle sensor LWS4 for attachment to the end of the steering-column shaft

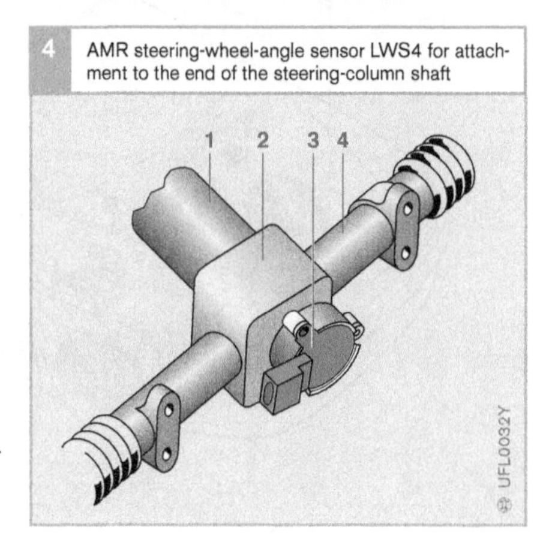

Fig. 4
1 Steering column
2 Steering box
3 Steering-wheel-angle sensor
4 Steering rack

Sensotronic brake control (SBC)

Sensotronic brake control (SBC) is an electrohydraulic brake system that combines the functions of a brake servo unit and the ABS (antilock braking system) equipment, including ESP (electronic stability program). The mechanical operation of the brake pedal is redundantly measured by the actuator unit and transmitted to the control unit. There, control commands are calculated according to specific algorithms and passed to the hydraulic modulator where they are converted into pressure modulating operations for the brakes. If the electronics fail, a hydraulic fallback system is automatically available.

Purpose and function

By utilizing its "brake-by-wire" capabilities, SBC can control the hydraulic pressure in the wheel brake cylinders independently of driver input. As a result, functions beyond those performed by ABS (antilock braking system), TCS (traction control system) and ESP (electronic stability program) can be implemented. One example is the convenient method of brake application for ACC (adaptive cruise control).

Basic functions

As with a conventional braking system, sensotronic brake control must be capable of

- reducing the speed of the vehicle,
- bringing the vehicle to a halt, and
- keeping the vehicle stationary when it is stopped.

As an active braking system, it also performs the tasks of

- operating the brakes,
- amplifying the brake force, and
- modulating the brake force.

SBC is an electronic control system with hydraulic actuators. Braking force distribution takes place electronically to each wheel in response to driving conditions. A vacuum source for the brake servo function is no longer required. The self-diagnosis capability enables an early warning function for detection of possible system faults.

SBC uses hydraulic standard wheel brakes. Because of the fully electronic pressure control, SBC can be easily networked with vehicle handling systems. It thus meets all the demands made of future braking systems.

1 SBC components in the car

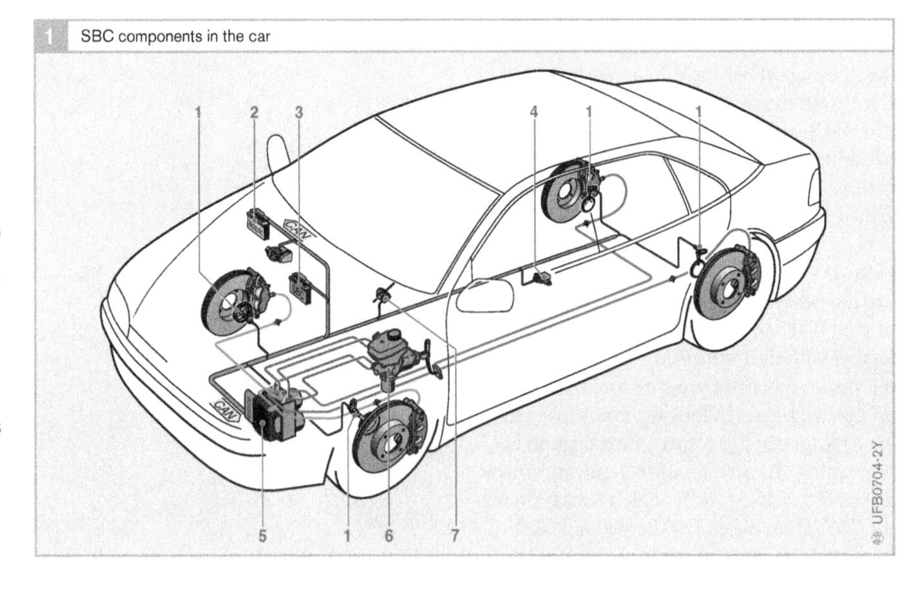

Fig. 1
1 Active wheel speed sensor with direction sensing
2 Engine management ECU
3 SBC ECU
4 Yaw rate and lateral acceleration sensor
5 Hydraulic modulator (for SBC, ABS, TCS and ESP)
6 Actuator unit with pedal travel sensor
7 Steering angle sensor

By using a high-pressure accumulator, SBC is capable of extremely rapid dynamic pressure increases and thus offers the potential for achieving short braking distances and excellent vehicle handling stability. Brake pressure modulation and active braking are silent and produce no brake pedal feedback. Consequently, SBC also satisfies demands for greater levels of comfort.

Braking characteristics can be adapted to the driving conditions (e.g. sharper response at high speeds or with more dynamic driving styles). "Duller" pedal characteristics allow the reduction of the braking effect, which is necessitated by physics, to be signaled to the driver before fading due to overheating occurs.

Additional SBC functions

The auxiliary functions provided by SBC make a significant extra contribution to safety and convenience.

Hill hold control

After hill hold control is activated by a significant increase in brake pressure while the vehicle is stationary, the vehicle remains braked without the need to keep the pedal down. The hill hold control is automatically released as soon as the driver has built up sufficient engine torque by depressing the accelerator. This allows the driver to start the car on a hill, for example, without activating the parking brake system. Likewise, in other situations in which the vehicle would roll out of position if not braked, the driver does not need to keep his or her foot on the brake at all times once hill hold control has been activated.

Enhanced brake assist function

If the driver abruptly releases the accelerator, an automatically regulated brake pressure build-up takes place that gently applies the brake pads. If panic braking follows, this allows the brake to "grab" more quickly and thus allows a shorter total braking distance.

If the system detects panic braking, the brake pressure is briefly increased until the optimum friction value is fully utilized. This results in a significant reduction of total braking distance for hesitant drivers. The highly dynamic braking force build-up of SBC exceeds that of conventional systems in this regard.

Soft stop assist

SBC provides comfortable braking that stops the car with no jerking by automatically reducing the pressure just before the car comes to a complete stop. If more deceleration is desired, this function is not activated and SBC minimizes the braking distance.

Traffic jam assist

When traffic jam assist is activated, SBC builds up a higher drag torque, which means that the driver does not have to constantly alternate between the accelerator and brake. The vehicle is automatically braked and, if necessary, brought to a complete stop and kept at a complete stop. This function can be activated at speeds of 50 - 60 km/h.

Brake wiping

"Brake wiping" is an operation whereby the film of water is regularly removed from the brake disks in wet weather. It results in shorter stopping distances in wet conditions. You can take the information for activating this function from the windshield wiper signal, for example.

2 Evolution of brake systems

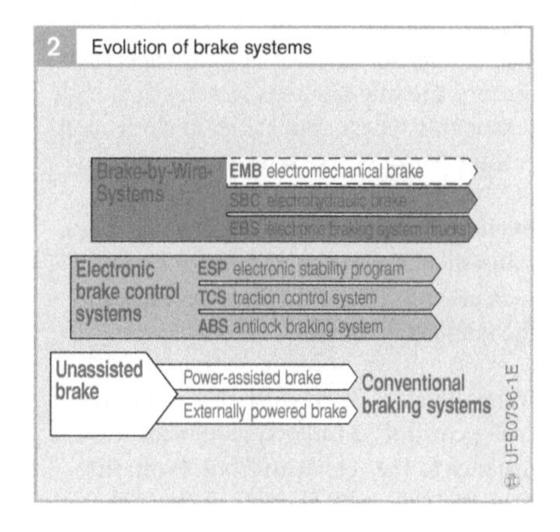

Design

The sensotronic brake control system consists of the following components:

- Actuator unit (Fig. 1, Item 6)
- Vehicle dynamics sensors (1, 4 and 7)
- Discrete (separate) ECU (3)
- Hydraulic modulator with add-on ECU (5).

Those components are interconnected by electrical control leads and hydraulic pipes. Fig. 1 shows where they are fitted in a car.

Actuator unit

The actuator unit consists of:

- Master cylinder with expansion tank,
- Pedal travel simulator and
- Pedal travel sensor.

Pedal travel simulation

The pedal travel simulator makes it possible to realize a suitable force-distance curve and an appropriate damping of the brake pedal. In that way, the driver obtains the same "brake feel" with the sensotronic brake control system as with a very well designed conventional braking system.

Sensors of the SBC systems

The SBC sensors consist of the vehicle dynamics sensors familiar from ESP and the actual SBC sensors.

Vehicle dynamics sensors

The ESP sensors consist of four wheel speed sensors, the yaw rate sensor, a steering angle sensor and, where applicable, a lateral acceleration sensor. These sensors provide the ECU with data relating to the speed and movement status of the wheels and driving states such as cornering. Control functions such as ABS, TCS and ESP are executed in the familiar manner.

If the vehicle is fitted with ACC (adaptive cruise control), a radar system measures the distance to the vehicle in front. From these data, the SBC control unit calculates the brake pressure to be applied, which is built up without pedal feedback.

SBC sensors

Four pressure sensors measure the pressure individually for each wheel circuit (Fig. 3). A pressure sensor measures the storage pressure of the high-pressure accumulator. The driver's braking request is calculated by the pedal travel sensor attached to the actuator unit and a pressure sensor that detects the brake pressure applied by the driver.

The pedal travel sensor consists of two redundant, independent angle sensors. Together with the pressure sensor for the driver's brake pressure, they allow a threefold detection of the driver's request, and the system can continue to work without errors even if one of these systems fails.

Method of operation

Normal operation

Fig. 3 illustrates the components of SBC in the form of a block diagram. An electric motor drives a hydraulic pump. This charges a high-pressure accumulator to a pressure between approx. 90 and 130 bar, which is monitored by the accumulator pressure sensor. The four independent wheel pressure modulators are supplied by this accumulator and adjust the required pressure individually for each wheel. The pressure modulators themselves each consist of two valves with proportional control characteristics and a pressure sensor.

In normal operation, the isolating valves interrupt the connection to the actuator. The system is in "brake-by-wire" mode. It electronically detects the driver's braking request and transmits it "by wire" to the wheel pressure modulators. The interaction of the engine, valves and pressure sensors is controlled by the electronics, which are installed in the add-on ECU in hybrid technology. They have two microcontrollers that monitor each other. The essential feature of these electronics is their extensive self-diagnosis, which monitors the plausibility of every system state at all times. In this way, the driver can be notified of any failures before critical states occur. If components fail, the system automatically provides the optimal remaining partial function to the driver. An extensive fault memory allows prompt diagnostics and repair in the event of a fault.

An intelligent interface with CAN bus establishes the connection to the discrete ECU. The following functions are integrated there:

- ESP (electronic stability program),
- TCS (traction control system),
- ABS (antilock braking system),
- Driver brake request calculation and
- SBC auxiliary functions (assist functions).

Braking in the event of system failure

For safety reasons, the SBC system is designed so that in the event of any serious errors (such as failure of the power supply), the system is switched to a state in which the vehicle can be braked even without active brake force support. When de-energized, the isolating valves establish a direct connection to the actuator (Fig. 3) and thus allow a direct hydraulic connection from the actuator unit to the wheel brake cylinders.

To maintain optimum function even if the system fails, the plunger pistons in the illustration serve as a medium separator between the active circuit of the SBC and the conventional front axle brake circuit. These prevent any gas that might escape from the high-pressure accumulator from reaching the brake circuit of the front wheels, which would diminish the instantaneous braking power in the event of a system failure.

3 Interaction of SBC functional modules

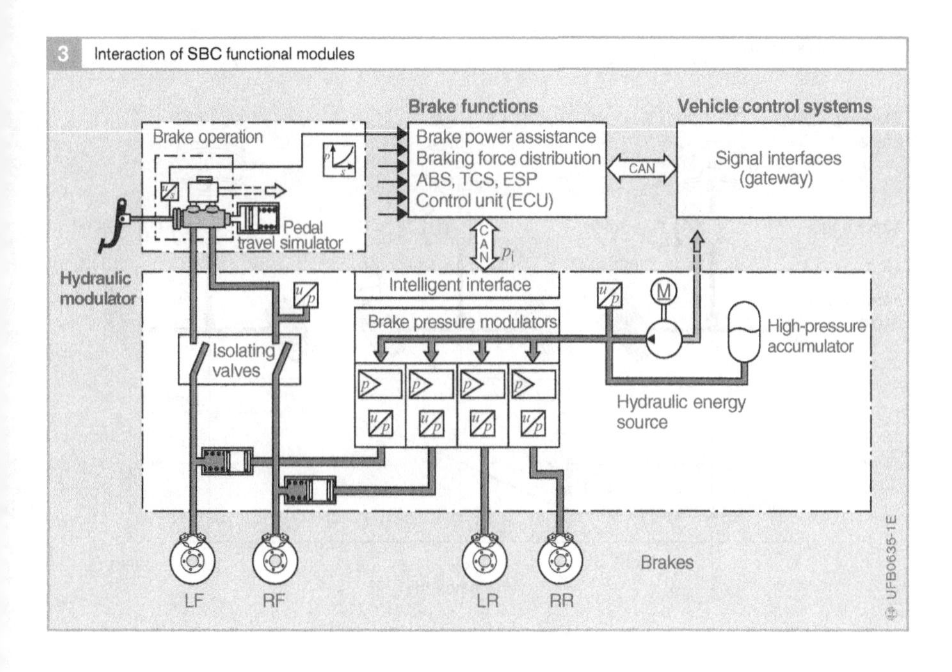

Active steering

The development of vehicle steering systems is characterized by the consistent introduction of hydraulic servo assist and the replacement of ball-and-nut-type steering in the car by the easier and more inexpensive rack-and-pinion steering. Recently, electromechanical power steering has been displacing hydraulic power steering in small and lightweight cars. By law, however, pure "steer-by-wire" technology is not yet permitted in motor vehicles. European Union safety regulations still require a mechanical connection between the steering wheel and the wheels of the vehicle.

All of these developments have the goal of making vehicle handling as easy as possible and to limit steering forces to a logical amount. The best possible feedback about the contact of the tires to the road is to be ensured. This has a decisive impact on the driver's ability to manage his or her task in the control loop between driver, vehicle and environment.

Purpose

The newly developed active steering can affect the steering forces and the steering angle set by the driver. It fulfills the wish for a direct steering ratio to improve handling at low speeds. It also meets requirements for ensuring comfort, drivability and straight-running stability at high speeds. Active steering is an initial step towards a "steer-by-wire" function. Although it does not let the car drive itself, it provides correction functions and added comfort and convenience.

Design

The primary difference between active steering and a "steer-by-wire" system is the fact that the steering train, and thus the driver's mechanical control of the steered front wheels, is maintained during active steering.

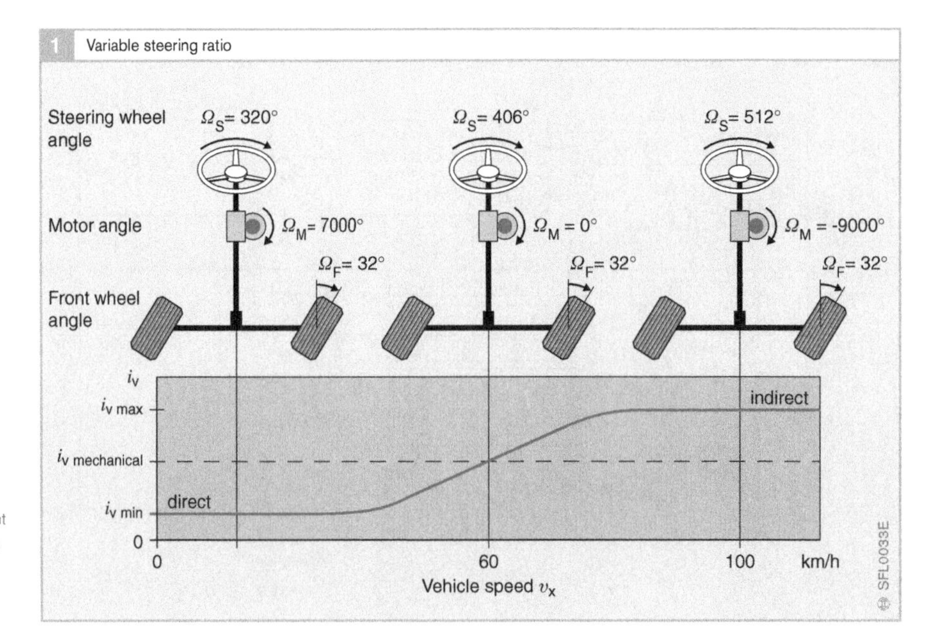

Fig. 1
Changing the ratio between the steering wheel angle and the median angle of the front wheels → Reducing the steering effort at low speeds and providing stability at high speeds

Mechanical system
The steering train consists, as usual, of the steering wheel, steering column, steering gear and tie rods. The special feature of the new active steering is a differential gearbox (Fig. 2). For this purpose, a planetary gear (6) is integrated with two input shafts and one output shaft in the steering gear. One input shaft is connected to the steering wheel, and the other drives an electric motor (4) via a worm gear pair (3) as a reduction stage. The connected ECU processes the necessary sensor signals, controls the electric motor and monitors the entire steering system.

The electric motor and differential gearbox allow steering intervention at the front axle to take place independently of the driver. At low speeds, the effective steering angle at the wheels is greater than the angle set on the steering wheel, as the system adds a part that is proportional to the steering angle. At high speeds, it subtracts a corresponding amount so that the wheel angle is smaller than that set by the driver (Fig. 1). When the electric motor is inactive, the steering wheel controls the vehicle's wheels directly, as with conventional steering systems.

Hydraulic system
The principle of differential steering usually requires a hydraulic servo assist to limit the forces applied by hand to a logical amount. This is accomplished using an "open center" steering valve specially adapted to the high performance requirements. The vector superimposition of the positioning rates of the drivers and engine can, in certain cases, cause significantly higher rack-and-pinion speeds than those of conventional steering systems. The geometric flow of the vane-type power steering pump with flow rate controller is designed for the theoretical maximum positioning rate. Regulation at the output side provides a highly dynamic and quiet energy supply for the active steering system.

2 Actuators

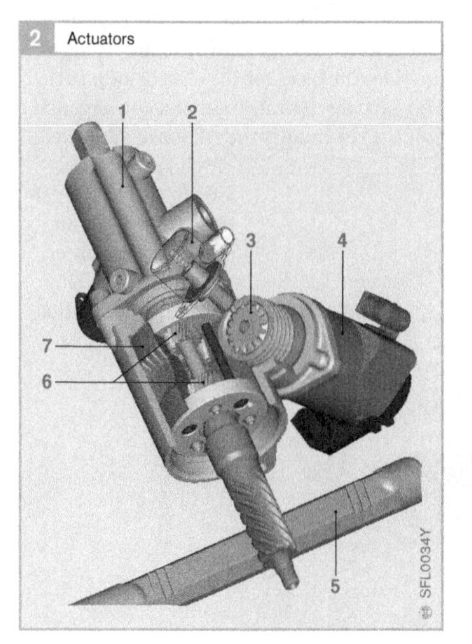

3 Components of active steering

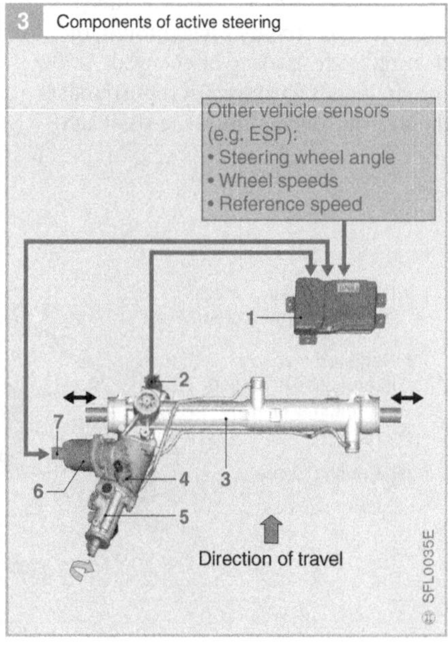

Fig. 2
1 Servotronic II valve
2 Electromagnetic block
3 Worm
4 Electric motor
5 Steering rack
6 Planetary gear
7 Worm gear

Fig. 3
1 Electronic control unit
2 Pinion angle sensor
3 Substructure
4 Block
5 Servotronic II valve
6 Actuator module
7 Motor angle sensor

Method of operation

Activation concept

The actuator adjustment at the front axle, which takes place independently of the driver, requires a complex activation concept that is implemented in an ECU with two processors that communicate with each other. One processor is responsible for activating the servomotor, the other for calculating the correct control angle. Both processors monitor each other to ensure that they are functioning properly. The movement status of the steering gear is detected using one angle sensor each on the steering pinion and servomotor. The steering wheel angle signal is also used as the setpoint specification from the driver. Sensors for yaw velocity, lateral acceleration and wheel speed, which are already in place for the driving stability systems (ESP), provide additional input signals for the active steering.

The ECU is networked to the system using powertrain CAN and the new chassis CAN at the required high data rate (see Fig. 3). Approximately 100 times a second, the necessary data are collected using sensors and evaluated by the ECU. The ECU then decides whether, and by what amount, the steering angle needs to be changed. At low speeds, it adds an amount proportional to the steering angle; at high speeds, it subtracts a corresponding amount. Therefore, from the driver's point of view, the impression given is of a steering ratio that is variable over the driving speed. The steering effort remains largely constant over a wide speed range. Except at extremely low speeds, a significantly greater steering angle than 180° is required.

Driving stabilization

To calculate the stabilizing steering intervention, the vehicle movement variables of yaw angular acceleration and lateral acceleration are returned and compared in the stabilization controller with the setpoint specified by the driver, which is made up of the steering wheel angle and driving speed (Fig. 4).

Setpoint value

The setpoint for the control angle to be set, which is calculated in the ECU of the active steering, can be separated into a part with open-loop control and a part with closed-loop control. The open-loop control part, known simply as the variable steering ratio, by definition is calculated from the reference variable only, the driver's steering angle. Additional information from the closed-loop control system known as the "vehicle" provides the basis for the closed-loop part. The partial setpoints meet at a convergence point. They modify the response of the vehicle closed-loop control system to the driver's steering input. These steering interventions are usually continuous and thus do not disturb the driver.

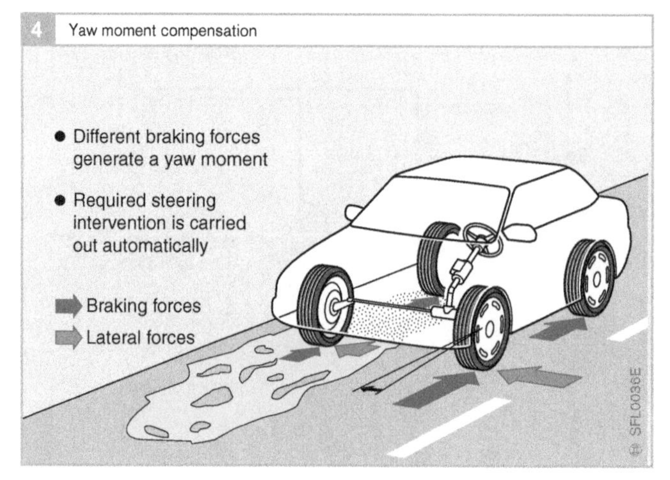

Fig. 4
Example: braking on non-uniform road surface

Cooperation with driving stability systems

Compared to conventional driving stability systems based on wheel slip control, driving stabilization via steering intervention at the front axle has different characteristics:

- The steering intervention is considerably less noticeable to the driver than the clearly audible braking intervention.
- The steering intervention is faster than a radial braking intervention, which requires a certain pressure build-up time.
- The braking intervention provides superior stabilization performance than the steering.

By combining active steering (steering intervention) and wheel slip control (braking intervention), optimum driving stabilization is attained.

Safety concept

If the servomotor has to be switched off due to a fault, this path is mechanically blocked. The planetary gear then rolls off internally while the worm gear is blocked, and the vehicle remains steerable with no restrictions and at a constant gear ratio. Thus active steering is possible even if the mechanical connection fails. This is a great advantage over pure "steer-by-wire" systems. In active steering, all relevant input signals are safeguarded by redundant sensors or measurements. Two different processors calculate the setpoint signal in the ECU. Although the conversion of the setpoint signal in the electromechanical converter takes place in one channel only, no unwanted actuation in the region relevant to vehicle dynamics is possible due to the selection of a BLDC motor.

The safety concept is expanded by adding an adapted shutoff concept. Function shutoffs range from temporary or long-term disabling of the driving stabilization to limited vehicle operation with substitute values all the way to complete shutoff of the entire system. The state transitions here are comparable to familiar, manageable disturbances such as those caused by crosswinds or rail grooves in the road.

Active steering requires no additional control elements because all of its subfunctions are activated automatically when the engine is started. If the internal combustion engine is not working (while the vehicle is being towed), the active steering, like conventional power steering, is disabled. Such situations are indicated by a lamp in the instrument cluster.

Benefits of active steering for the driver

- Driving errors are corrected or compensated for in such a way that the driver is not surprised by the vehicle's reaction.
- The steering ratio, which depends on the driving situation, makes it easier to maneuver the car at very low speeds, as fewer complete turns of the steering wheel are required with the same amount of effort.
- Greater convenience and comfort at high speeds, as the driver no longer needs to fear losing control of the car by accidentally applying too much steering force.
- The steering lead is another comfort and convenience feature. It allows more agile activation in response to the steering command.

Occupant protection systems

In the event of an accident, occupant protection systems are intended to keep the accelerations and forces that act on the passengers low and lessen the consequences of the accident.

Vehicle safety

Active safety systems help to prevent accidents and thus make a preventative contribution to road traffic safety. One example of an active driving safety system is the antilock braking system (ABS) with electronic stability program (ESP) from Bosch, which stabilizes the vehicle even in critical braking situations and maintains steerability in the process.

Passive safety systems help to protect the occupants against serious or even fatal injuries. An example of passive safety are airbags, which protect the occupants after an unavoidable impact.

Seat belts, seat belt pretensioners

Function

The function of seat belts is to restrain the occupants of a vehicle in their seats when the vehicle impacts against an obstacle.

Seat belt pretensioners improve the restraining characteristics of a three-point inertia reel belt and increase protection against injury. In the event of a frontal impact, they pull the seat belts tighter against the body and hold the upper body as closely as possible against the seat backrest. This prevents excessive forward displacement of the occupants caused by inertia (Fig. 1).

Method of operation

In a frontal impact with a solid obstacle at a speed of 50 km/h, the seat belts must absorb a level of energy comparable to the kinetic energy of a person dropping in free fall from the fifth floor of a building.

Due to a loose belt ("seat belt slack"), seat belt stretch and the film-reel effect, three-point inertia reel belts provide only limited protection in frontal impacts against solid obstacles at speeds of over 40 km/h because they can no longer safely prevent the head and body from impacting against the steering wheel or the instrument panel. Without a restraint system, an occupant experiences extensive forward displacement (Fig. 2).

In an impact, the *shoulder belt tightener* eliminates the seat belt slack and the "film-reel effect" by rolling up and tightening the belt webbing. At an impact speed of 50 km/h, this system achieves its full effect within the first 20 ms of impact; this supports the airbag, which needs approx. 40 ms to inflate completely. After that, an occupant continues to move forwards by a certain amount, thereby expelling the gas (N_2) from the airbag so that the occupant's kinetic energy is dissipated in a relatively gradual manner. This protects occupants from injury

1 Occupant protection systems with seat belt pretensioners and front airbags

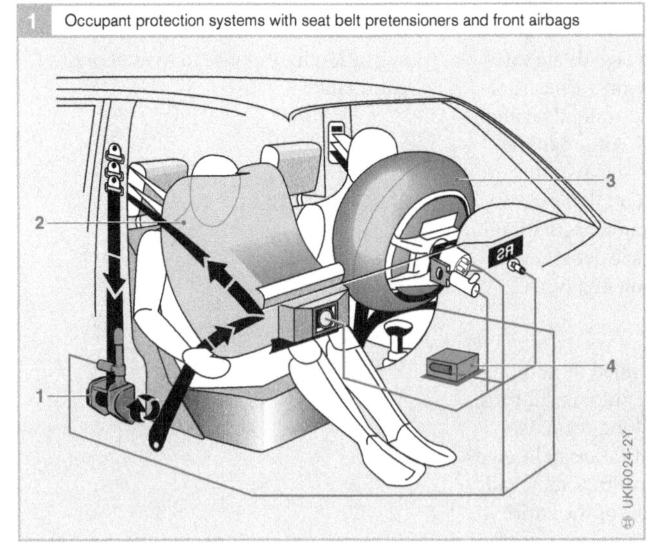

Fig. 1
1 Seat belt pretensioner
2 Passenger front airbag
3 Driver front airbag
4 ECU

because it prevents impact with rigid parts of the vehicle structure.

A prerequisite for optimum protection is that the occupants' forward movement away from their seats must be minimal as they decelerate with the vehicle. Activation of the seat belt pretensioners takes care of this virtually from the moment of impact, and ensures restraint of occupants as early as possible. The maximum forward displacement with tightened seat belts is approx. 2 cm and the duration of mechanical tightening is 5-10 ms.

On activation, the system electrically fires a pyrotechnic propellant charge. The rising pressure acts on a piston, which turns the belt reel via a steel cable in such a way that the belt is held tightly against the body (Fig. 3).

Variants

In addition to the described shoulder belt tighteners for rewinding the belt reel shaft, there are variants which pull the seat belt buckle back (buckle tighteners) and simultaneously tighten the shoulder and lap belts. *Buckle tighteners* further improve the restraining effect and the protection to prevent occupants from sliding forward under the lap belt ("submarining effect"). The tightening process in these two systems takes place in the same period of time as for shoulder belt tighteners.

A larger degree of tightener travel for achieving a better restraining effect is provided by the combination of two seat belt pretensioners for each (front) seat, which, on the Renault Laguna for instance, consist of a shoulder belt tightener and a *belt buckle tightener.* The belt buckle tightener is activated either only in an impact above a certain degree of severity, or with a certain time lag (e.g. approx. 7 ms) relative to activation of the shoulder belt tightener.

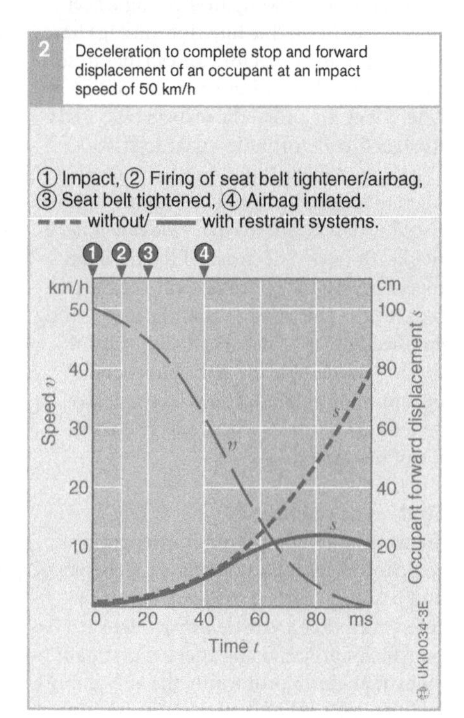

2 Deceleration to complete stop and forward displacement of an occupant at an impact speed of 50 km/h

Fig. 2
① Impact
② Triggering of seat belt pretensioner/airbag
③ Belt tensioned
④ Airbag filled
- - Without passenger restraint systems
— With passenger restraint systems

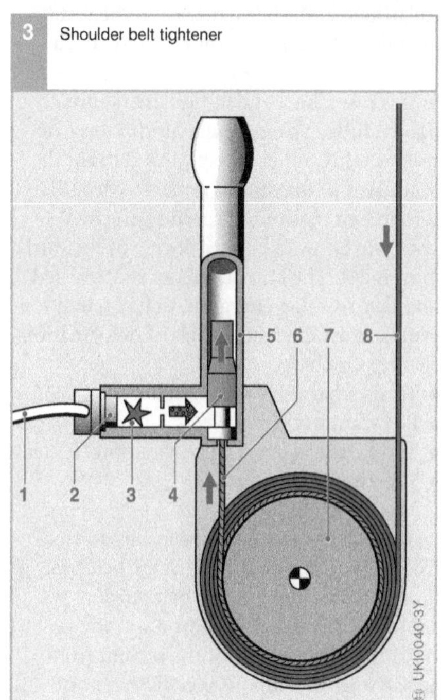

3 Shoulder belt tightener

Fig. 3
1 Firing cable
2 Firing element
3 Propellant charge
4 Piston
5 Cylinder
6 Wire rope
7 Belt reel
8 Belt webbing

Apart from pyrotechnical seat belt pretensioners, there are also mechanical versions. In the case of a mechanical tightener, a mechanical or electrical sensor releases a pretensioned spring, which pulls the seat belt buckle back. The sole advantage of these systems is that they are cheaper. However, their deployment characteristics are not so well synchronized with the deployment of the airbag as pyrotechnical seat belt pretensioners, which, of course, have the same electronic impact-sensing equipment as the front airbags.

In order to achieve optimum protection, the response of all components of the complete occupant protection system, comprising seat belt pretensioners and airbags for frontal impacts, must be adapted to one another. Seat belts and seat belt pretensioners provide the greater part of the protective effect since they absorb 50-60% of impact energy alone. With front airbags, the energy absorption is about 70% if deployment timing is properly synchronized.

A further improvement, which prevents collarbone and rib fractures and the resulting internal injuries to more elderly occupants, can be achieved by *belt force limiters*. In this case, the seat belt tighteners initially tighten fully (using the maximum force of approx. 4 kN, for example) and restrain the occupants to maximum possible effect. If a certain belt tension is exceeded, the belt gives and allows a greater degree of forward movement. The kinetic energy is converted into deformation energy so that acceleration peaks are avoided. Examples of deformation elements include:

- Torsion bar (belt reel shaft)
- Rip seam in the belt
- Seat belt buckle with deformation element
- Shearing element

DaimlerChrysler, for example, has an electronically controlled single-stage belt force limiter, which reduces the belt tension to 1-2 kN by firing a detonator a specific period after deployment of the second front airbag stage and after a specific extent of forward movement is reached.

Further developments

The performance of pyrotechnical seat belt pretensioners is constantly being improved. "High-performance tighteners" are capable of retracting an extended belt length of about 15 cm in roughly 5 ms. In the future there will also be two-stage belt force limiters consisting of two torsion bars with staggered response or a single torsion bar combined with an extra deformation plate in the retractor.

Front airbag

Function

The function of front airbags is to protect the driver and front passenger against head and chest injuries in a vehicle impact with a solid obstacle at speeds of up to 60 km/h. In case of a frontal impact of two vehicles, the front airbags provide protection at relative speeds of up to 100 km/h. In a serious accident, a seat belt pretensioner cannot keep the head from striking the steering wheel. In order to fulfill this function, airbags have different filling capacities and shapes to suit varying vehicle requirements, depending on where they are fitted, the vehicle type, and its structure deformation characteristics.

In a few vehicle types, front airbags also operate in conjunction with "inflatable knee pads", which safeguard the "ride down benefit", i.e. the speed decrease of the occupants together with the speed decrease of the passenger cell. This ensures that the upper body and head describe the rotational forward motion needed for the airbag to provide optimum protection, and is of particular benefit in countries where seat belt usage is not mandatory.

Method of operation

To protect driver and front passenger, pyrotechnical gas inflators inflate the driver and passenger airbags using dynamic pyrotechnics after a vehicle impact detected by sensors. In order for the affected occupant to enjoy maximum protection, the airbag must be fully inflated before the occupant comes into contact with it. On contact with the upper body, the airbag partly deflates in order to "gently" absorb impact energy acting on the occupant with noncritical (in terms of injury) surface pressures and declaration forces. This concept significantly reduces or even prevents head and chest injuries.

The maximum permissible forward displacement before the driver's airbag is fully inflated is approx. 12.5 cm, corresponding to a period of approx. 10 ms + 30 ms = 40 ms after the initial impact (at 50 km/h with a solid obstacle) (see Fig. 2). It takes 10 ms for electronic firing to take place and 30 ms for the airbag to inflate (Fig. 4).

In a 50 km/h crash, the airbag takes approx. 40 ms to inflate fully and a further 80-100 ms to deflate through the deflation holes. The entire process takes little more than a tenth of a second, i.e. the bat of an eyelid.

Impact detection

Optimum occupant protection against the effects of frontal, offset, oblique or pole impact is obtained (as mentioned above) through the precisely coordinated interaction of electronically detonated pyrotechnical front airbags and seat belt pretensioners. To maximize the effect of both protective devices, they are activated with optimized time response by a common ECU (trigger unit) installed in the passenger cell. This involves the electronic control unit using one or two electronic linear acceleration sensors to measure the deceleration occurring on impact and calculate the change in velocity. In order to be able to better detect oblique and offset impacts, the deployment algorithm can also take account of the signal from the lateral acceleration sensor.

The impact must also be analyzed. The airbag should not trigger from a hammer blow in the workshop, gentle impacts, bottoming out, driving over a curbstone or a pothole. With this goal in mind, the sensor signals are processed in digital analysis algorithms whose sensitivity parameters have

4 "Dynamic" inflation of a driver airbag

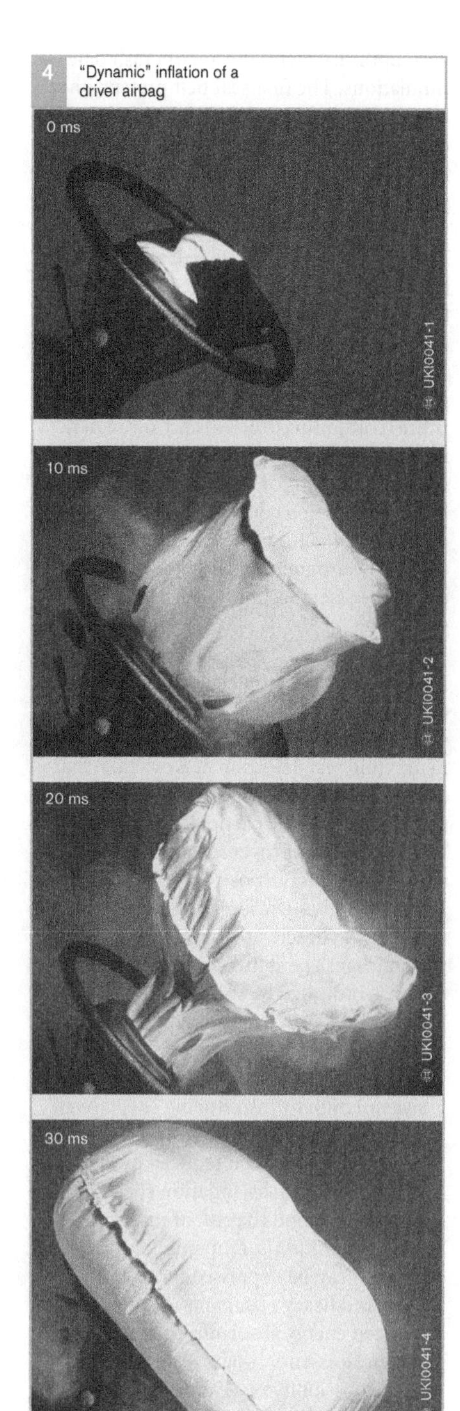

been optimized with the aid of crash data simulations. The first seat belt pretensioner trigger threshold is reached within 8-30 ms depending on the type of impact, and the first front airbag trigger threshold after approx. 10-50 ms.

The acceleration signals, which are influenced by such factors as the vehicle equipment and the body's deformation characteristics, are different for each vehicle. They determine the setting parameters which are of crucial importance for sensitivity in the analysis algorithm (computing process) and, ultimately, for triggering the airbag and seat belt pretensioner. Depending on the vehicle manufacturer's production concept, the deployment parameters and the vehicle's equipment level can also be programmed into the ECU at the end of the assembly line ("end-of-line programming").

In order to prevent injuries caused by airbags or fatalities to "out-of-position" occupants or to small children in child seats with automatic child seat detection, it is essential that the front airbags are triggered and inflated in accordance with the particular situations. The following improvement measures are available for this purpose:

1. *Deactivation switches.* These switches can be used to deactivate the driver or passenger airbag. The status of the airbag function is indicated by special lamps.
2. In the USA, where approximately 160 fatalities have been caused by airbags, attempts are being made to reduce aggressive inflation by introducing *"depowered airbags"*. These are airbags whose gas inflator power has been reduced by 20-30%, which itself reduces inflation speed, inflation severity and the risk of injury to "out-of-position" occupants. "Depowered airbags" can be depressed more easily by large and heavy occupants, i.e. they have a reduced energy absorption capacity. It is therefore essential – above all with regard to the possibility of severe frontal impacts – for occupants to fasten their seat belts.
   In the USA, the "low-risk" deployment method is currently preferred. This means that in "out-of-position" situations, only the first front airbag stage is triggered. In heavy impacts, the full gas inflator output can then be brought into effect by triggering both inflator stages. Another way of implementing "low-risk" deployment with single-stage inflators and controllable deflation vents is to keep the deflation valve constantly open.
3. *"Intelligent airbag systems"*. The introduction of more and improved sensing functions and control options for the airbag inflation process, with the accompanying improvement in protective effect, is intended to result in a gradual reduction in the risk of injury. Such functional improvements are:

- Impact severity detection by improvements in the deployment algorithm or the use of one or two upfront sensors, refer to "restraint system electronics", RSE (Fig. 5) These are acceleration sensors installed in the vehicle's crumple zone (e.g. on the radiator crossmember) which facilitate early detection of impacts that are difficult to detect centrally, such as ODB (offset deformable barrier) crashes, pole or underride impacts. They also allow an assessment of the impact energy:
- Seat belt usage detection
- Occupant presence, position and weight detection
- Seat position and backrest inclination detection
- Use of front airbags with two-stage gas inflators or with single-stage gas inflators and pyrotechnically triggered gas discharge valves (see also "low-risk" deployment method)
- Use of seat belt pretensioners with occupant-weight-dependent belt force limiters
- CAN bus networking of the occupant protection system for communication and synergy utilization of data from "slow" sensors (switches) in other systems (data on vehicle speed, brake operation, seat

belt buckle and door switch status) and for activation of warning lamps and transmission of diagnostic data.

For transmission of emergency calls after a crash and for activation of "secondary safety systems" (hazard warning signals, central locking release, fuel supply pump shutoff, battery disconnection etc.) the "crash output" is used (Fig. 6).

Side airbag

Function

Side impacts make up approx. 30% of all accidents. This makes the side collision the second most common type of impact after the frontal impact. An increasing number of vehicles are therefore being fitted with side airbags in addition to seat belt pretensioners and front airbags. Side airbags, which inflate along the length of the roof lining for head protection (inflatable tubular systems, window bags, inflatable curtains) or from the door or seat backrest (thorax bags, upper body protection) are designed to cushion the occupants and protect them from injury in the event of a side impact.

Method of operation

Due to the lack of a crumple zone, and the minimum distance between the occupants and the vehicle's side structural components, it is particularly difficult for side airbags to inflate in time. In the case of severe impacts, therefore, the time needed for impact detection and activation of the side airbags must be approx. 5-10 ms and the time needed to inflate the approx. 12 l thorax bags must not exceed 10 ms.

Bosch offers the following option to meet the above requirements: an instrument cluster ECU, which processes the input signals of peripheral (mounted at suitable points on the body), side-sensing acceleration sensors, and which can trigger side airbags as well as the seat belt pretensioners and the front airbags.

5 "Restraint system electronics" (RSE) electronic impact protection system

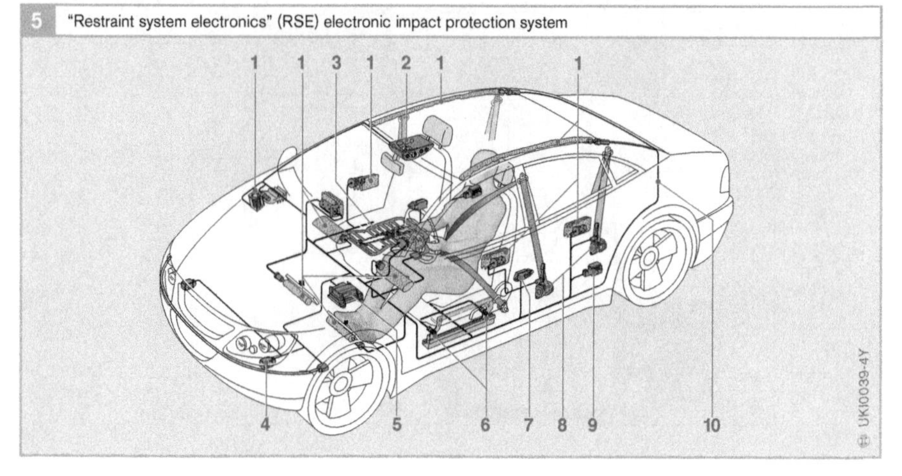

Fig. 5
1 Airbag with gas inflator
2 iVision™ passenger compartment camera
3 OC mat
4 Upfront sensor
5 Central electronic control unit with integrated rollover sensor
6 iBolt™
7 Peripheral pressure sensor (PPS)
8 Seat belt pretensioner with propellant charge
9 Peripheral acceleration sensor (PAS)
10 Bus architecture (CAN)

Components

Acceleration sensors

Acceleration sensors for impact detection are integrated directly in the control unit (seat belt pretensioners, front airbag), and mounted in selected positions on both sides of the vehicle on supporting structural components such as seat crossmembers, sills, B and C-pillars (side airbags) or in the crumple zone at the front of the vehicle (upfront sensors for "intelligent airbag systems"). The precision of these sensors is crucial in saving lives. Nowadays, those acceleration sensors are surface micromechanical sensors consisting of fixed and moving finger structures and spring pins. A special process is used to incorporate the "spring-mass system" on the surface of a silicon wafer. Since the sensors have only a low working capacitance (≈ 1 pF), it is necessary to accommodate the evaluation electronics in the same housing in the immediate proximity of the sensor element so as to avoid stray capacitance and other forms of interference.

6 Central combined airbag 9 ECU (block diagram)

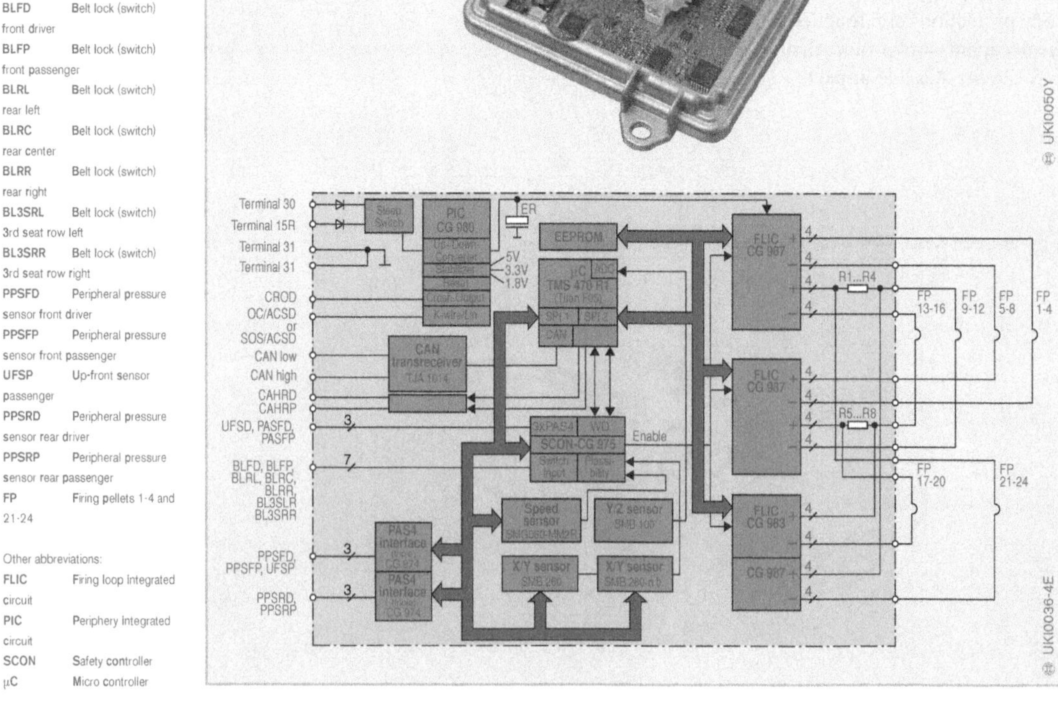

Fig. 6

Terminal 30 Direct battery positive, not fed through ignition switch

Terminal 15R Switched battery positive when ignition switch in "radio", "ignition on" or "starter" position

Terminal 31 Body ground (at one of the device mounting points)

CROD Crash output digital

OC/ACSD Occupant classification/automatic child seat detection

SOS/ACSD Seat occupancy sensing/automatic child seat detection

CAN low Controller area network, low level

CAN high Controller area network, high level

CAHRD Crash active head rest driver

CAHRP Crash active head rest passenger

UFSD Up-front sensor driver

PASFD Peripheral acceleration sensor front driver

PASFP Peripheral acceleration sensor front passenger

BLFD Belt lock (switch) front driver

BLFP Belt lock (switch) front passenger

BLRL Belt lock (switch) rear left

BLRC Belt lock (switch) rear center

BLRR Belt lock (switch) rear right

BL3SRL Belt lock (switch) 3rd seat row left

BL3SRR Belt lock (switch) 3rd seat row right

PPSFD Peripheral pressure sensor front driver

PPSFP Peripheral pressure sensor front passenger

UFSP Up-front sensor passenger

PPSRD Peripheral pressure sensor rear driver

PPSRP Peripheral pressure sensor rear passenger

FP Firing pellets 1-4 and 21-24

Other abbreviations:

FLIC Firing loop Integrated circuit

PIC Periphery Integrated circuit

SCON Safety controller

µC Micro controller

Combined ECUs for seat belt pretensioners, front and side airbags and rollover protection equipment

The following functions are incorporated in the central ECU, also referred to as the trigger unit (current list):

- Crash sensing by acceleration sensor and safety switch or by two acceleration sensors without safety switch (redundant, fully electronic sensing).
- Rollover detection by yaw rate and low g, y and z acceleration sensors (refer to the section on "Rollover sensing").
- Prompt activation of front airbags and seat belt pretensioners in response to different types of impact in the vehicle longitudinal direction (e.g. frontal, oblique, offset, pole, rear-end).
- Activation of rollover protection equipment.
- For the side airbags, the ECU operates in conjunction with a central lateral sensor and two or four peripheral acceleration sensors. The peripheral acceleration sensors (PAS) transmit the triggering command to the central ECU via a digital interface. The central ECU triggers the side airbags provided the internal lateral sensor has confirmed a side impact by means of a plausibility check. Since the central plausibility confirmation arrives too late in the case of impacts into the door or above the sill, peripheral pressure sensors (PPS) inside the door cavity are to be used in the future to measure the adiabatic pressure changes caused by deformation of the door. This will result in rapid detection of door impacts. Confirmation of "plausibility" is now provided by PAS mounted on supporting peripheral structural components. This is now unquestionably faster than the central lateral acceleration sensors.
- Voltage transformer and energy accumulator in case the supply of power from the vehicle battery fails.
- Selective triggering of the seat belt pretensioners, depending on monitored belt buckle status: firing only takes place if key is in the ignition switch. At present, proximity-type seat belt buckle switches are used, i.e. Hall-effect IC switches which detect the change in the magnetic field when the buckle is fastened.
- Setting of multiple triggering thresholds for two-stage seat belt pretensioners and two-stage front airbags depending on the status of belt use and seat occupation.
- Watchdog (WD): Airbag triggering units must meet high safety standards with regard to false activation in non-crash situations and correct activation when needed (crashes). For this reason, the ninth-generation airbag triggering unit (AB 9), introduced in 2003, incorporate three independent, intensive monitoring hardware watchdogs (WDs):
  WD1 uses its own independent oscillator to monitor the 2-MHz system clock.
  WD2 monitors the realtime processes (time base 500 µs) for correct and complete sequence. For this reason, the safety controller (SCON; refer to the AB 9 block diagram) sends the microcomputer 8 digital messages, to which it must respond by sending 8 correct replies to the SCON within a time window of (1 ± 0.3) ms.
  WD3 monitors the "background" processes such as the "built-in self-test" routines of the ARM core for correct operation. The microcomputer's response to the SCON in this case must be provided within a period of 100 ms.
  On AB 9 sensors, analyzer modules and output stages are linked by two serial peripheral interfaces (SPIs). The sensors have digital outputs whose signals can be transmitted directly via SPIs. Signal changes can then be detected by line connections on the printed circuit board, or else they have no effect and a high level of functional reliability is achieved. Deployment is only permitted if an independent hardware plausibility channel also detects an impact and enables the output stages for a limited period.
- Diagnosis of internal and external functions and of system components.

- Storage of fault types and duration with crash recorder; readout via the diagnostic or CAN bus interface.
- Warning lamp activation.

Gas inflators

The pyrotechnical propellant charges of the gas inflators for generating the airbag inflation gas (mainly nitrogen) and for actuating seat belt pretensioners are activated by an electrically operated firing element.

The gas inflator in question inflates the airbag with nitrogen. The driver airbag built into the steering wheel hub (volume approx. 60 l) or, as the case may be, the passenger airbag fitted in the glove box space (approx. 120 l) is inflated in approx. 30 ms after detonation.

AC firing

In order to prevent inadvertent triggering through contact between the firing element and the vehicle system voltage (e.g. faulty insulation in the wiring harness), AC firing is used. This involves firing by alternating-current pulses at approx. 80 kHz. A small capacitor with a capacitance of 470 nF incorporated in the firing circuit in the firing element plug electrically isolates the firing element from the DC current. This isolation from the vehicle system voltage prevents inadvertent triggering, even after an accident when the airbag remains untriggered and the occupants have to be freed from the deformed passenger cell by emergency services. It may even be necessary to cut through the (permanent +) firing circuit wires in the steering column wiring harness and short-circuit them according to positive and ground.

7 Force sensing "iBolt" (functional principle)

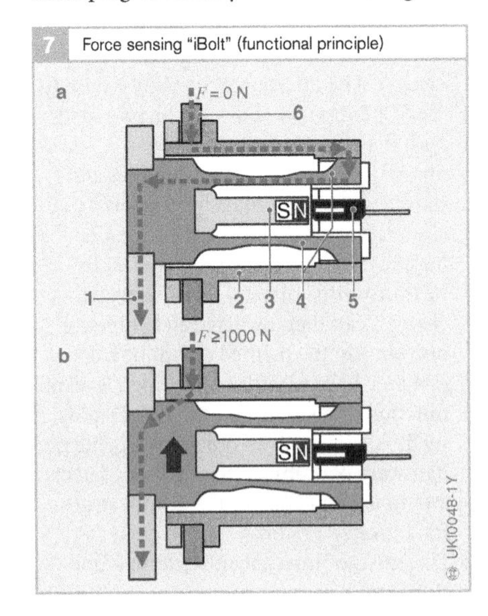

Fig. 7
a Initial position
b In function, i.e. in overload stop

1 Sliding base
2 Sleeve
3 Solenoid holder
4 Double flexing rod (spring)
5 Hall-effect IC
6 Seat frame

Passenger compartment sensing

Occupant classification mats ("OC mats"), which measure the pressure profile on the seat, are used to distinguish whether the seat is occupied by a person or by an object. In addition, the pressure distribution and the pelvic bone spacing are used to indicate the occupant's size and thus indirectly the occupant's weight. The mats consist of individually addressable force sensing points which reduce their resistance according to the FSR principle force sensing resistor) as pressure increases.

In addition, *absolute weight measurement* using four piezo-resistive sensors or wire strain gauges on the seat frame is also under development. Instead of using deformation elements, the Bosch strategy for weight measurement involves the use of "iBolts" ("intelligent" bolts) for fixing the seat frame (seat cradle) to the sliding base. These force sensing "iBolts" (Fig. 5 and 7) replace the four fixing bolts otherwise used.

They measure the weight-dependent change in the gap between the bolt sleeve and the internal bolt with integral Hall-element IC connected to the sliding base. Four different concepts are under consideration for detecting "out-of-position" situations:

- Determining the position of the occupant's center of gravity from the weight distribution on the seat detected by the four weight sensors.
- Using the following optical methods:

- "Time of flight" (TOF) principle. This system sends out infrared light signals and measures the time taken for the reflected signals to be received back, which is dependent on the distance to the occupant. The time intervals being measured are of the order of picoseconds!
- "Photonic mixer device" (PMD) method. A PMD imaging sensor sends out "ultrasonic light" and enables spatial vision and triangulation.
- "iVision" passenger compartment stereo video camera using CMOS technology (the option favored by Bosch, see system diagram of "restraint system electronics, RSE"). This detects occupant position, size and restraint method and can also control convenience functions (seat, mirror and radio settings) to suit the individual occupant.

No unified standard for passenger compartment sensing has yet been able to establish itself. Jaguar, for example, uses occupant classification mats combined with ultrasonic sensors.

Rollover protection systems

Function

In the event of an accident where the vehicle rolls over, open-top vehicles such as convertibles, off-road vehicles etc., lack the protecting and supporting roof structure of closed-top vehicles. Initially, therefore, rollover sensing and protection systems were only installed in convertibles and roadsters without fixed rollover bars (Fig. 8).

Now engineers are developing rollover sensing for use in closed passenger cars. If a car turns over, there is the danger that non-belted occupants may be thrown through the side windows and crushed by their own vehicle, or the arms, heads or torsos of belted occupants may protrude from the vehicle and be seriously injured.

To provide protection in such cases, already existing restraint systems such as

8 Quick activation of retractable head restraints during a convertible rollover test

Fig. 8
a Rollover begins
b Head restraints are triggered
c Vehicle rolls over
d Vehicle hits the ground
(Source: Mercedes-Benz)

seat belt pretensioners and head airbags are activated. In convertibles, the extendable rollover bars or the extendable head restraints are also triggered.

Method of operation

The earlier sensing concepts (from mid-1989) were based on an omnidirectional sensing function. In other words, a rollover in any direction from the horizontal should be detectable. For this purpose, manufacturers used either all-around-sensing acceleration sensors that were AND-wired to an omnidirectional tilt sensor or level gauge (water level principle) and gravitation sensors (sensor closes a spring-assisted reed switch when contact with the ground is lost).

Current sensing concepts no longer trigger the system at a fixed threshold but rather at a threshold that conforms to a situation and only for the most common rollover situation, i.e. about the longitudinal axis. The Bosch sensing concept involves a surface micromechanical yaw sensor and high-resolution acceleration sensors in the vehicle's transverse and vertical axes (y and z axes).

The yaw rate sensor is the main sensor, while the y and z-axis acceleration sensors are used both to check plausibility and to detect the type of rollover (slope, gradient, curb impact or "soil-trip" rollover). On Bosch systems, these sensors are incorporated in the airbag triggering unit.

Deployment of occupant protection systems is adapted to the situation according to the type of turnover, the yaw rate and the lateral acceleration, i.e. systems are triggered after between 30 and 3000 ms by automatic selection and use of the algorithm module appropriate to the type of rollover.

Outlook

In addition to front airbag shutoff using deactivation switches, soon there will also be an increasing number of child seats with standardized anchor systems ("ISOFIX child seats"). Switches integrated in the anchoring locks initiate an automatic passenger airbag shutoff, which must be indicated by a special lamp.

For further improvement of the deployment function and better advance detection of the type of impact ("pre-crash" detection), microwave radar, ultrasound or lidar sensors (optical system using laser light) will be used to detect relative speed, distance and angle of impact for frontal impacts (Fig. 9).

In connection with pre-crash sensing, reversible seat belt pretensioners are being developed. They are electromechanically actuated, i.e. they take longer to tighten, and must be triggered earlier, i.e. 150 ms before initial impact, by pre-crash sensing alone (prefire function).

A further improvement in the restraining effect will be provided by airbags integrated in the thorax section of the seat belt ("air belts", "inflatable tubular torso restraints" or "bag-in-belt" systems), which will reduce the risk of broken ribs in older occupants.

The same path for improving protective functions is being pursued by engineers developing "inflatable headrests" (adaptive head restraints for preventing whiplash trauma and cervical injuries), "inflatable carpets" (prevention of foot and ankle injuries), two-stage seat belt pretensioners and "active seats". In the case of "active seats", an airbag made of thin steel sheet (!) is inflated to prevent occupants sliding forwards under the lap belt ("submarining effect").

To reduce wiring harness size and complexity, firing circuit networking is being developed. The "Safe-by-Wire" bus (originally developed by Philips) is an example of a product for such applications. More recently, a consortium of companies, including Bosch, has been formed with the aim of developing a line production "Safe-by-Wire" firing bus. The current designation for the "Safe-by-Wire" bus is the "ASRB2.0" bus, short for "Automotive safety restraints bus 2.0". The DSI bus (developed by Motorola for TRW) also continues to be used. However, it is still entirely uncertain whether a firing bus concept will become established.

Signals from "slow" sensors or switches (e.g. the seat belt buckle or ISOFIX switches) can also be transmitted by the firing bus.
Efforts are currently underway in the USA to standardize the "ASRB2.0" bus concept. Standardization is imperative in order to ensure market penetration and the potential usability of standardized firing elements with standardized bus device electronics. Efforts are underway to integrate the receiver electronics in the firing elements, without increasing diameter and while maintaining a maximum cap extension of 5 mm. This would increase the usability of standard gas inflators.

In addition to the "firing bus", there will also be a "sensor bus" for networking the signals of "fast" sensors. This will make it possible to combine inertial sensors, for instance, in a "sensor cluster". The overall picture of vehicle dynamics can then be made available via CAN to the evaluation chips of various vehicle systems. Conceivable sensor buses include TT CAN time-triggered CAN), TTP time-triggered protocol) and FlexRay, the option currently favored by Bosch. The requirements of a sensor bus regarding transmission reliability and speed are extremely high.

The first phase of legally required measures for improving pedestrian protection can be expected in 2005. Therefore, OEMs urgently need to develop solutions for their new models to meet the pedestrian trauma limits which will then be in force and which in most cases will be achievable by passive design features (body shape, use of impact-absorbing materials). Enactment of the second stage of the legislation (in approximately 2010) providing for even lower trauma limits will then require active safety features, i.e. pedestrian impact will have to be detected and protective actuators actuated.

Pedestrian impact sensing will initially be implemented by *deformation or force sensors* in the fender and possibly the front of the hood, e.g. in the form of

- Fiber-optic cables which utilize the "microbending" effect
- Film pressure sensors (as in occupant classification mats)
- Acceleration sensors or knock sensors on the fender crossmembers

9 Pre-crash traffic situation

Fig. 9
150 ms before impact: "Prefire" (triggering of reversible seat belt pretensioner)
10 ms before impact: "Preset" (determining the trigger thresholds of the airbags)

At a later date contactless sensors will be used to reliably distinguish between a pedestrian and an object. These might, for instance, be:

- Ultrasonic sensors or
- External stereo video cameras

The *protective actuators* consist of A-pillar airbags and hoods which can be raised by approx. 10 cm so that, if impacted by a pedestrian's head, they are not depressed far enough to come in contact with the rigid engine components due to the greater clearance. As a result, the trauma suffered is less severe.

In Europe, 7000 pedestrians are killed every year. That Figure represents 20% of the total number of road accident fatalities. In Japan, for example, there are 17,000 pedestrian deaths a year. For this reason, legislators in Japan are deliberating whether to make safety features for pedestrians a legal requirement as in Europe.

The following additional improvements for softer cushioning of occupants are also likely:

1. Airbags with active ventilation system: These airbags have a controllable deflation valve to maintain the internal pressure of the airbag constant even if an occupant falls against it and to minimize occupant trauma. A simpler version is an airbag with "intelligent vents". These vents remain closed (so that the airbag does not deflate) until the pressure increase resulting from the impact of the occupant causes them to open and allow the airbag to deflate. As a result, the airbag's energy absorption capacity is fully maintained until the point at which its motion-damping function comes into effect.
2. Adaptive, pyrotechnically triggered steering column release. This allows the steering wheel to move forward in a severe impact so that the occupant can be more softly cushioned over a greater distance of travel.
3. Networking of passive and active safety features. The first example of the synergetic use of sensors in different safety systems will be implemented in ROSE II (rollover sensing II). ROSE II will utilize the signals available on the CAN from the speed vector sensor for improved detection of soil trip rollover situations. The speed vector sensor is part of the ESP system and is used to measure the deviation of the vehicle motion vector from the vehicle's longitudinal axis. The ESP, on the other hand, can utilize the signals from the ROSE II low-g acceleration sensors (y and z axes) for improved detection of unstable dynamic handling situations.

Piezoelectric acceleration sensors

Application

Piezoelectric bimorphous bending elements and two-layer piezoceramic elements are used as acceleration sensors in passenger-restraint systems for triggering the seat-belt tighteners, the airbags, and the roll-over bar.

Design and operating concept

A piezo bending element is at the heart of this acceleration sensor. It is a bonded structure comprising two piezoelectric layers of opposite polarities ("bimorphous bending element"). When subjected to acceleration, one half of this structure bends and the other compresses, so that a mechanical bending stress results (Fig. 1).

The voltage resulting from the element bend is picked off at the electrodes attached to the sensor element's outside metallised surfaces.

The sensor element shares a hermetically-sealed housing with the initial signal-amplification stage, and is sometimes encased in gel for mechanical protection.

1 Bending element from a piezoelectric acceleration sensor

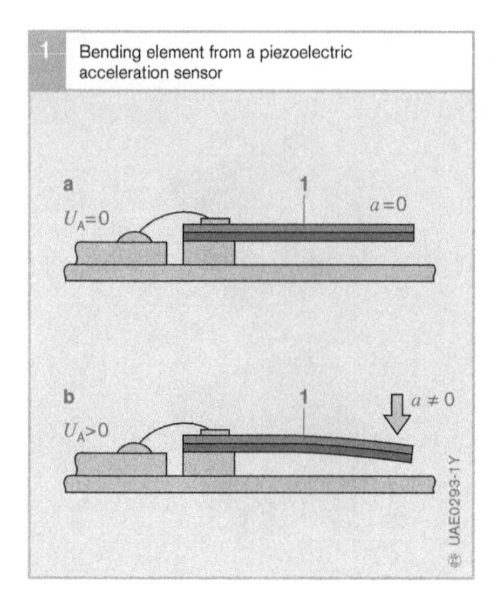

Fig. 1
a Not subject to acceleration
b Subject to acceleration a
1 Piezoceramic bimorphous bending element
U_A Measurement voltage

2 Piezoelectric acceleration sensor (dual sensor for vertical mounting)

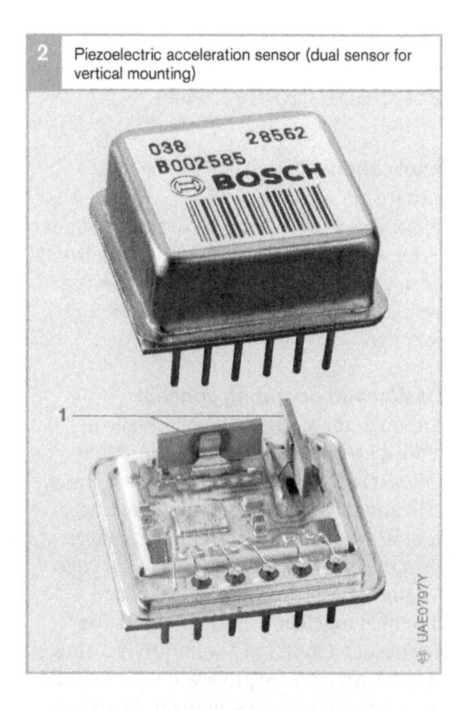

Fig. 2
1 Bending element

For signal conditioning, the acceleration sensor is provided with a hybrid circuit comprised of an impedance converter, a filter, and an amplifier. This serves to define the sensitivity and useful frequency range. The filter suppresses the high-frequency signal components. When subjected to acceleration, the piezo bending elements deflect to such an extent due to their own mass that they generate a dynamic, easy-to-evaluate non-DC signal with a maximum frequency which is typically 10 Hz.

By "reversing" the actuator principle and applying voltage, the sensor's correct operation can be checked within the framework of OBD "on-board diagnosis". All that is required is an additional actuator electrode.

Depending upon installation position and direction of acceleration, there are single or dual sensors available (Fig. 2). Sensors are also on the market which are designed specifically for vertical or horizontal mounting (Fig. 2).

Surface micromechanical acceleration sensors

Application

Surface micromechanical acceleration sensors are used in passenger-restraint systems to register the acceleration values of a frontal or side collision. They serve to trigger the seatbelt tightener, the airbag, and the rollover bar.

Design and operating concept

Although these sensors were initially intended for use with higher accelerations (50...100 g), they also operate with lower acceleration figures when used in passenger-restraint systems. They are much smaller than the bulk silicon sensors (typical edge length: approx. 100...500 µm), and are mounted together with their evaluation electronics (ASIC) in a waterproof casing (Fig. 1). An additive process is used to build up their spring-mass system on the surface of the silicon wafer.

The seismic mass with its comb-like electrodes (Figs. 2 and 3, pos. 1) is spring-mounted in the measuring cell. There are fixed comb-like electrodes (3, 6) on the chip on each side of these movable electrodes. This configuration comprising fixed amd movable electrodes corresponds to a series circuit comprising two differential capacitors (capacity of the comb-like structure: approx. 1 pF). Opposed-phase AC voltages are applied across the terminals C_1 and C_2, and their superimpositions picked-off between the capacitors at C_M (measurement capacity), in other words at the seismic mass.

Since the seismic mass is spring-mounted (2), linear acceleration in the sensing direction results in a change of the spacing between the fixed and movable electrodes, and therefore also to a change in the capacity of C_1 and C_2 which in turn causes the electrical signal to change. In the evaluation electronics circuit, this change is amplified, and then filtered and digitalised ready for further signal processing in the airbag ECU. Due to the low capacity of approx. 1 pF, the evaluation electronics is situated at the sensor and is

1 Surface micromechanical acceleration sensors for airbag triggering (Example)

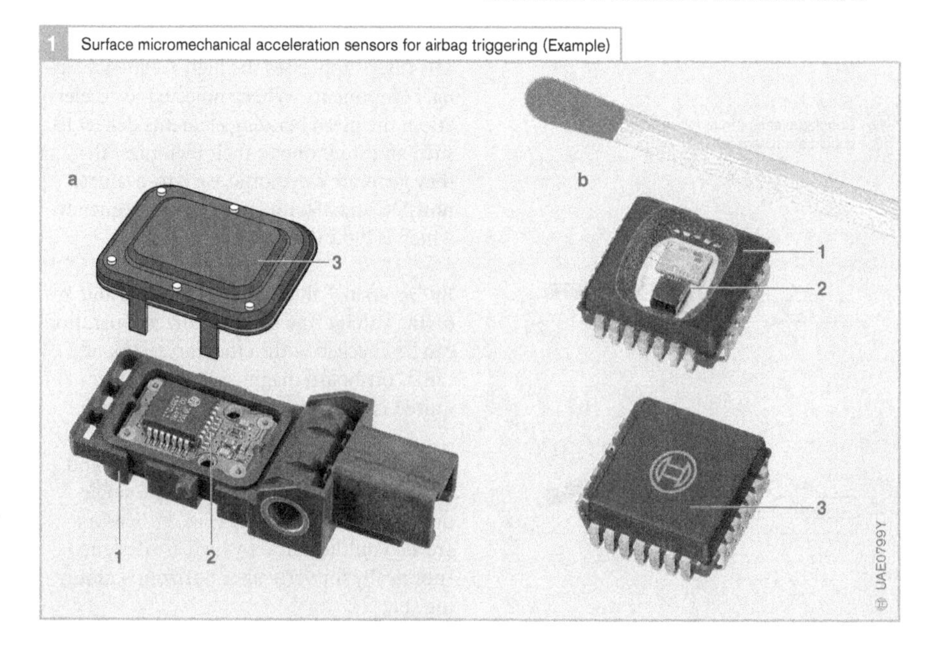

Fig. 1
a Side-airbag sensor
b Front-airbag sensor
1 Casing
2 Sensor and evaluation chip
3 Cover

either integrated with the sensor on the same chip, or is located very close to it. Closed-loop position controls with electrostatic return are also available.

The evaluation circuit incorporates functions for sensor-deviation compensation and for self-diagnosis during the sensor start-up phase. During self-diagnosis, electrostatic forces are applied to deflect the comb-like structure and simulate the processes which take place during acceleration in the vehicle.

Dual micromechanical sensors (4) are used for instance in the ESP Electronic Stability Program for vehicle dynamics control: Basically, these consist of two individual sensors, whereby a micromechanical yaw-rate sensor and a micromechanical acceleration sensor are combined to form a single unit. This reduces the number of individual components and signal lines, as well as requiring less room and less attachment hardware in the vehicle.

2 Comb-like structure of the sensor measuring element

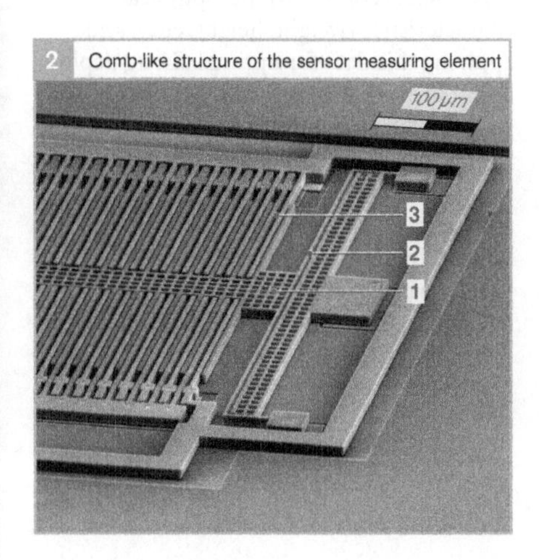

4 Lateral-acceleration sensor combined with yaw-rate sensor (dual sensor)

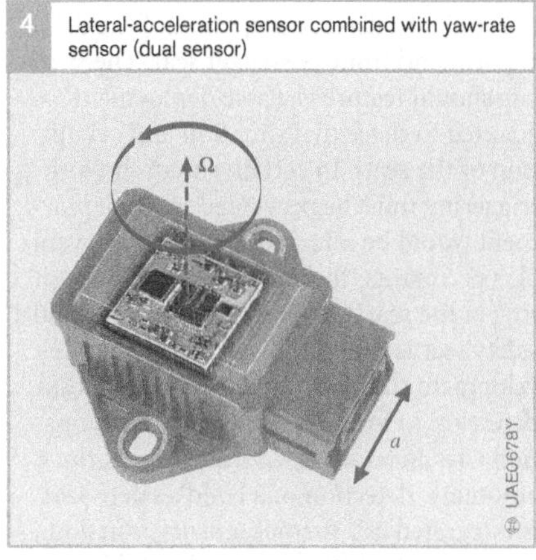

Fig. 2
1 Spring-mounted seismic mass with electrode
2 Spring
3 Fixed electrodes

Fig. 4
a Acceleration in sensing direction
Ω Yaw rate

3 Surface micromechanical acceleration sensor with capacitive pick-off

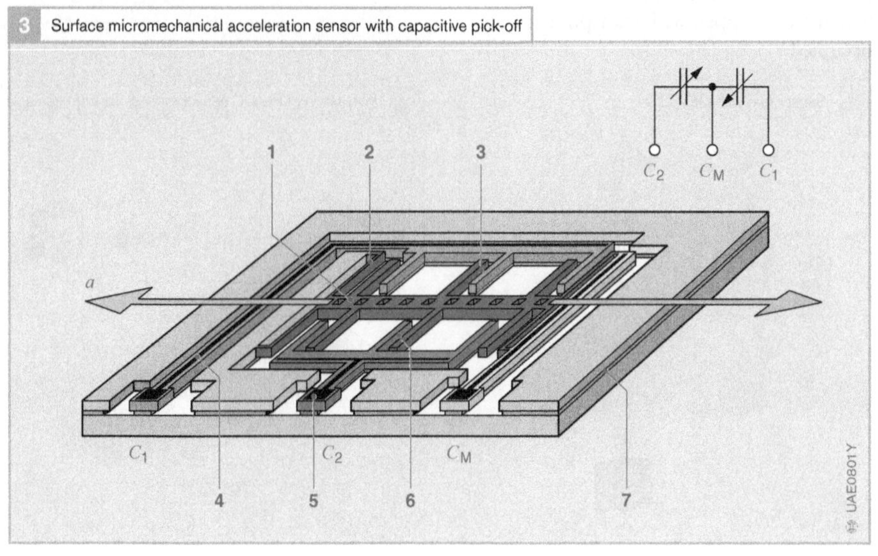

Fig. 3
1 Spring-mounted seismic mass with electrodes
2 Spring
3 Fixed electrodes with capacity C_1
4 Printed Al conductor
5 Bond pad
6 Fixed electrodes with capacity C_2
7 Silicon oxide
a Acceleration in sensing direction
C_M Measuring capacity

Seat occupancy sensing

Assignment

Following introduction of the airbag for the front-seat passenger, safety and actuarial considerations made it necessary to detect whether the front-seat passenger's seat is occupied or not. Otherwise, when an accident occurs and both front airbags are deployed, unnecessary repair costs result if the passenger seat is unoccupied.

The development of the so-called "Smart Bags" marked an increase in the demand for the ability to detect occupation of the driver-seat and front-passenger seat. The smart bag should feature variable deployment adapted to the actual situation and occupation of the seats. In certain situations, airbag triggering must be prevented when deployment would be injurious to one of the vehicle's occupants (for instance, if a child is sitting in the seat next to the driver, or a child's safety seat is fitted). This led to further development of the "simple" seat-occupation detection to form the "intelligent" **O**ccupation **C**lassification (OC). In addition, the automatic detection of a child's safety seat is integrated as a further sensory function. It can detect whether or not the child seats, which are equipped with transponders, are occupied.

Design and construction

A so-called sensor mat and ECU incorporated in the vehicle's front seats (Figs. 1 and 2) registers the information on the person in the seat and sends this to the airbag ECU. These data are then applied when adapting the restraint-system triggering to the current situation.

Operating concept

Measuring concept

This relies upon the classification of passengers (OC) according to their physical characteristics (weight, height, etc.), and applying this data for optimal airbag deployment. Instead of directly "weighing" the person concerned, the OC system primarily applies the correlation between the person's weight and his/her anthropometric[1]) characteristics (such as distance between hipbones). To do so, the OC sensor mat measures the pressure profile on the seat surface. Evaluation indicates first of all whether the seat is occupied or not, and further analysis permits the person concerned to be allocated to a certain classsification (Fig. 3).

[1]) The study of human body measurements, especially on a comparative basis.

1 Sensor mat with OC-ECU

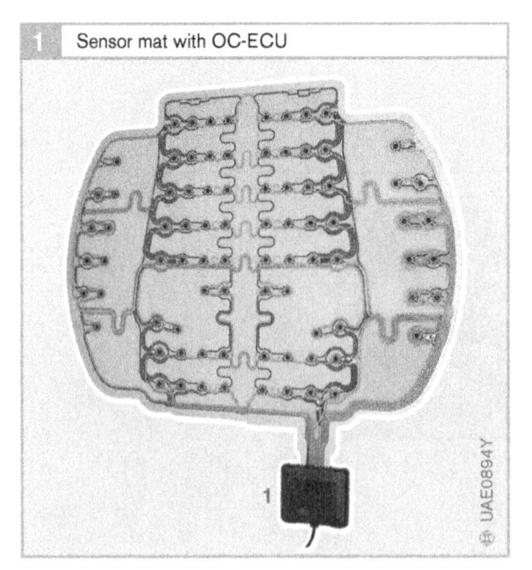

2 Installation of the OC sensor mats in the front seats

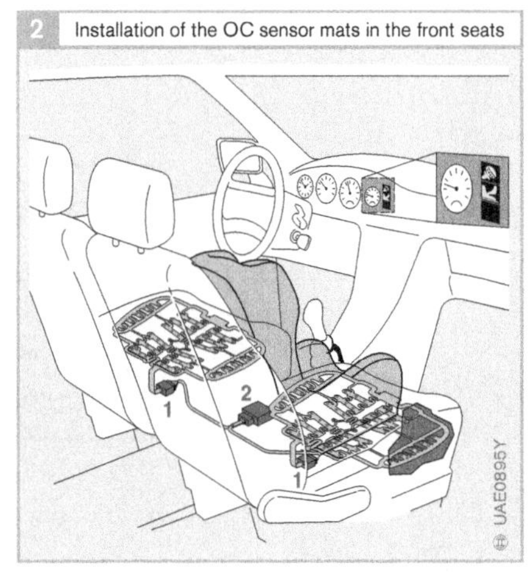

Fig. 1
1 ECU

Fig. 2
1 OC-ECU
2 Airbag ECU

Sensor technology

Basically, the OC sensor mat comprises pressure-dependent FSR resistance elements (FSR: Force-Sensitive Resistance), the information from which can be selectively evaluated. A sensor element's electrical resistance drops when it is subjected to increasing mechanical load. This effect can be registered by inputting a measuring current. The analysis of all sensor points permits definition of the size of the occupied seat area, and of the local points of concentration of the profile.

A standalone sending antenna and two receive antennas in the OC sensor mat serve to implement the ACSD function. During the generation of a sending field, transponders in the specially equipped child's seats are excited so that they impose a code on the sending field by means of modulation. The data received by the receive antenna and evaluated by the electronic circuitry is applied in determining the type of child's seat and its orientation.

ECU

The ECU feeds measuring currents into the sensor mat and evaluates the sensor signals with the help of an algorithm program which runs in the microcontroller. The resulting classification data and the information on the child's safety seat are sent to the airbag ECU in a cyclical protocol where, via a decision table, they help to define the triggering behaviour.

Algorithm

Among other things, the following decision criteria serve to analyse the impression of the seating profile:

Distance between hip-bones:
A typical seating profile has two main impression points which correspond to the distance between the passsenger's hip-bones.

Occupied surface:
Similarly, there is a correlation between the occupied surface and the person's weight.

Profile coherence:
Consideration of the profile structure.

Dynamic response:
Change of the profile as a function of time.

3 Seat profile of the human body (a), with assignment of the distance between hip-bones to the person's weight (b)

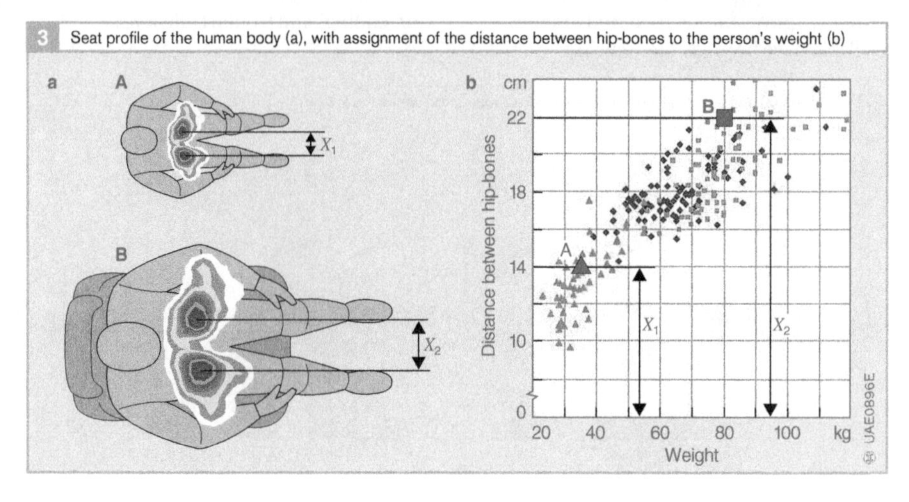

Fig. 3
- a Seating profile
- b Diagram
- A Child with distance between hip-bones X_1
- B Adult with distance between hip-bones X_2

Driving assistance systems

On average, someone dies every minute somewhere in the world as a result of a traffic accident. Bosch pursues the aim of reducing the frequency and the severity of accidents by developing active and passive driving assistance systems.

Critical driving situations

Driving assistance systems aim to make the vehicle capable of perceiving its surroundings, interpreting them, identifying critical situations and assisting the driver in performing driving maneuvers. The object is, at best, to prevent accidents completely and, at worst, to minimize the consequences of an accident for those concerned.

In critical driving situations, fractions of a second are often decisive in determining whether an accident occurs or not. According to various studies, approximately 60% of rear-end collisions and virtually a third of frontal impacts could have been avoided if the driver had reacted only half a second earlier. Every second accident at an intersection could be prevented by faster driver reaction.

At the end of the 1980s, when the vision of highly efficient and partially automated road traffic was presented as part of the "Prometheus" project, the electronic components for this task were not in existence. However, the highly sensitive sensors and extremely powerful microcomputers now available have brought the "sensitive" vehicle a step nearer to realization. Sensors scan the vehicle's surroundings and systems generate warnings based on the objects identified or immediately perform the required driving maneuvers. All of this is done more quickly – by those decisive fractions of a second – than even a fully attentive and experienced driver would be capable of.

Accident causes, measures

In 2001, over 96,000 people were killed in road traffic accidents in Europe, the USA and Japan. The resulting cost to the various national economies totaled more than €400 billion (Fig. 1). The fact that many motorists are overburdened by the complexities of road traffic situations is shown by recent statistics. In Germany, for example, there were 2.37 million road accidents in 2001, 375,345 of which resulted in personal injury. In nine out of ten cases, the cause was human error.

A statistical analysis of accident causes outside built-up areas in Germany (Fig. 2) shows more than a third of the total number are caused by drivers changing lanes or unintentionally failing to stay in their lane. Systems which can see the "blind spot" and lane change alarms offer a means of reducing the number of accidents caused in this way. About another third of accidents result from "rear-ending" and frontal collisions.

1 Fatalities (a) and economic cost (b) resulting from road traffic accidents in 2001

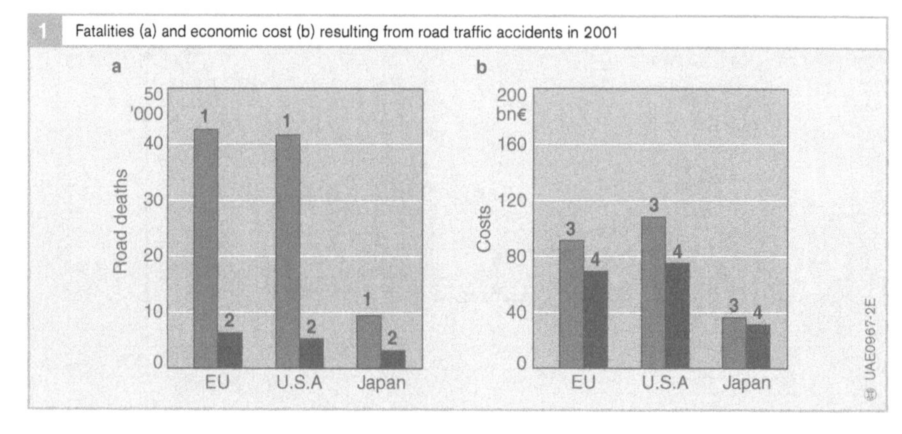

Fig. 1
1 Total fatalities (in thousands)
2 Ratio of pedestrians
3 Personal injuries
4 Property damage

Collision warning systems can represent a first line of defense in combating such accidents. A second level can be provided by collision avoidance systems which actively intervene in the control of the vehicle. A first step in this direction has already been made with the development of ACC (adaptive cruise control).

Accidents involving pedestrians and at intersections have a high level of complexity. Only networked sensor systems with scenario interpretation capabilities can master such complicated accident situations. This is one of the issues currently occupying researchers.

Application areas

Driving assistance systems with multiple applications (Fig. 3) fall into these two categories:

- Safety systems aimed at preventing accidents
- Comfort and convenience systems with the long-term goal of "semi-automated driving".

A further distinction is made between:

- Active systems which intervene in vehicle dynamics
- Passive, i. e. informational systems, which do not intervene in vehicle control.

Safety and convenience

Passive safety (Fig. 3, lower left quadrant) consists of features for lessening the consequences of accidents, such as pre-crash functions and pedestrian protection features.

Systems for *driver assistance* without intervention into vehicle control (bottom right quadrant) represent a precursor to vehicle handling. Such systems merely warn the driver or recommend maneuvers. Examples: The *parking assistant* first measures the length of the parking space and indicates to the driver whether it will be easy or difficult to park in the space, or if the space is too small. In the next stage, the system offers the driver recommendations for steer-

2 Causes of accidents outside built-up areas in Germany in 2000

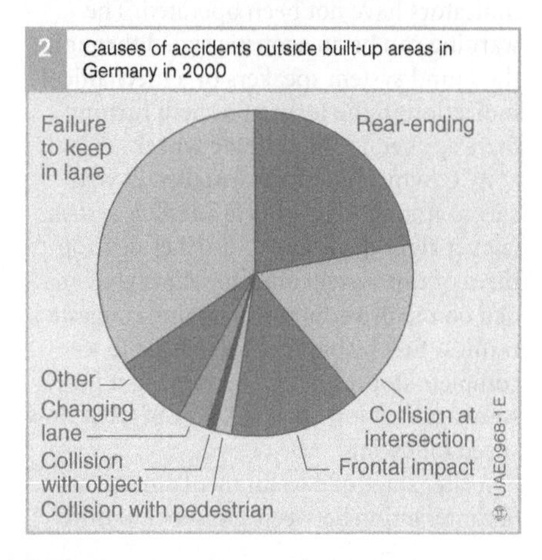

Fig. 2
1 Changing lanes
2 Failure to stay in lane
3 Rear-ending
4 Frontal impact
5 Collision with object
6 Collision with pedestrian
7 Collision at intersection
8 Other

3 Safety and convenience functions based on all-around visibility sensors

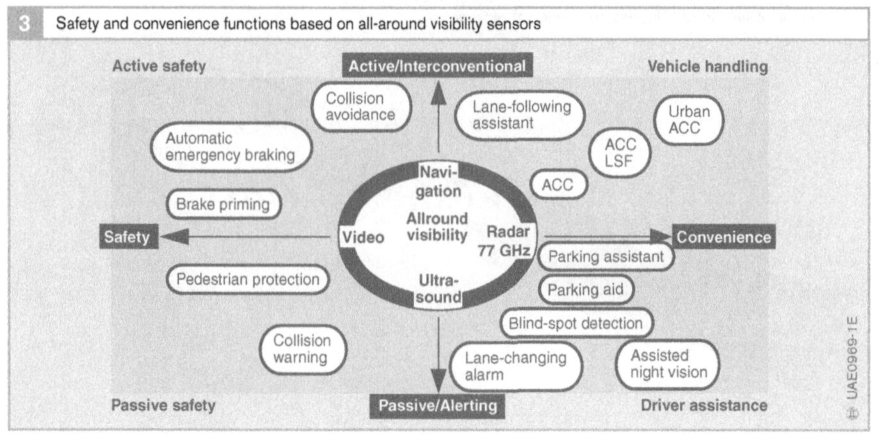

ing the vehicle during the parking maneuver based on the measurement it has taken of the parking space. The long-term aim for Bosch in the development of parking systems is the autonomous parking assistant – a system that actively intervenes in vehicle handling and automatically maneuvers the vehicle into the parking space. Detection of potentially dangerous objects in the blind spot is performed by close-range sensors (ultrasound sensors, radar sensors or lidar sensors). *Video sensors* can be effectively used to improve visibility for the driver at night. *Lane change alarms* use a video camera to extrapolate the course of the lane in front of the vehicle and warn the driver if the vehicle leaves its lane and the direction indicators have not been operated. The warning can be an acoustic signal through the sound system speakers or a mechanical indication in the form of a small turning force applied to the steering wheel.

ACC, which is already installed in vehicles, is among these *vehicle handling systems* (upper right quadrant). A further development of this system aims to relieve the burden on the driver in slow-moving congested traffic – first by braking the vehicle to a complete stop, and then by moving it forward again at low speed (ACC LSF: ACC low speed following).

A later stage of development aims to utilize interaction between a number of different sensors to enable complete linear control even in city areas (ACC Stop & Go) and at high vehicle speeds. The basis for this is a complex fusion of radar and video data. By combining the linear control system with a (similarly video-based) lateral control (lane-following assistance) system, an autonomous vehicle handling system is conceivable in theory. The lane-following assistance system is a further development of the lane-changing alarm.

The *active safety* functions (top left quadrant of diagram) encompass all features intended to prevent accidents. The high demands placed on them regarding functionality and reliability extend from the simple parking assistance brake, which automatically brakes the vehicle before it hits an obstruction, to computer-aided driving maneuvers for the purpose of avoiding collisions. Intermediate stages are represented by "predictive safety systems" (PSS). They extend from pre-pressurizing the brakes when a potential hazard is detected, to brief sharp application of the brakes, through to automatic emergency braking, which always triggers full braking force if the vehicle computer detects that a collision is unavoidable.

Comfort and convenience systems and driver support systems (such as ParkPilot and adaptive cruise control, or "ACC") are the foundations on which Bosch will be

4 Vehicle all-around visibility, detection range of sensors

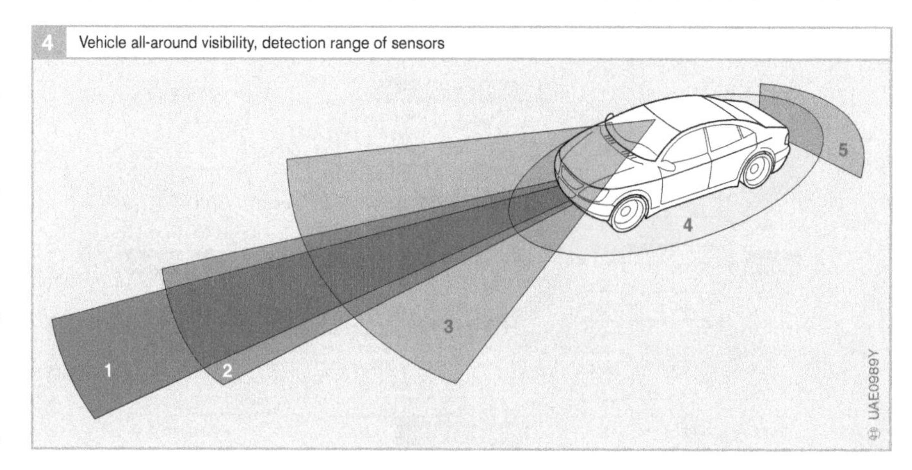

Fig. 4
1 77 GHz long-range radar
Long range ≥150 m
Horizontal aperture angle: ±8°
2 Infrared night vision range ≤150 m
Horizontal aperture angle: ±10°
3 Video medium range ≤80 m
Horizontal aperture angle: ±22°
4 Ultrasound ultra-close-range ≤3 m
Horizontal aperture angle: ±60° (individual sensor)
5 Video rear-end
Horizontal aperture angle: ±60°

developing safety systems to full production maturity over the next few years. In the medium term the aim of these systems is to reduce the severity of accidents, while the longer-term objective is to prevent accidents altogether.

Electronic all-around visibility

Using "electronic all-around visibility", numerous driving assistance systems are achievable – both for *warning* and for *active intervention* purposes. Fig. 4 shows the protection areas covered by present and future all-around visibility sensors.

Close range

Due to the limited availability of sensors, only a few driving assistance systems have been able to become established on the marketplace so far. One of them is "ParkPilot", which uses ultrasound technology to monitor the close-range area. Ultrasound sensors integrated in the vehicle's fenders ensure that the driver is given an acoustic or optical warning if the vehicle approaches an obstruction. The sensors have a range of ≤ 1.5 m. The next (fourth) generation of ultrasonic sensors will have a range of ≈ 2.5 m. These sensors will then be suitable for the future, more demanding, advanced functions of *parking space measurement* and *parking assistant.*

The close-range system now in widespread use has been well accepted by users and is already standard equipment on some models.

Long range

For long-range applications, ACC systems with long-range radar (LRR) sensors are already in use. They have an operating frequency of 76.5 GHz and a range of ≈ 120 m. A very narrow-beam radar lobe scans the area in front of the vehicle in order to determine the distance to the vehicle in front. The driver specifies the required speed and safety distance. If a slow-moving vehicle is detected in front by the sensors scanning the vehicle's surroundings, the ACC automatically applies the brakes to maintain the distance previously specified by the driver. As soon as the scanned area is clear of vehicles, ACC accelerates the vehicle again up to the preset cruising speed. In this way, the vehicle integrates harmoniously within the traffic flow. It not only allows the driver to reach his/her destination in a more relaxed state, it also increases the level of attention that can be devoted to traffic conditions. The ACC data can also be used to warn the driver if the vehicle is approaching too close to the vehicle in front.

The current Bosch ACC version meets these requirements by automatically intervening in the braking and engine management systems at speeds of > 30 km/h. At lower speeds, the system is deactivated.

The second generation (ACC2, beginning in 2004) features a doubling of the horizontal scanning range to ± 8° and a substantial reduction in size. The device is thus the smallest radar distance control system with integrated ECU.

ACC is the first driving assistance system which not only warns the driver but also actively intervenes in vehicle dynamics. The current version of ACC is designed particularly for use on expressways. Due to its larger beam width, ACC2 will be able to assess the traffic situation better, in particular when negotiating curves or filtering in, and will also be usable on highways with tight curves.

Virtual safety shield

Ultrasonic sensors with extended scanning range or short-range sensors can form a "virtual safety shield" around the vehicle. This shield can be used to implement a number of functions. Firstly, the signals from this safety shield warn the driver of potentially dangerous situations; secondly, it acts as a data source for safety, comfort and convenience systems. Even the driver's "blind spot" can be monitored by these sensors.

Video sensors

Video sensors play a major role in driving assistance systems because they specifically assist with the interpretation of visual information (object classification, Fig. 5). In the near future, Bosch will be able to offer this type of sensor for use in vehicles. This will open the way for a wide range of new functions.

Rear-end video sensors (in their simplest form) can assist the ultrasound-based ParkPilot system during *parking maneuvers.*

Greater benefit is offered by a *rear-end camera* if the detected objects can be interpreted by image-processing software and the driver is alerted in critical situations. Such a situation might be one where an intended lane change would be dangerous because a vehicle is approaching fast in the outside lane.

The use of a *front-end camera* is necessary to implement functions for *night vision enhancement,* for instance. To this end, the system illuminates the road ahead of the vehicle with infrared light. A display shows the image recorded by an infrared-sensitive camera. Visibility is enhanced for the driver without dazzling oncoming vehicles, and obstructions and hazards can be identified more quickly in the dark.

A *day-and-night sensitive front-end camera* allows several assistance functions. For example, Bosch is currently developing systems for lane detection and road sign recognition based on this technology.

The "*lane detection system*" can identify the lane boundaries and the lane direction ahead. If the vehicle is about to move out of its lane unintentionally, the system alerts the driver. At a later stage, Bosch is planning to expand lane detection into a *lane-following assistance system* which can move the vehicle back to its lane by actively turning the steering wheel. Combined with ACC, this will make an ideal system for relieving driver stress in stop-and-go traffic.

Another function that makes use of data from the video sensor is *"road sign recognition".* This system is capable of recognizing ("reading") road signs (e.g. speed limit or no-overtaking signs). The instrument cluster then displays the last road signs recognized.The front-end camera also assists the ACC sensor by providing the capability of not only measuring the distance from an object, but of classifying it as well. By combining the video system with long-range radar, there are synergetic benefits: the visible range of the ACC is extended significantly, and object detection is even more reliable.

Video technology will initially be used in driving assistance systems that provide information. Current video sensors are still a long way from imitating the capabilities of the human eye in terms of resolution, sensitivity and light intensity response. However, advanced methods of image processing combined with recently developed dynamic imaging sensors already demonstrate the enormous potential of these sensors.

CMOS technology with nonlinear luminance conversion will be capable of covering a very wide range of brightness dynamics and will be far superior to conventional CCD sensors. Since the brightness of images in the automotive environment is not controllable, the dynamic range of conventional imaging sensors is inadequate; for this reason, highly dynamic imagers are required.

The video signals from the camera in a video system are transmitted to an image processor which extracts individual image features (Fig. 5). This information can also be sent over a data bus to other ECUs or information units (HMI: **h**uman **m**achine **i**nterface), where it can be used as the basis for initiating intervention in vehicle control or for driver information.

Sensor data fusion

To ensure that assistance functions are as resistant to fault as possible but remain capable of detecting and classifying several objects simultaneously, the signals of multiple sensors must be combined and analyzed. Sensor data fusion allows systems to create an overall, realistic picture of the vehicle's surroundings. In this way, information about the vehicle's surroundings is much more reliable than if it was detected by individual sensors.

Future driving assistance systems will incorporate the functions of an increasing number of sensors and actuators and will have more complex connections to other vehicle systems. Bosch is developing all the components and functions based on the "Cartronic" system architecture, which serves to network all control and regulation tasks in the vehicle. It consists of a clearly structured function architecture and modular software with open, standardized interfaces.

Summary and outlook

The development of sensors for detecting the vehicle's surroundings is progressing at a fast pace. New functions are quickly being integrated according to their relevance to safety and convenience.

Close-range sensors represent the next milestone in this development. They will form a "virtual safety shield" around the vehicle and provide the signals that will firstly alert the driver to potentially critical situations, and secondly act as a data source for active safety, comfort and convenience systems. At the same time (and as the high-performance sensor chips become available for volume production), video sensors will enter the world of automotive technology and open up a multiplicity of opportunities. This will open up the way for information systems that are based on highly complex image analysis using high-performance processors and which can be used as data sources by several driving assistance systems.

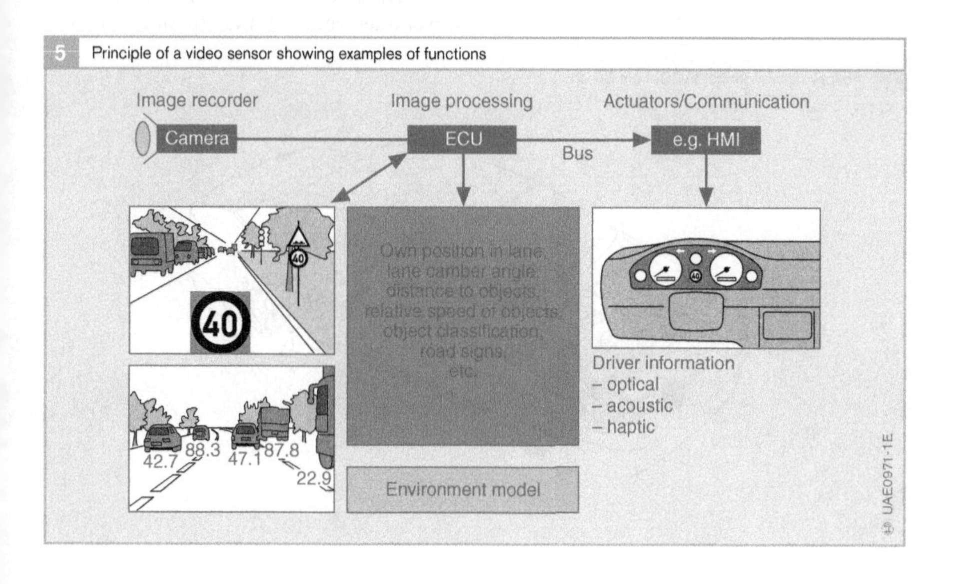

5 Principle of a video sensor showing examples of functions

Adaptive cruise control (ACC)

ACC (Adaptive cruise control) simplifies the task of driving a car because it relieves the driver of the mentally demanding task of keeping a check on the car's speed, thus allowing driving behind slower vehicles to be relaxed and safe.

System overview

Benefits and application area

It is that "following slower vehicles" function of ACC in particular that is perceived by the driver as a major gain in convenience and as a substantial mental relief. The side effects of this function also include improved road safety due to greater distances between vehicles and greater relaxation on the part of the driver.

The main area of application for ACC (Fig. 1) is on expressways and multilane trunk roads with light to relatively high traffic densities. Although use of the system in traffic jams and urban conditions is highly desirable, at present this remains an objective for future systems, as the technical difficulties associated with such a function demand considerable further development of sensor capabilities (refer to the chapter on "Future Developments").

Function

For every adaptive cruise control (ACC) system, the minimum function is cruise control function, namely keeping the speed constant at the setting selected by the driver. That function, referred to below as Set Speed Control, is employed primarily when there is no vehicle in front forcing the driver to adopt a slower speed than the set speed. Thus it is also effective if the car in front of the car with ACC is traveling faster.

The essential difference in function between ACC and standard cruise control is that a car with ACC will safely follow a vehicle that is traveling at a slower speed than the cruising speed to which the driver has set the ACC (Fig. 2).

If the vehicle in front is traveling at a constant speed, a car fitted with ACC will follow it at the same speed and a virtually constant distance. That is because the distance between the two vehicles is – at least within a broad speed range – virtually proportional to their speed. This "constant time gap" is (regardless of speed) equal to the time required for the most forward point on the car with ACC to reach the current position of the rearmost point on the vehicle in front.

The changeover between these two main functions is automatic without the need for driver intervention. If the situation should

1 Adaptive cruise control (ACC) from Bosch

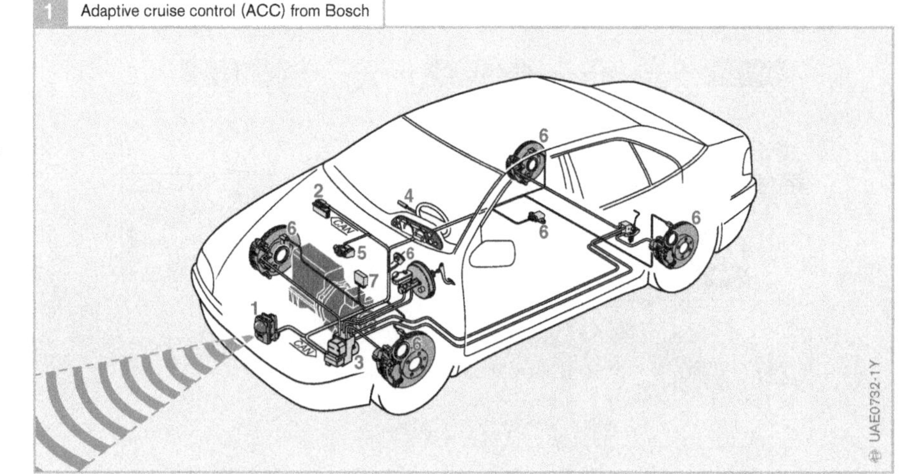

Fig. 1
1 ACC sensor & control unit
2 Engine management ECU
3 Active brake intervention via ESP
4 Control and display unit
5 Engine control intervention via ME Motronic (gasoline engines) or EDC (diesel engines)
6 Sensors
7 Transmission shift control (optional)

2 ACC function. The main application of the ACC is driving in low to high-density traffic

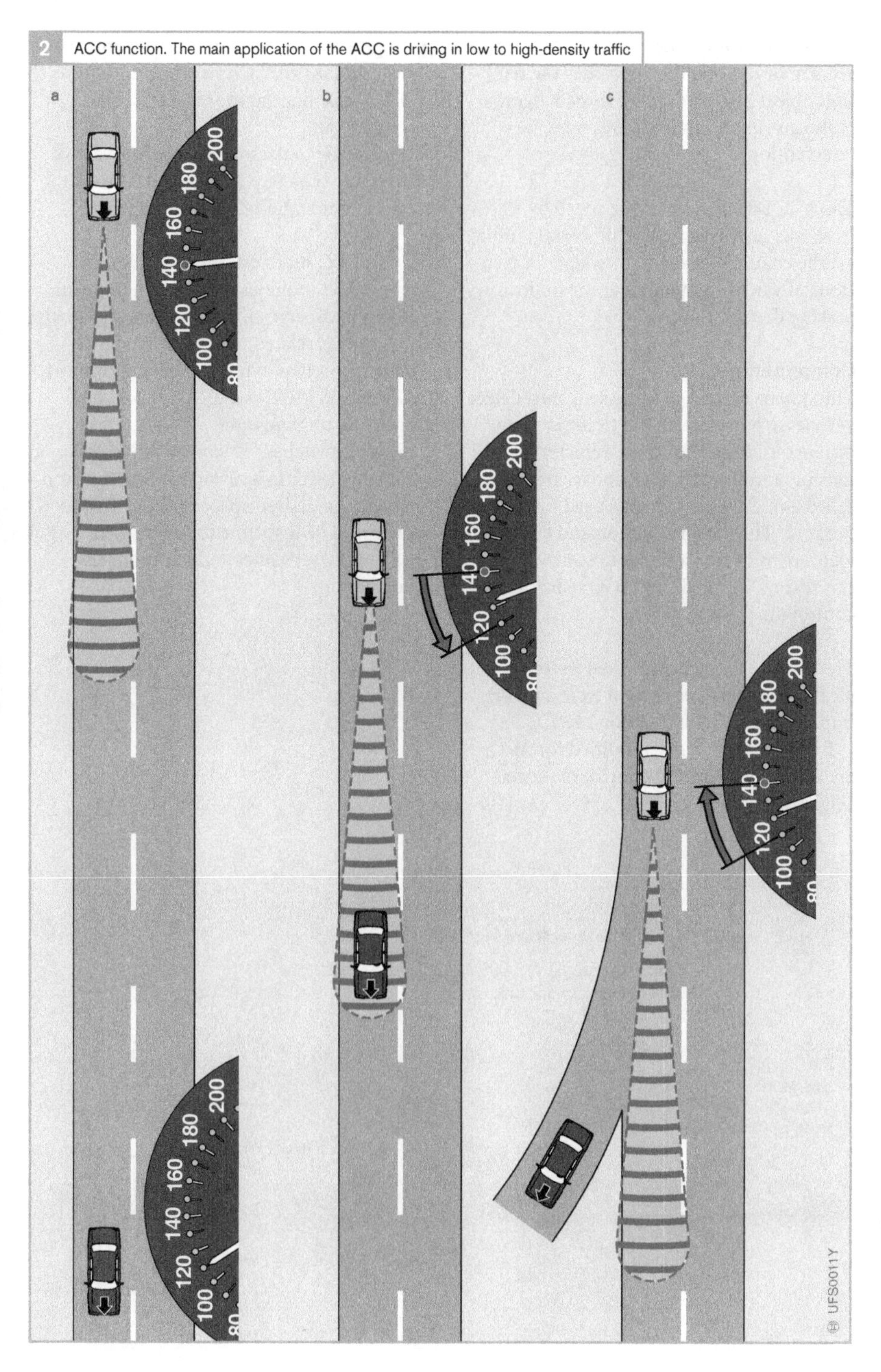

Fig. 2
- a Approaching a vehicle in front while driving at a constant speed (desired cruising speed)
- b Braking and driving behind a slower vehicle
- c After the vehicle in front turns, the ACC vehicle accelerates returns to the desired cruising speed originally set

change, for example because another vehicle veers in or out and thus becomes the relevant object being followed, this changeover is also automatic and requires no driver intervention.

The ACC system adjusts the speed by electronically accelerating, within certain limits, via the engine management system or electronically activates the brake system for braking deceleration.

Components

The system needs a ranging sensor to detect vehicles driving ahead and to measure the distance and speed of these vehicles. In Europe, a millimeter wave radar (frequently called a microwave radar) is used for this purpose. The sensor function and controller logic are built into a single unit which, for this reason, is called the "ACC sensor & control unit" (ACC-SCU).

The increased range and wider detection angle of the second generation were a further improvement to the function of ACC.

Existing subsystems, modified for ACC, are used to change and control the speed (Fig. 3):

- Engine management with electronic torque control, such as Motronic with ETC (gasoline engine) or EDC (diesel engine) and
- Electronic brake modulation with active pressure build-up (generally based on the electronic stability program, ESP).

So that ACC functions reliably (even in curves) ESP provides, in addition to the deceleration capacity, important sensor signals for variables related to vehicle dynamics. Furthermore, for complete driving comfort, ACC should ideally be combined with an automatic transmission.

Special switches for control and display allow the driver to activate the function and set both the desired speed and the desired time gap. The instrument cluster then displays the set values and additional ACC information.

3 Basic structure and components of the ACC control system

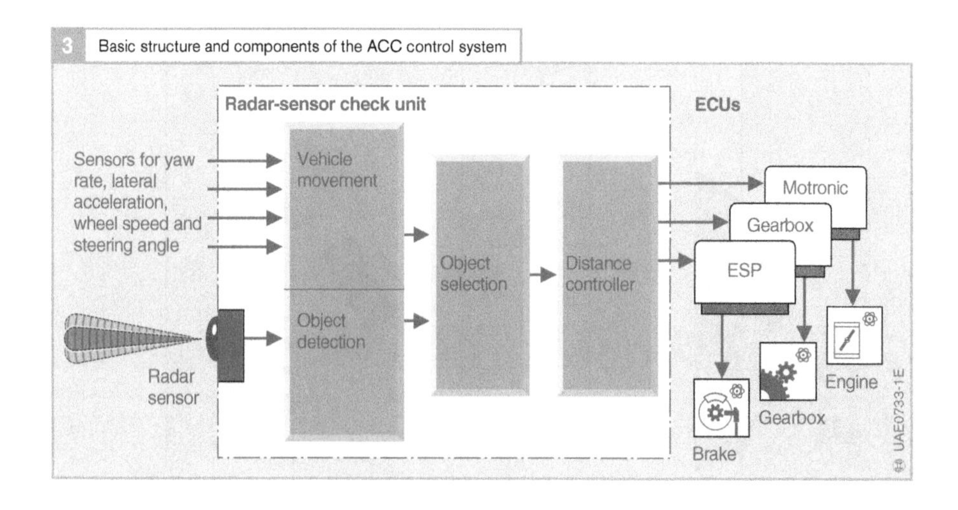

Ranging radar

Physical measuring principles

Reflection

The radar (**ra**dio **d**etecting **a**nd **r**anging) transmits an electronic wave using an antenna. This wave bounces off of an object in the radar beam's path, returns and is received.

Radar echoes are generated by all electrically conductive materials, particularly by all vehicles that make up road traffic. Therefore, radar is especially well suited as a distance measuring principle. Furthermore, radar offers advantages in unfavorable weather conditions (such as fog or rain) because it uses a longer wavelength than optical methods.

Other distance measuring principles (such as optical distance measuring devices) require surfaces with good reflective properties. Objects with optical reflectors that are dirty or not readily visible cannot be reliably detected.

Propagation time measurement

For all radar methods, the distance measurement is based on a direct or indirect propagation time measurement for the time between when the radar signal is transmitted and when the signal echo is received. With direct reflection, this time τ is equal to (twice) the distance d to the reflector and the speed of light c:

$$\tau = 2d/c$$

For a distance of $d = 150$ m and $c \approx 300{,}000$ km/s, the propagation time for $\tau \approx 1.0$ µs.

Doppler effect

For an object moving relative to the radar sensor with a relative speed v_{rel}, the signal echo undergoes a frequency shift f_D compared to the emitted signal. For the differential speeds listed here, this shift is (Fig. 1):

$$f_D = -2f_C \cdot v_{rel}/c$$

where

f_C Carrier frequency of the signal

At the radar frequencies commonly used for ACC, $f_C = 76.5$ GHz, there is a frequency shift of $f_D \approx -510 \cdot v_{rel}$/m, and thus 510 Hz at a relative speed of –1 m/s (proximity).

1 Use of the ranging radar (Doppler effect)

Fig. 1
d Distance
f_C Carrier frequency
f_D Differential frequency
v_1 Driving speed of vehicle 1
v_2 Driving speed of vehicle 2
v_{rel} Relative speed

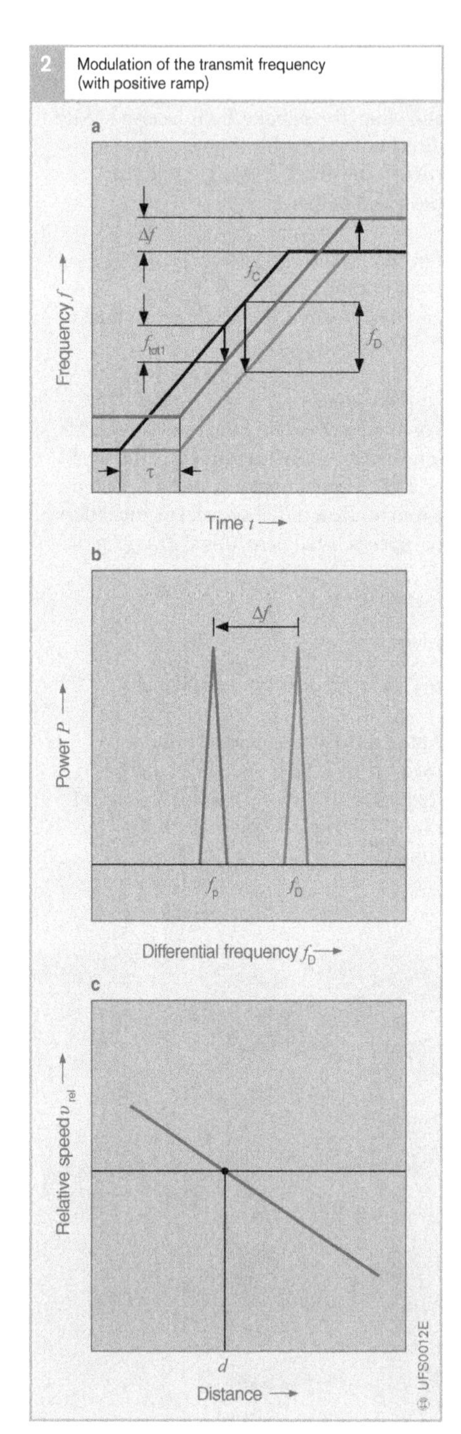

Fig. 2
a Positive addition of the Doppler shift to the differential frequency
b Effect of the Doppler shift
c "Distance" as a function of "relative speed"

f_C Carrier frequency (modulated transmit frequency)
f_D Differential frequency
Δf Doppler shift
$f_{tot1} = f_D - \Delta f$ Total frequency shift (ascending ramp)
f_P Positive frequency shift due to Doppler effect
τ Propagation time

Frequency modulation

Direct propagation time measurement requires much effort; therefore, in most cases, indirect propagation time measurement is used. One of these methods is known as FMCW (frequency modulated continuous wave). Rather than comparing the times between the transmitted signal and received echo, the FMCW radar compares the frequencies of the transmitted signal and received echo. The prerequisite for a meaningful measurement is a modulated transmit frequency.

For this purpose, the transmit frequency is modulated using a VCO (voltage controlled oscillator) in a linear ramp form with the gradient $m = df/dt$ (Fig. 2a). While the received signal returns after the propagation time $\tau = 2d/c$, the transmit frequency has changed in the meantime by the differential frequency $f_D = \tau \cdot m$. Therefore, the propagation time, and thus the distance, can be measured indirectly by ascertaining the differential frequency between the transmitted and received signals. The differential frequency, in turn, can be ascertained using a mixer, followed by low-pass filtering. To determine the frequency, the signal is digitized and converted into a frequency spectrum using an FFT (fast Fourier transform). A peak in the spectrum at f_D (Fig. 2b) corresponds to a distance of

$$d = f_D \cdot c/2m$$

However, the differential frequency information contains not only the propagation time information, but also the Doppler shift, which results from the frequency difference caused by the propagation time: $f_{tot\,1} = f_D - \Delta f$. This means that there will, at first, be ambiguity in evaluation. In addition to a single differential frequency, a linear combination of distance and relative speed values must be considered, which form a straight line in the diagram "distance" as a function of "relative speed" (Fig. 2c).

This ambiguity can be resolved by applying multiple FMCW modulation cycles with different gradients.

When the transmit frequency is modulated with a different ramp gradient, there is also an ambiguity between "distance" and "relative speed". However, in the diagram "distance" as a function of "relative speed" mentioned previously, this ambiguity expresses itself in the form of another straight line.

If, for example, a mirror image of the first ramp with negative gradient is used, the relationship shown in Fig. 3a results. According to this, a negative gradient leads to a negative addition of the Doppler shift Δf to the differential frequency shifted by the propagation time f_D (Fig. 3b).

In the diagram "distance" as a function of "relative speed", the straight line for the negative ramp now intersects the straight line that belongs to the first ramp with a positive gradient. The intersection of the two straight lines now provides the correct value for the "distance" and the "relative speed" (Fig. 3c).

However, the method must also be able to be used if there is more than one target. To do so, the method must be expanded by adding more modulation cycles so that an unambiguous assignment of "target frequencies" to "objects" is possible.

Determining the angle

To determine the angle at which the radar locates an object, multiple radar "lobes" are transmitted and evaluated.

Each radar beam has a characteristic "antenna diagram". For a defined target, the amplitude of the signal echo depends characteristically on the angle at which the radar hits the target (Fig. 4).

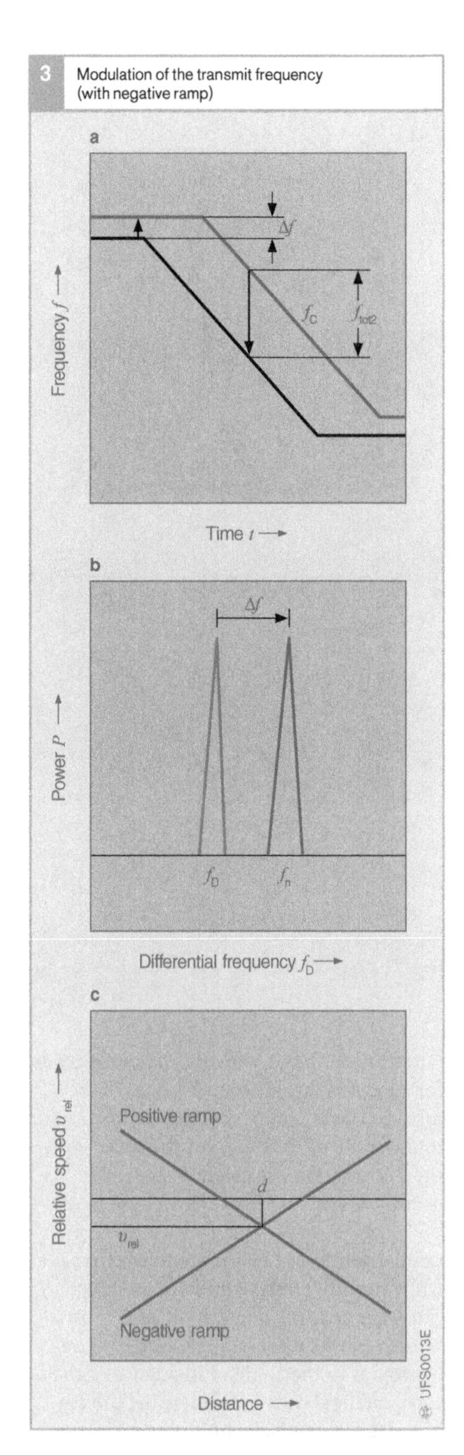

Fig. 3
a Negative addition of the Doppler shift to the differential frequency
b Effect of the Doppler shift
c "Distance" as a function of "relative speed"

f_C Carrier frequency (modulated transmit frequency)
f_D Differential frequency
Δf Doppler shift
$f_{tot2} = f_D + \Delta f$ Total frequency shift (descending ramp)
f_p Positive and
f_n Negative frequency shift due to Doppler effect
$d \sim f_p + f_n$ Distance
$v_{rel} \sim f_p - f_n$ Relative speed

4 Antenna diagrams of various radar beam lobes

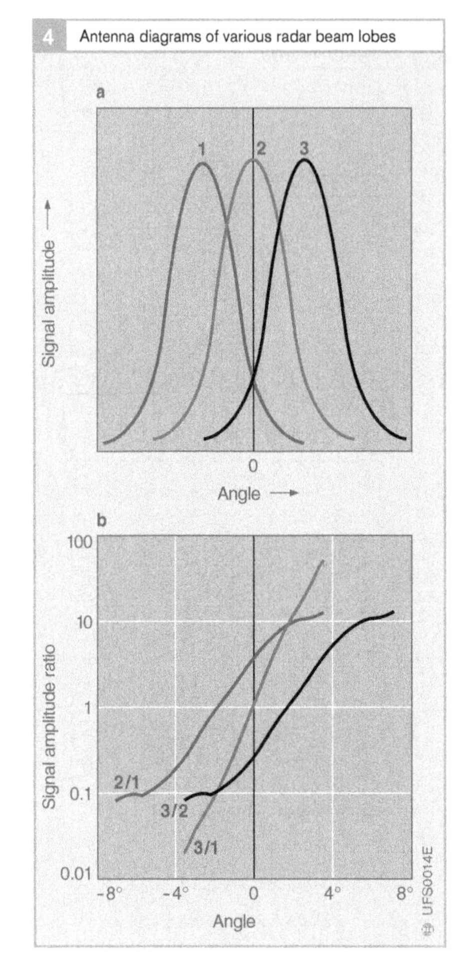

Fig. 4
a Overlapping area of the antenna diagram
b Illumination range of the radar beam lobes

1 Left beam
2 Center beam
3 Right beam
2/1 Signal amplitude ratio of beam lobe 2 to beam lobe 1 etc.

On the other hand, the reflective properties for a located target are unknown. Therefore, no direct conclusion can be made as to the angle of incidence of the wave from the information provided by the radar beam.

On the contrary, these prerequisites are met when multiple radar lobes are used. Because the antenna diagram of each lobe is sensitive in a different angle range, a conclusion as to the angle of incidence of the wave can be drawn by comparing the amplitudes for a signal echo in the various radar lobes. The quality of the amplitude comparison depends on the overlapping area of the individual antenna diagrams (Fig. 4a).

To "illuminate" an angle range of 8°, three radar lobes arranged in parallel with a full width at half maximum of +/– 2° are used (Fig. 4b).

Radar modules

Function

The actual measuring unit of the ACC-SCU is the radar transceiver(RTC).
It has the following tasks:

- Generating high-frequency radar radiation in the 76-77 GHz range
- Dividing and emitting three radar beam lobes simultaneously
- Subsequently receiving the echoes of this radiation reflected by objects
- Preparing these echoes for the downstream digital electronic signal processing

In addition, an electronic circuit is included for highly accurate stabilization of the transmit frequency and for generating a linear frequency modulation.

A Bosch radar unit has the following specifications (Table 1):

Table 1

1 Characteristic data of a Bosch radar unit

	First generation	Second generation
Range	2 to 120 m	2 to 150 m
Measurable relative speed	–50 to +50 m/s	–50 to +50 m/s
Angle range	±4°	±8°
Separability	0.85 m; 1.7 m/s	0.85 m; 1 m/s
Measuring rate	10 Hz	10 Hz
Frequency range	76-77 GHz	76-77 GHz
Mean transmission power	approx. 1 mW	approx. 5 mW
Bandwidth	approx. 200 MHz	approx. 200 MHz

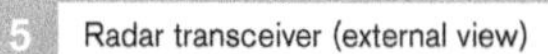

Design and method of operation

These are the basic components of the radar transceiver (RTC) (Fig. 5 and 6):

- High-frequency oscillator (Gunn oscillator) for generating the radar radiation
- Distribution network for antenna feed and reception mixing
- Frequency control electronics with reference oscillator
- Signal preamplifier

5 Radar transceiver (external view)

6 Radar transceiver (block diagram)

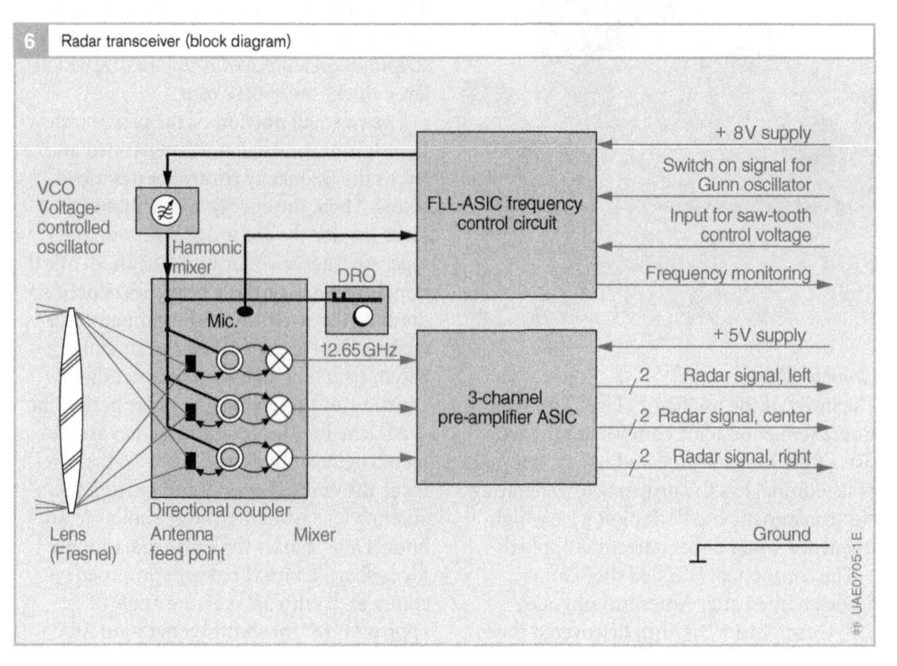

7 Gunn oscillator (structure)

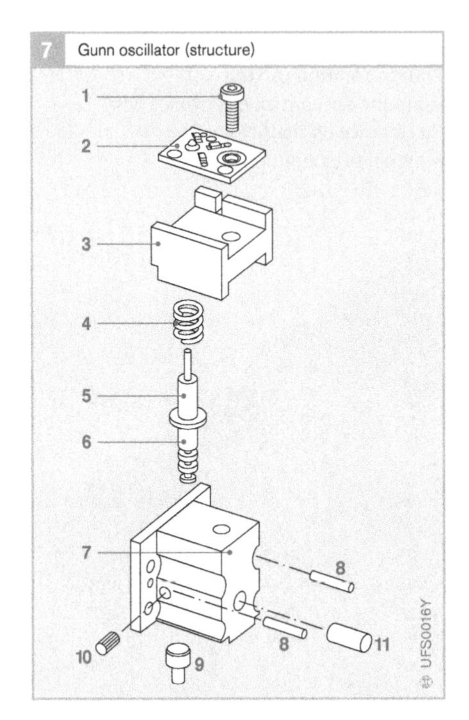

Fig. 7
1 Mounting screw
2 Auxiliary printed circuit board
3 Insulator
4 Helical spring
5 Ferrite sleeve
6 Bias choke
7 Oscillator body
8 Locating pins
9 Gunn semiconductor component
10 Frequency tuning pin
11 Power tuning pin

8 Distribution network (surrounding area)

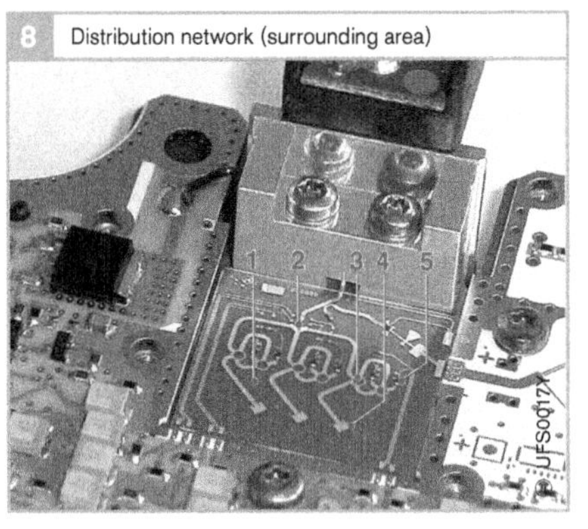

Fig. 8
1 Microstripline circuit on fused quartz
2 Wilkinson power splitter with two surface resistances
3 Directionally selective separation of transmitted and received signals
4 Three antenna patches
5 Seven mixer diodes

Gunn oscillator

The heart of the oscillator (Fig. 7) is an electronic semiconductor component that consists of gallium arsenide and which, because of its doping, has the property of generating electromagnetic oscillations at a very high frequency when direct current is applied.

The component is called the "Gunn diode", named after American physicist I. G. Gunn. In 1963, Gunn discovered the effect that in certain semiconductors (gallium arsenide), oscillations in the microwave range occurred at large electric field strengths. The Gunn diode is packaged in a ceramic housing and built into an oscillator block consisting of aluminum.

The contact pin to the power supply contains filter structures and a resonator disk which, together with the rectangular inner cross section of the block, forms a hollow resonator. The oscillating behavior of the diode is, in large part, determined by the geometric dimensions of the components surrounding it. These facts place extremely high demands on the precision with which these components are manufactured.

The frequency can be varied within the range of 76-77 GHz by modifying the applied voltage (hence the technical name VCO: voltage controlled oscillator). The high-frequency energy generated is fed to a distribution network via a rectangular hollow conductor integrated into the oscillator block.

Distribution network and antenna feed

The distribution network (Fig. 8) is an electrical stripline circuit realized using gold striplines on a piece of fused quartz, which has a thickness of 0.17 mm.

First, a small portion of the power transferred from the oscillator is separated and fed to the frequency controller described below. Then, the energy is distributed to three (or, for the second generation, four) separate transceiver branches with identical structure. Each of these branches contains a double-ring structure and terminates in a rectangular element called the antenna patch, each of which has the function of emitting and receiving one radar beam lobe.

Attached to the antenna patches are elements made of dielectric material that prefocus the emitted energy and beam it to an antenna lens which, in a way similar to an optical lens, causes the radar radiation to focus sharply into three superimposed cones, each with an aperture angle of approx. 4° (8° for second generation ACC).

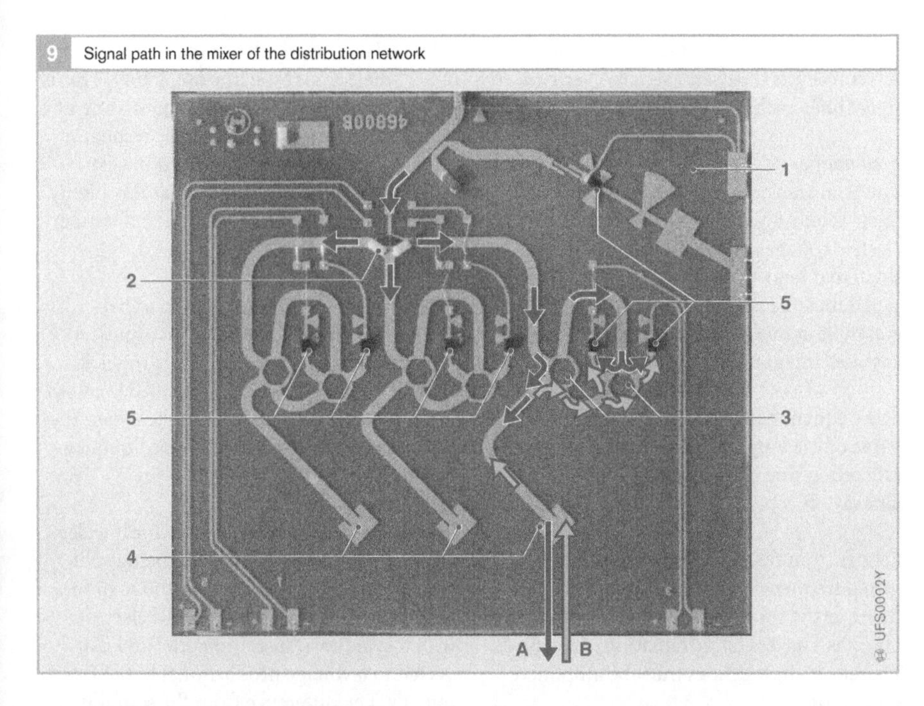

9 Signal path in the mixer of the distribution network

Fig. 9
A Transmit beam
B Receive beam

1 Microstripline circuit on fused quartz
2 Wilkinson power splitter with two surface resistances
3 Directionally selective separation of transmitted and received signals
4 Three antenna patches
5 Seven mixer diodes

Greater focusing would be possible only with larger antenna lenses. Because any two beam cones are approximately half superimposed, the total scanning range is 8° ($\triangleq \pm 4°$) or 16° for second generation ACC, beginning at the installation point of the radar at the front of the vehicle.

The Bosch system is a monostatic radar system, meaning that it also uses the same antenna arrangement in the opposite direction to receive the radar echo. A system of this type needs less installation space than a bistatic system with separate arrangements. Thus it is better suited to installation in vehicles.

Mixer
The first ring structure of the distribution network separates the power fed to it so that the antenna patches (Fig. 9) receive only about half of the power. The other half is fed to a second ring structure. Simultaneously, the energy received by the antenna patches as radar echo is also conducted there.

Each second ring structure of the distribution network forms, together with the antiparallel switched diodes, makes up a mixer in which the transmission and reception power are combined into a single electrical signal. The frequency of this signal corresponds to the difference between the emitted and received frequencies.

This electrical signal is the actual useful radar signal. Its frequency, in the range from 20-200 kHz, contains the information about the distance in the longitudinal direction and the relative speed of the detected objects. The differences in amplitude of the three branches are used to determine the angle.

The electronic circuit of the radar transceiver receives the useful signals over two signal lines each.

Preamplifier
The total transmission power of the ACC radar is only about 1 mW. Therefore, the electric voltages of the useful signal are so small that before it is processed further, the signal must be amplified by a factor of several million in a specially designed, three-channel, integrated amplifier circuit.

The frequency-dependent amplification curve of the amplifier circuit ensures accurate processing of echo signals, even from faraway objects.

Echoes from faraway objects return higher mixed frequencies and low voltage amplitudes, as the more distant the reflecting object is, the less radar radiation is received. Therefore, these echoes must be amplified even more.

Frequency control electronics
Because all important information is contained in the frequency of the useful signal, fluctuations of the transmission frequency would falsify measurement results, as would deviations from the time linearity of the transmitted frequency ramp.

The Bosch ACC radar is thus equipped with fast frequency control electronics which, approximately every millionth of a second, compare the emitted frequency with the current setpoint value and adjust it accordingly.

This also ensures conformity with legal and/or telecommunications regulations, which prescribe the frequency range from frequency range from 76-77 GHz for operation of long-range automotive radar systems.

To do this, in addition to the main oscillator, another oscillator is included as a reference oscillator with a rated frequency of 12.65 GHz, and which has the structure of a DRO (**d**ielectric **r**esonator **o**scillator). This is an electronic oscillator circuit consisting of a power transistor and a dielectric resonator element for frequency stabilization (like quartz clock circuits, this oscillator is highly stable with regard to service life and temperature).

The energy of the DRO is fed to a "harmonic mixer" located on the distribution network. There it is mixed, at six times its basic frequency ($6 \cdot 12.65 = 75.9$ GHz), with a small portion of the power taken from the main oscillator, resulting in mixed frequencies from 100-1100 MHz.

This signal is the input variable for the electronic frequency control. After being split again (the frequencies are still too high for further processing with standard electronics), it is fed to a "discriminator" and converted to a voltage proportional to the frequency. The difference from the setpoint value, which is also in the form of a voltage, is measured. In case of deviations, the supply voltage of the oscillator is modified until the setpoint value is again reached.

The setpoint value itself is a changing variable. The signal processing unit sets this variable to attain the change of the transmit frequency of 200 MHz during one millisecond; this change is required for the measurement.

Furthermore, fixed, programmed maximum and minimum values ensure that even if the frequency control fails, the frequency cannot leave the permitted range.

ACC sensor and control unit

Mechanical design

Requirements

The installation point for the radar sensor portion of the ACC electronics is, of necessity, the front area of the vehicle. This results in the following requirements:

- Temperature-resistant in a range from –40 °C to ≥ +80 °C
- Tightly sealed against splash water and steam jet
- Resistant to vehicle vibrations in rough road conditions
- Resistant to stone impact
- Smallest possible dimensions.

Furthermore, it must also be possible to adjust the radar sensor, as a highly accurate match of the center position of the radar beam cone with the center axle of the vehicle is required for drivability during ACC control.

Components

The Bosch ACC electronics unit (Fig. 1) contains not only the actual radar sensor, but also all of the electronics necessary for vehicle control. Therefore, it is called the ACC-SCU (ACC sensor and control unit).

This unit does not require an extra installation space or extra effort for wiring, as all signals of the ACC control are exchanged over the vehicle data bus (CAN) with the engine and brake electronics and the gauge.

Lens

Generally, two essential physical parameters determine the size of the radar devices (Fig. 2):

- The outer dimensions of the antenna system
- The associated focal length, i. e. the distance between the beam source and the bottom of the lens.

1 ACC sensor and control unit (ACC-SCU) (second generation)

2 ACC sensor and control unit (ACC-SCU) (cross-section diagram, first generation)

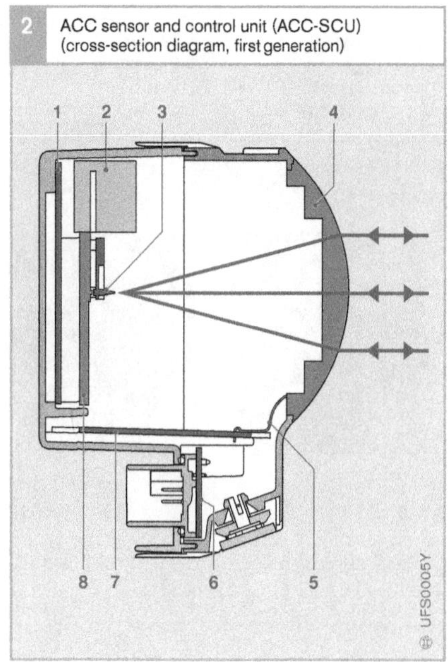

Fig. 2
1 Printed circuit board 1
2 Oscillator block
3 Beam sources (dielectric rod antenna)
4 Lens
5 Contact of the lens heater
6 Printed circuit board 3
7 Printed circuit board 2
8 Radar transceiver unit

For a given frequency, the diameter of a lens results from the desired beam focusing. The most complete illumination of the lens possible can be attained only if the distance between the beam source and lens is correct.

The lens of the Bosch radar is made of a special dielectric plastic that is temperature resistant and resistant to stone impact.
It is part of the plastic top section of the housing, which is thus tightly sealed to the outside.

Optionally, an electric heater can be integrated into the lens to prevent malfunctions as a result of being coated with ice and snow (wet snow, in particular, causes a noticeable damping of radar beams).

Electronic components

The basic electronics of the ACC-SCU consist of three printed circuit boards and the radar transceiver unit (Fig. 2 and 3):

Radar transceiver unit
The radar transceiver unit is placed directly on printed circuit board 1 with short connections that are less prone to malfunctions.

Printed circuit board 1
Printed circuit board 1 contains all of the components required for digital signal processing (calculating positions and speeds from raw radar data). At its heart is a digital signal processor.

Printed circuit board 2
In addition to another processor (16-bit microcontroller), which performs all of the calculations for vehicle control, printed circuit board 2 has a voltage regulator and additional switching and control components.

Printed circuit board 3
Printed circuit board 3 contains the plug and the driver components for the connec-

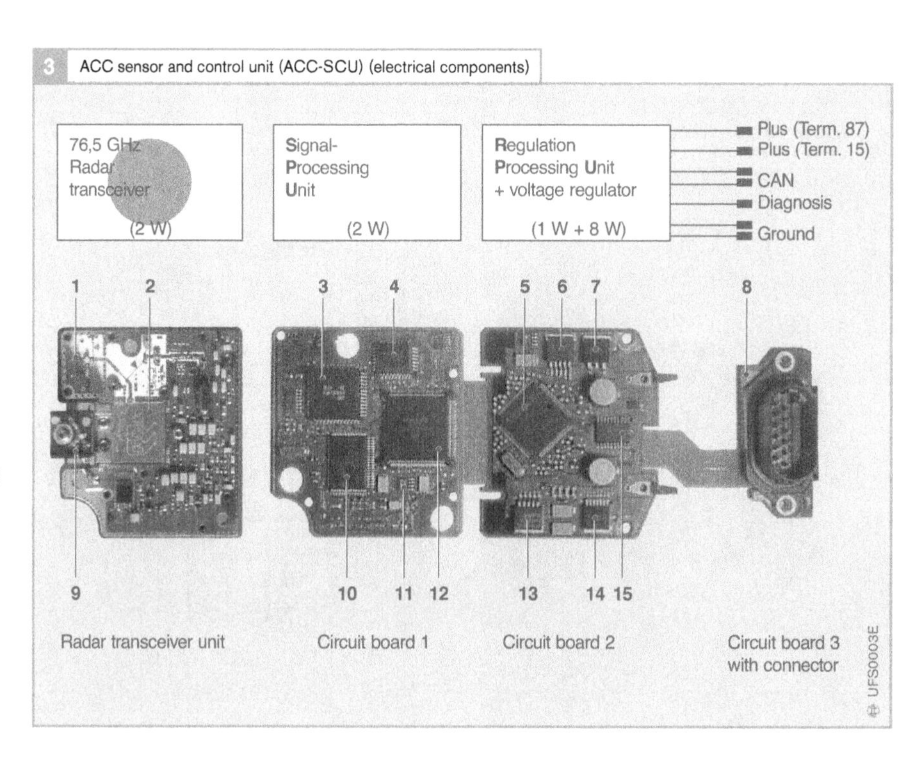

Fig. 3
Dual printed circuit board concept for first-generation ACC, single printed circuit board concept for second-generation ACC

1 Dielectric resonator oscillator (DRO)
2 Dielectric rod antenna
3 SRAM
4 Flash
5 16-bit microcontroller
6 Terminal, 5V (digital)
7 Switch, 3A
8 MQS plug with CAN transceiver
9 Gunn oscillator
10 ASIC CC610
11 Switching controller, 4.1 V
12 DSP 56002
13 Regulator, 8V
14 Terminal, 5V (analog)
15 K-line interface

tion to the vehicle electrical system and the vehicle CAN bus, as well as interference suppression chokes and capacitors.

Due to the integrated connections, all three conductors form one electrical unit. These flexible connections allow the entire printed circuit board assembly to be folded and thus arranged in a space-saving way in the overall unit (Fig. 4).

Housing and adjustment device

The Bosch radar sensor controller unit has a die cast aluminum housing. The way in which the electronic printed circuit boards are installed in the housing ensures optimum dissipation of heat losses.

On the outside of the housing, three pivot eyes with plastic ball joint shells as fastening elements. They hold three ball head screws with threads that grip the plastic elements of the holder.

5 ACC adjustment in horizontal and vertical directions (front view)

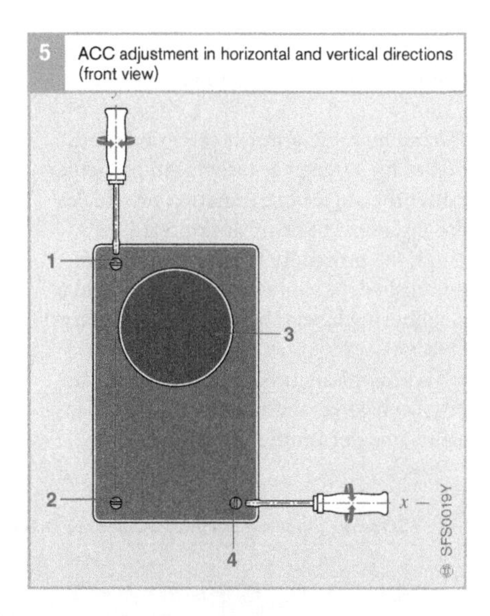

Fig. 5
1 Adjusting screw 1 for vertical adjustment
2 Screw 2 (fixed bearing)
3 Lens
4 Adjusting screw 3 for horizontal adjustment
x Axis for vertical adjustment
y Axis for horizontal adjustment

These screws are arranged at an angle that allows the unit to pivot in two levels (Fig. 5). In a vertical direction, this occurs by turning on screw 1 (the x-axis is the pivot axis) and in a horizontal direction by turning on screw 3 (the y-axis is the pivot axis). Here, screw 2 serves as the fixed bearing and does not move.

To compensate for any inaccuracy in installation, this fixture allows the unit to be adjusted in the vehicle in a way similar to a headlight. The holder is adapted to the specific vehicle, and there are also versions with angular gears for installation positions in which the adjusting screws cannot be accessed from the front.

4 ACC sensor and control unit (ACC-SCU) (mechanical structure)

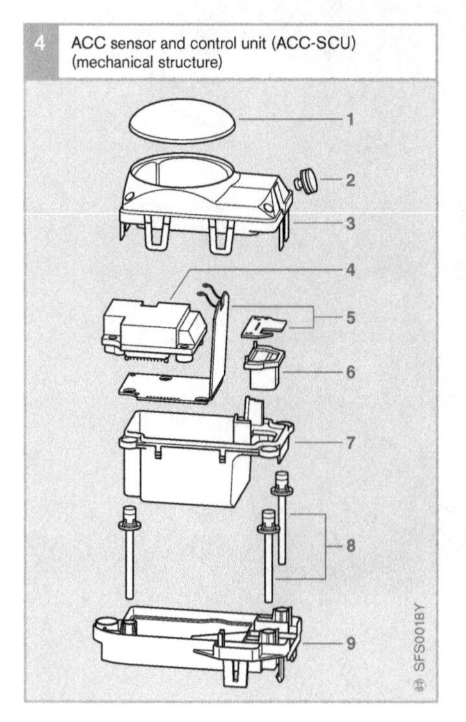

Fig. 4
1 Lens
2 Pressure compensation element
3 Top section of housing
4 Radar transceiver
5 Fixed flexible printed circuit board
6 12-pin contact strip for MQS contact points
7 Housing base
8 Adjusting screws (for the specific vehicle)
9 Sensor holder

Adjustment

The adjustment is made in two steps:

- Determining the longitudinal axis of the vehicle (driving axis)
- Aligning the radar axis parallel to the longitudinal axis of the vehicle.

The vehicle axis can be determined using common wheel alignment check procedures. The accuracy of ACC sensor adjustment, meaning the accuracy with which the sensor is aligned to the vehicle axis, is very important for the correct function of the ACC.

Horizontal misalignment can impair location of the target, because this in particular causes the angle determination of vehicles driving ahead to be misinterpreted. As a result, the proximity behavior would be diminished, possibly causing a vehicle in a neighboring lane to be selected as the target object.

Vertical misalignment of the sensor can negatively affect the sensor range and cause errors in determining the angle.

The requirements for the accuracy of the alignment are determined by the lane prediction, angle evaluation and plausibility algorithms (calculation processes). Misalignment of the sensor has the same effect on these functional components as an offset error. When the horizontal misalignment is about 0.3° or more, the impaired function becomes noticeable to the driver. Therefore, the required alignment accuracy should be significantly less than this value.

6 ACC sensor and control unit (ACC-SCU) (cross-section view)

Fig. 6
1 Lens
2 Radar transceiver
3 Beam sources (dielectric rod antenna)

Electronics hardware

Digital electronics

Functions

The digital electronics assume the following functions:

- Digitizing radar signals
- Carrying out FFT (fast Fourier transform)
- Calculating the distance, relative speed and angle of the radar targets
- Carrying out distance and cruise control, course prediction and self-diagnosis
- Exchanging data via CAN with the electronic stability program (ESP), Motronic, transmission control and gauge
- Making diagnostics via plug possible
- Controlling the lens heater, under certain conditions
- Monitoring voltages and signals

Design and operating concept

The digital electronics (Fig. 7) can be divided into the RPU (regulation processing unit) and the SPU (signal processing unit).

The heart of the SPU is a DSP (digital signal processor). This highly integrated electronic component was originally used in car audio applications. It is exceptionally well suited to carry out many arithmetic operations (such as multiplication and division) quickly. Thus it is the ideal component to perform the necessary calculations of detection, distance, speed, angle and tracking.

For the arithmetic operations just mentioned, the signals must be made available in digital form. This task is assumed by a highly complex CC610 circuit. Bosch developed this circuit specifically for ACC signal processing.

7 ACC radar sensor (PU processing unit, block diagram)

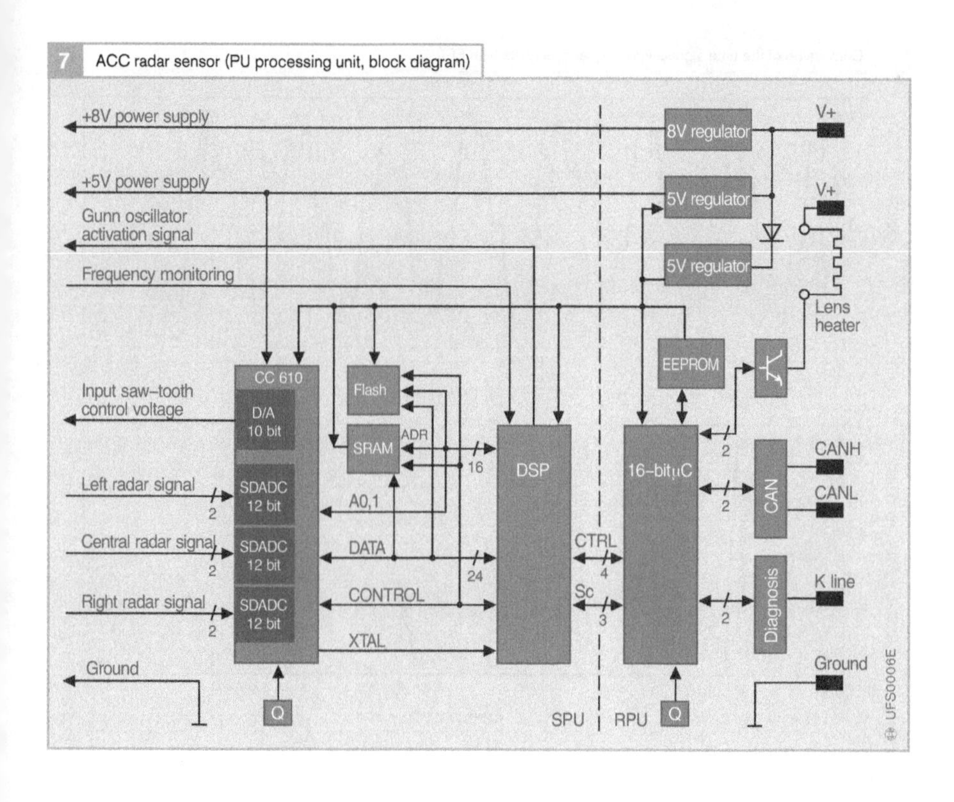

An additional integrated DAC (digital-to-analog converter) outputs a staircase-shaped voltage ramp. It serves as the setpoint value for the FLL regulation (frequency locked loop, refer to the section on "Frequency control") and generates a linear modulation of the transmit frequency. While this voltage is ramping up, the three mixed radar signals are amplified by the preamplifier and digitized with a resolution of 12 bits, filtered and subjected to a FFT (fast Fourier transform). The FFT allows the time signals to be converted to frequency signals very quickly (Fig. 8).

The DSP controls the chronological sequence of the modulation and retrieves the results from the CC610 control circuit via the parallel interface. The data are stored temporarily in an SRAM with fast read/write access. Once the two double ramps for frequency modulation have been run through, the arithmetic operations described else where are carried out. The duration of one cycle is 80-100 ms. The program required for this sequence is stored in a separate SPU Flash EEPROM.

The objects with the attributes "distance", "speed", "angle" etc. are transmitted to the RPU via the serial interface. The function of the RPU is described in a separate section.

The single-chip controller of the RPU contains all switching modules such as the CPU core, RAM, CAN controller, ADC (analog/digital converter), counter, digital interfaces to the EEPROM (programmable, non-volatile read/write memory), the SPU, the diagnosis module and the clock generating oscillator.

The program stored on an integrated flash EEPROM can also be modified by the vehicle manufacturer in the vehicle if additional interfaces are provided.

8 Conversion of the time signals into frequency signals with FFT

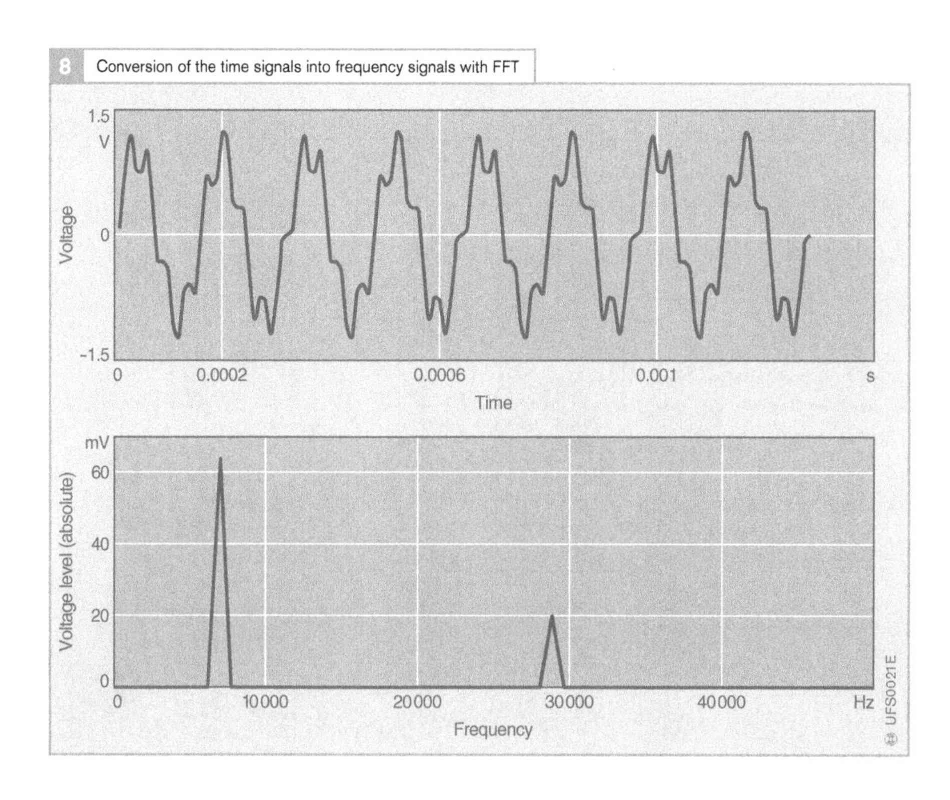

Fig. 8
a Time signals
b Frequency signals

The analog/digital converters monitor the voltages. If, for example, the measured supply voltage falls below a certain value, the ACC function is no longer permitted. Likewise, the stabilized supply voltages that are generated are tested for certain limit values. In case of an error, the ACC function is disabled, a "disabled" message sent to the display, and a fault code stored in the EEPROM.

The CAN interface module allows reliable digital communication with the partner ECUs in the vehicle. The CAN bus has become established as the standard serial data transfer method in motor vehicles. The usual transfer rates are 250 to 500 Kbit/s.

These high transfer rates require special precautions to be taken. These include suitable filter components, which prevent interference from harmonic waves that could, for example, impair radio reception in the vehicle.

To allow diagnosis in the workshop, it is necessary to store any faults that occur. An EEPROM is installed in the ACC-SCU for this purpose. Part of it is dedicated as fault storage memory. In conjunction with a test device, its contents can be read out and interpreted for workshop diagnosis. In addition, the vehicle manufacturer can store other typical data for the vehicle in the EEPROM.

The diagnosis module is the bidirectional interface to the diagnostic tester. If the diagnostic tester sends the "read fault memory" command, this message is interpreted by the controller of the RPU. The controller reads out the data from the EEPROM and converts them into a protocol that the diagnostic tester can understand. The diagnosis module also has a protective function to protect the sensitive controller from the rough operating environment of the vehicle.

The control function for the lens heater activates the heating wire in the lens in cold weather. Once the lens surface is heated, no snow or ice can settle on it. Both ice and snow can, to a certain extent, dampen the radar beams and limit the intended range. Although a monitoring function would ensure that the ACC function would be disabled in such a case, this would limit the availability of the system under these extreme ambient conditions. Pulse width modulation allows the lens heater to be flexibly activated depending on the temperature and the supply voltage.

Voltage regulators

The digital and analog components need a constant voltage supply in order to work without errors. This is provided by several voltage regulators. The battery voltage supplied by the battery and generator would destroy the sensitive components. Due to the threat of malfunction, any voltage peaks of ±100 V would have to be filtered out, as would a superimposed alternating voltage of ±2 V. The ACC-SCU must also withstand polarity reversal of the battery or starting of the vehicle with a 24V car battery.

The division into two voltage regulators is necessary to dissipate the heat loss generated. They supply the analog and digital components of the RPU, SPU and RTC with voltage.

The Gunn oscillator is supplied by an 8V voltage regulator.

Composite system

System architecture

Purpose of the system architecture

Because ACC is a function that includes multiple subsystems, the *system architecture* plays a key role. Only with a suitable system architecture can the subfunctions be linked in a way that provides harmonious and reliable overall function.

One particular challenge for the system architecture is the fact that the participating subsystems are often developed by different suppliers – frequently, competitors – and sometimes vary even in the same vehicle model.

Structure of the composite system

An overview of the basic structure of the ACC control and integration of the system into the vehicle is provided in Fig. 1 and 3 in the "System overview" chapter. The ACC sensor & control unit detects vehicles driving ahead and calculates a setpoint acceleration of the vehicle[1]). This setpoint acceleration is then converted into suitable command signals for the partner systems of the engine and braking system. Therefore, the ACC system is not an autonomous system, but is based on networking of the various participating partner systems.

Method of operation of the composite system

Fig. 1 provides an overview of the participating partner systems necessary for the overall function of ACC:

- The ACC setpoint values for linear acceleration are implemented by the engine management and braking systems. Conversely, ACC needs information from these partner systems about the status of the vehicle, such as vehicle speed, vehicle acceleration, rotational motion of the vehicle, current engine torque etc.

1) In this case, acceleration means the full range of accelerations – including negative acceleration – that cause vehicle deceleration.

1 ACC in the composite system with the participating partner systems

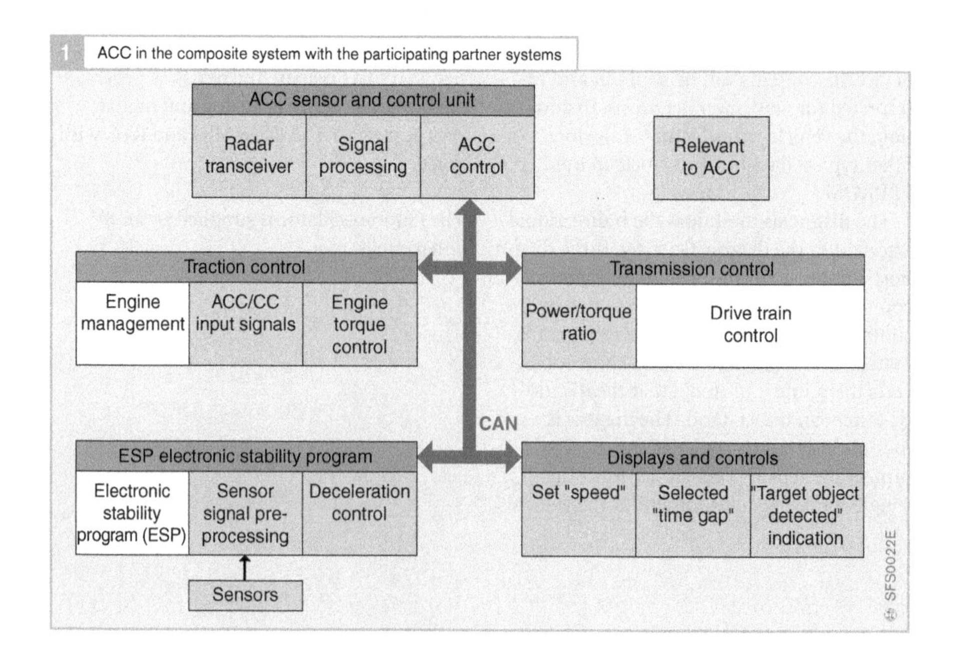

- The display and control elements can also be accessed via CAN. ACC needs the driver command information (set speed, selected time gap) and provides information to the driver (such as whether a target has been detected). The control elements are also needed by the conventional cruise control and thus are usually evaluated by the engine management system.
- The transmission control is not used by the ACC as an actuator system. However, information from the transmission control is needed about the current force/moment ratio of the transmission.

For data transmission, the CAN (controller area network) control unit bus is used. This network connects the individual systems. Frequently, other devices are also connected or can be reached via gateway functions.

In addition to the type of data transmission, the convention of the data content of the network signals is defined. This results from the functional division. Therefore, the content of the individual interface signals can vary depending on the partner systems present.

Traction control

The ACC system needs a way to intervene into engine management in order to implement a setpoint torque command using the engine management system.

Most of the currently used engine management systems offer this possibility (such as EGAS systems, Motronic ME7, electronic diesel engine control EDC). They use the existing internal interface to the conventional cruise control. Thus it is possible to control the drive train based on engine torques without knowledge of engine characteristic fields.

Brake control

If the drag torque deceleration of the drive train is no longer sufficient to implement the setpoint deceleration requested by the ACC control system, the active brake is triggered. Two actuator versions are used for this purpose:

Active brake boosters

A brake booster with electronically controlled movement of the pedal linkage makes it possible to brake automatically instead of "manually" using the driver's foot. The prerequisites for this are a suitable diaphragm design and an additional pneumatic valve with proportional control characteristics. The brake light switch on the brake pedal still controls the brake lights. An additional "release switch" indicates when the driver applies the brakes manually.

If there is a pressure sensor for pressure measurement on the main cylinder outlet (as is normally the case with the ESP vehicle dynamics control system), the active braking for ACC is often controlled via a pressure or torque interface.

Hydraulic brake actuators

In the TCS and ESP vehicle safety systems, electronically controlled brake actuators are already in widespread use – but only in situations where the vehicle's stability is threatened, not in normal driving conditions. With improved activation techniques, the hydraulic actuators (generally, motor pumps and valves) can be activated in a way that also allows convenient brake activation.

For the brake light activation, in addition to the existing signal of the brake light switch activated by the brake pedal, an additional switching signal is generated, so that the brake light is illuminated even for active (automatic) braking.

The braking system of the SBC (sensotronic brake control) requires no additional hardware whatsoever. As a "brake-by-wire" system, SBC is ideal for ACC. The ACC request is merely an additional setpoint branch for this system.

Curve sensors

Familiar ACC systems use ESP sensor signals to measure the vehicle's movement. To do so, CAN typically transmits the measured quantities from an ESP control unit, which is also in the vehicle, to the ACC control unit. In this way, the extra costs for a sensor system exclusive to the ACC can be avoided.

The following ESP sensors are available for ACC (the chapter on "Sensors" contains a detailed description of the applications, design and method of operation of these sensors):

Yaw rate sensors

The yaw rate sensors measure the rotational motion of the vehicle around its vertical axis. The physical measuring principle is based on the measurement of the Coriolis force. Under the influence of a rotational motion, there is a change of the translational oscillation movement of a mass excited to translational oscillation.

Steering wheel angle sensor

The physical measuring principle is based on the angle measurement at the steering column. Depending on the required task, these sensors provide sliding contact or non-contact measurement.

Acceleration sensor

The physical measuring principle is based on measurement of the excursion of a mass supported by elastic bearings when subjected to forces of inertia parallel or transverse to the vehicle axis.

Wheel speed sensor

From the wheel speed sensor signals, the corresponding ECU derives the wheels' rotation rates (wheel speed). The following types of sensors are used for this purpose:

- *Passive (inductive) speed sensors* with a gear connected to the wheel hub.
- *Active speed sensors* with a *multi-pole ring* on which magnets with opposite polarity are arranged. Correspondingly, the measuring element detects the change of the magnetic flow.

Safety concept

Purpose of the safety concept

The objective of the ACC safety concept is to avoid critical driving situations and driving states in the event of errors of the ACC system. At the same time, the limitations on availability that originate from the safety measures are to be minimized.

The safety concept must guarantee fail-safe behavior of the ACC control unit and allow accurate workshop diagnosis by doing the following:

- Shutting off the radar emissions
- Disabling the ACC control system
- Storing a fault memory entry.

Doing so requires reliable and differentiated detection of all possible error states, and initiation of a fault response suited to the specific type of fault.

Structure of the safety concept

Generally recognized methods for monitoring safety-related systems are diversity and redundancy.

In *diversified* information processing, all calculations on a control computer of different types are reconstructed using different software.

To achieve *redundancy*, twice the amount of the same hardware and software is used.

With the increasing complexity of ECU functions in today's vehicles, bit-accurate matching of the results of diversified calculation methods is not attainable. Instead of a simple query regarding consistency, a complex plausibility algorithm must be developed that tolerates deviations within defined limits. However, complete error detection is no longer provided in this case.

Furthermore, in the development of ECUs, diversity and redundancy result in a conflict of objectives with cost and size reduction.

For these reasons, a monitoring concept was developed for the ACC ECU that is based on the computing structure specific to ACC while, at the same time, taking into account the complexity of the tasks and the specific safety requirements. As a result, the ACC ECU, with its dual-processor structure and the associated internal communication, meets the safety requirements with regard to redundant hardware structures and monitoring units.

The monitoring concept of the ACC ECU is divided into three logical levels that are located in the two controller units and in the external partner ECUs:

"Component monitoring" level

The *component monitoring* level is composed of two parts, which are independent of each other, in the two controllers. It is limited to the discovery of faults in the periphery of the controller. No monitoring of the computational logic is associated with this.

Examples of component monitoring are:

- Monitoring the radar transceiver
- Detecting misalignment of the sensor
- Detecting "sensor blindness"
- Monitoring the power supply
- Monitoring the CAN data bus
- Monitoring the lens heater.

"Function monitoring" level

Likewise, the *function monitoring* level is implemented into each of the two controllers independently. Each controller carried out tests of its own computational logic.

In addition, there are tests located outside the ACC ECU that are executed by partner ECUs. In this process, they check the ACC messages for consistency and plausibility. This detects faults of the ACC function that lead to implausible CAN signals or an irregular CAN transmission cycle.

Examples of function monitoring are:

- Internal hardware tests of the processor
- Internal checksum tests of the processor
- Testing of CAN checksums
- Testing of CAN message counters
- CAN timeout monitoring.

"Mutual control" level

The *mutual control* level includes the interaction of the two controllers in a shared monitoring structure. The primary difference in contrast to the function monitoring is that the monitoring and the function to be monitored do not run on the same hardware; rather, there is a mutual control between the two controllers.

Examples of mutual control are:

- Checksum testing of the internal communication
- "Timing" monitoring of the internal communication
- Calculation and mutual checking of testing tasks.

Method of operation of the safety concept

The error messages of the individual monitoring functions are centrally evaluated in the ECU. The fault response takes place in a differentiated manner according to the severity of the faults that occur and the momentary driving situation.

The possible responses are as follows:

- ACC control continues with no restrictions, no fault display, fault code storage for workshop diagnosis
- An ACC deceleration intervention ends, followed by fault display and fault code storage for workshop diagnosis
- ACC control aborts immediately with fault display and fault code storage for workshop diagnosis.

Furthermore, a distinction is made between reversible and irreversible faults:

- Reversible faults disable the ACC control for the duration of the fault detection only
- Irreversible faults disable the ACC control for the duration of the driving cycle.

Therefore, in all fault events, ACC is again available if no more faults are detected after the next "ignition on". The sole exception: after misalignment of the sensor is detected, the ACC function must be re-enabled in the workshop.

Most entries into fault code storage for workshop diagnosis fall into one of the following groups:

- ECU faults (which require the ECU to be replaced)
- Deviations of the operating voltage
- Overheating
- Misalignment of the sensor
- Hardware errors of the CAN bus
- Errors in communication with the partner ECUs
- Receiving an error signal from a partner ECU.

In the case of an ACC error shutoff, the vehicle can continue to be used with no restrictions of other functions. The vehicle does not need to be immediately brought into the workshop for service.

In principle, there are only a few components in the ACC ECU that can be detected by only one monitoring function. In most cases, depending on the nature of the fault, various monitoring systems may be activated.

The following example demonstrates how the monitoring levels complement each other. Let us assume a fault in the power supply for the controller units:

The component monitoring system provides a voltage test for this purpose by feeding the voltage into a monitoring path and comparing it to fault thresholds. However, this requires that the controller to be monitored still works correctly, despite the assumed voltage deviation.

However, if the assumed fault leads to a malfunction of one of the two controller units, this can be detected by the mutual control of the internal communication.

However, in this example the most likely scenario is total failure of both controllers, which will be detected by function monitoring in the partner ECUs as a CAN timeout.

▶ History of radar

Technology borrowed from the animal world
RADAR (**Ra**dio **D**etection **a**nd **R**anging) is a system that uses radio waves to locate distant objects and is traditionally employed primarily in aviation and shipping. It has also been widely used for military applications since the development of radar-assisted air defences in the Second World War. More recent areas of application include space exploration, weather forecasting and, now, motor vehicles where it is used to measure the distance between vehicles for the ACC (**A**daptive **C**ruise **C**ontrol) system.

The idea for RADAR came from the sonar (**So**und **Na**vigation and **R**anging) system which uses sound echoes to determine the distance and position of objects, and which itself was copied from the navigation techniques of certain animals. Bats, for example, make high-pitched sounds with frequencies in the *ultrasonic range* of 30...120 kHz. The echoes that bounce back off solid objects are picked up by the bat's highly sensitive ears. That information then helps the bat to find its way around and to locate its prey.

RADAR functions in a similar manner but by using *radio signals* instead of sound. Measurement of distance by RADAR is based on timing the interval between transmission of an *electromagnetic wave* signal and reception of the signal echo that is reflected back by an object in its path.

While radar systems used in aviation and shipping operate at frequencies between 500 MHz and 40 GHz, the frequency band approved for ACC is 76...77 GHz.

Stages in the development of RADAR
The development of electromagnetic detecting and ranging equipment with long-range capabilities was an enormous challenge to the designers. Only a minute part of the energy originally transmitted was reflected back by the target. For that reason a very high-energy signal that is concentrated in as narrow a beam as possible has to be produced. This demands highly sensitive transmitters and receivers using signals with a wavelength that is shorter than the dimensions of the target.

The development of radar technology was marked by the following milestones and personalities:

1837 Morse: The telegraph becomes widely established. Here, electrical currents are used for the first time in communicating over longer distances.

1861/1876 Reis and **Bell:** Replacement of the telegraph by the telephone provides a much more direct and user-friendly method of telecommunication

1864 Maxwell, Hertz and **Marconi:** Existence of "radio waves" is theoretically and experimentally confirmed. Radio waves are reflected off metal objects in precisely the same way as light is reflected by a mirror.

1922 Marconi: The pioneer of radio provides the impetus for the continuation of earlier research into radio ranging

1925 Appleton and **Barnett:** The principle of radio-wave reflection is used to demonstrate the existence of conductive layers in the atmosphere
Breit and **Tuve:** Development of pulse modulation which enables precise measurement of distances

1935 Watson-Watt: Invention of radar

1938 Ponte: Invention of the magnetron (velocity-modulated electron tube for generating high-frequency oscillations)

Control and display

Function

Control and display elements are the immediate interface between the ACC system and the driver. Operation and interpretation should be as simple, unambiguous and intuitive as possible (i.e. an operation or situation should be able to be understood immediately).

Design and operating concept

Particularly for the control and display elements in the driver information area, a large amount of design leeway exists that is used differently by different vehicle manufacturers (see Fig. 1 for an example).

Therefore, the following describes the typical elements and their function without necessarily reflecting any one particular arrangement. Since controls are frequently acknowledged by a display function, control and display are grouped together in each case.

Activation

Although ACC is used frequently, it still must be actively switched on by the driver. In some vehicle models, it must first be enabled using a master switch. In other models, it is in a passive standby mode as soon as the ignition key is in the "ignition on" position.

The required conditions for activation include:

- The driving speed is higher than the minimum desired cruising speed.
- The brake pedal is not depressed.
- The handbrake is released.
- No faults are detected in the ACC-SCU or ACC system.

As soon as the requirements for activation have been met and the driver presses a button provided for this purpose, ACC takes effect.

An important prerequisite for beginning this regulation is, of course, the desired cruising speed and desired time gap, so that the driver receives immediate feedback about the configured desired settings and can modify them if necessary. Therefore, it is absolutely necessary for these values to be displayed, at least at activation.

1 Driver information area with display elements for ACC (example)

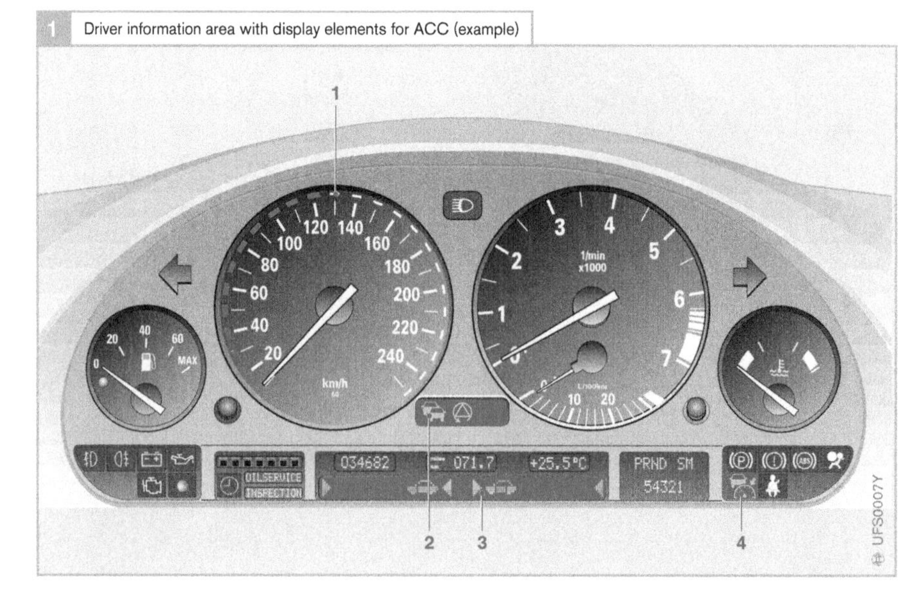

Fig. 1
1 Speedometer; LEDs for displaying the desired cruising speed ("ACC active")
2 Relevant target object detected ("ACC active")
3 "Selected distance" display with vehicle symbols (lights up for 6 seconds after ACC is activated or entries are made) or "ACC inactive" error message or "Clean sensor" prompt
4 "Standby"

To clearly and unambiguously distinguish ACC from other functions, the ISO (International Organization for Standardization) has defined a symbol for it (Fig. 2). This symbol can be used both as a readiness indicator and an activation indicator.

Setting and display of the desired cruising speed

All previously existing operating concepts combine the activation and setting of the desired cruising speed, meaning that as soon as the driver uses the switch for setting the desired cruising speed, ACC is also simultaneously activated out of standby mode (Fig. 3).

Although frequently, the same switches are used as for conventional cruise control, the setting process is substantially different. Specifically, real-world experience has shown that drivers find it more useful to have larger increments for ACC. For example, instead of increments of approximately 1 km/h used for conventional cruise control, increments of 5 or 10 km/h haven proven useful for ACC.

These larger increments make it easier to adjust the desired cruising speed over wider ranges, for example when changing from a construction zone to "open-road" driving on the expressway and vice versa.

There are four functions for setting ACC:

1. Accepting the actual speed as the desired cruising speed (**Set**).
2. Accepting the next higher increment than the actual speed (**Set +**).
3. Accepting the next lower increment than the actual speed (**Set –**).
4. Accepting the stored desired cruising speed (**R Resume**).

2 ISO symbols for ACC activation

Fig. 2
a ACC function
b ACC malfunction

3 ACC control elements on the steering wheel (example)

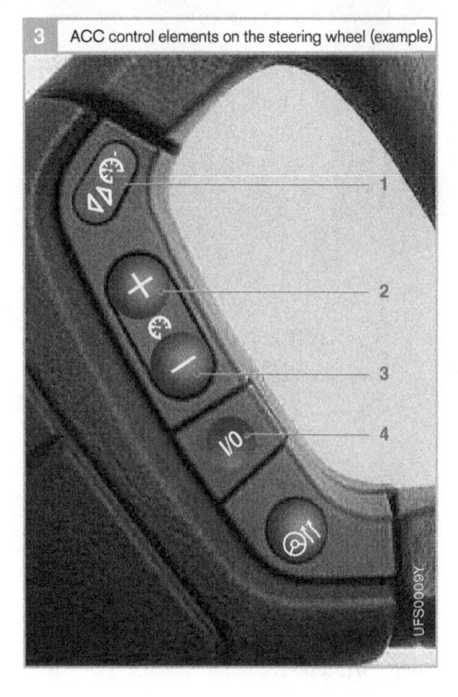

Fig. 3
1 "Resume": Call up the last saved desired cruising speed ("ACC passive") Select and display the setpoint distance for three distance increments ("ACC active")
2 "+" button: Activate the speed displayed by the speedometer ("ACC passive") Select the desired cruising speed in increasing increments of 10 km/h ("ACC active")
3 "–" button: Similar function to the "+" button, but selects the desired cruising speed in decreasing increments of 10 km/h
4 "I/O" button: Switch the ACC system on and off in "off" status and switching "ACC active" to "ACC passive"

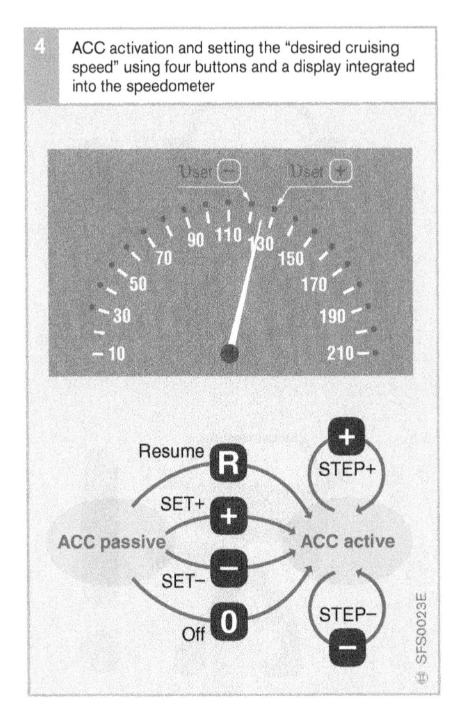

4 ACC activation and setting the "desired cruising speed" using four buttons and a display integrated into the speedometer

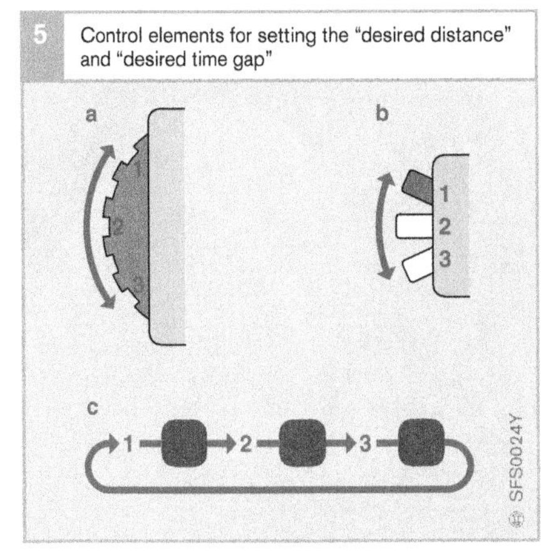

5 Control elements for setting the "desired distance" and "desired time gap"

Fig. 5
a Small knob
b Step switch
c Button for stepping through a program sequence

1 "Green" area, long distance
2 "Yellow" area, medium distance
3 "Red" area, short distance

Some combination of these is offered depending on the operating concept. After it is initially set, the desired cruising speed can be adjusted by holding down and/or tapping, in the increments mentioned above (**Step** +)/(**Step** –).

The "Set" and "Step" functions are combined, but in a manner that often differs depending on the vehicle manufacturer (Fig. 4).

These are typical combinations:

Step + with **Set** or with **Set** +

Step – with **Set, Set** – or with **Resume**

The display is integrated with the speedometer (Fig. 4) or appears in a separate field as a digital value.

Setting the desired distance and desired time gap

The desired distance and desired time gap depend on personal preference, but also on traffic and weather conditions. For this variation range, all manufacturers offer at least three different settings in the range of 1.0 to 2.0 s (time gap).

Various operating philosophies also exist for this configuration option:

- Continuous adjustment using a small knob (Fig. 5a)
- Step switch (Fig. 5b)
- Button for stepping through a program sequence, such as long, medium, short, long, medium etc. (Fig. 5c).

When changing the time gap, the driver receives feedback about the selected setting. Fig. 6 shows two possibilities of how this can be displayed.

"Object detected" display

In addition to the absolutely necessary displays of the "desired cruising speed" and "desired distance", the "object detected" display has also proven useful. It informs the driver when the ACC sensor has found a relevant object (such as a car driving ahead of it).

If a detected object is moving slower than the currently set desired cruising speed, it is classified as a control object.

Fig. 7 shows examples of possible configurations.

Other display functions

A display that the driver would rather not see is the error message, which appears in case of a function shutoff or failed activation.

In addition to plain text messages, the ISO symbol can also be used to display this message.

In addition to "real" errors in the various ECUs used in the ACC composite system, temporary errors can also cause a shutoff.

Specifically, if the sensor's view is impaired, for example by a thick coat of wet snow, there is a shutoff with an indication of the impairment.

Deactivation

As with conventional cruise control, ACC is deactivated by pressing an off button or using the brake pedal. Deactivation also occurs in case of impermissible operational states of the vehicle or if the vehicle falls below the minimum control speed.

A partial deactivation takes place after interventions of the TCS or ESP slip control systems. In this case, only the brake is activated; no acceleration takes place. This provides the opportunity to finish a deceleration maneuver. To continue driving with ACC, the driver must manually reactivate its full functionality.

6 Display of settings for "desired distance" and "desired time gap"

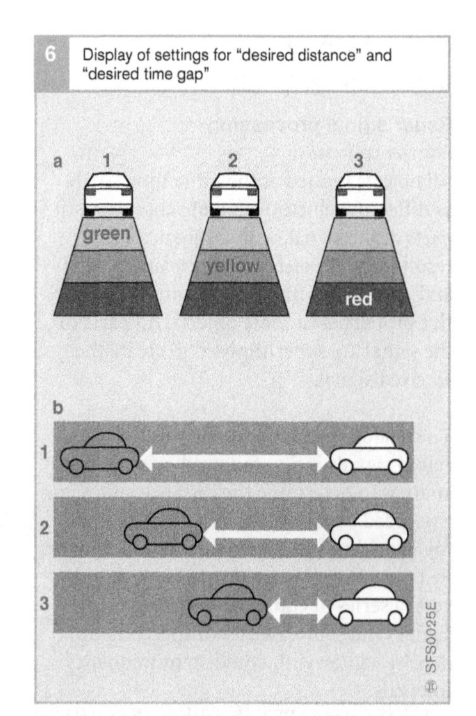

Fig. 6
a Perspective view in direction of travel
b Side view

1 "Green" area, large distance
2 "Yellow" area, medium distance
3 "Red" area, short distance

7 Versions of the "Object detected" display

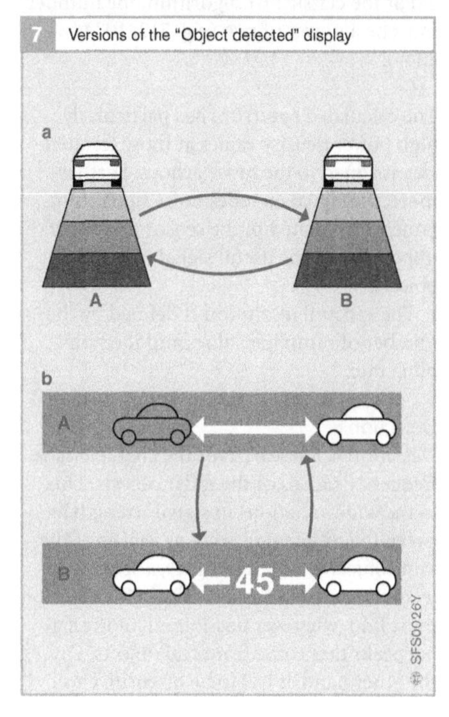

Fig. 7
a Perspective view in direction of travel
b Side view

A No relevant object
B Relevant object detected

Detection and object selection

Radar signal processing

Fourier transform

All objects located at the same time (such as different vehicles) generate characteristic parts of the signal, with frequencies that result from the distance and relative speed and amplitudes that result from the reflective properties of these objects. All parts of the signal are superimposed to create the received signal.

The analog-digital conversion of the received signals is first followed by a spectral analysis to determine the distance and relative speed of the objects. To do so, a powerful algorithm (calculation process) known as FFT (fast Fourier transform) is used. It converts a series of equidistant sampled time signal values into a series of spectral power density values, with equidistant frequency intervals.

For the classic FFT algorithm, the number must be a power of two (e.g. 512, 1024, 2048).

The calculated spectrum has particularly high power density values at those frequencies assigned to the radar echoes. Furthermore, the signal includes noise signal components generated in the sensor and superimposed over the useful signal of the target objects.

The spectral resolution is defined by the number of sampling values and the sampling rate.

Detection

Detection is the search for the characteristic frequency signals of the radar objects. Due to the wide variations in signal strength between the different objects, as well as of the same object at different times, a special detector is used. On the one hand, this detector must find, wherever possible, all of the signal peaks that come from real objects. On the other hand, it has to be insensitive to those parts of the signal caused by background noise or interference signals. For - example, the noise signal generated in the radar itself is not constant in the spectrum, but depends on time and frequency.

For each spectrum, a noise analysis is carried out first. A threshold value curve is defined depending on the spectral distribution of the noise power. Only peak signal values that are above this threshold are interpreted as target frequencies.

Object detection

Although the echo signals in the individual modulation cycles contain the information about the distance and relative speed of the objects, they cannot be unambiguously assigned to the objects. Only by linking the detection results of the modulation cycles can the results for distance and relative speed of the objects be obtained.

A found target frequency is composed of one part that depends on the distance and one part that depends on the relative speed. Therefore, to determine the distance and relative speed, target frequencies from multiple modulation ramps must fit together.

For the "multiple ramp FMCW measurement principle" a target frequency must be found for a physically existing radar object in each modulation ramp. This target frequency must be calculated from the distance and relative speed of the object (refer to the chapter on "Ranging radar").

The assignment is difficult when the spectra contain a large number of target frequencies.

The angle of a radar target to the axis of the radar is calculated from the comparison of the amplitude values for the same object in three adjacent radar beams.

Tracking

Tracking compares the measured data of the currently detected objects with the data from the previous measurement.

An object which, at the last measurement, was measured at the distance d with a relative speed v_r, has continued to move in the time Δt between the previous measurement and the new measurement and should now be measured at the expected distance

$$d_e = d + v_r \cdot \Delta t$$

Considering the fact that the measured object can also accelerate or decelerate, there is a margin of uncertainty regarding the distance d_e in which the newly measured distance value may be expected.

If the new measurement does indeed find an object within the expected range for distance and relative speed, the conclusion can be drawn that it is the same vehicle. Since the previously measured object was again found in the current measurement, the measured data are filtered with consideration of the "historical" measured data.

If, however, a previously measured object is *no longer* found in the current measurement (for example, because it is outside the radar beam or generates too little signal echo), the predicted object data continue to be used.

Additional measures for object tracking are necessary if multiple echo signals from various distances are generated by the same object. This is typically the case for trucks. These vehicles must be combined into *one* object.

Furthermore, the signal echoes are analyzed to detect "blindness" and malfunctions of the radar components.

Object selection

To select the most significant objects, in a *first step*, the lateral (sideways) position d_{yc} *(course offset)* relative the vehicle's own predicted course is determined. As shown in Fig. 1, it is calculated on the one hand from the lateral offset d_{yv} relative to the vehicle axis. In doing so, the lateral offsets relative to the sensor axis x_S are transformed using the *sensor offset* $d_{ySensor}$ to the center axis of the vehicle x_F.

On the other hand, using a description of the predicted course $d_{yvCourse} = k_y \cdot d^2/2$ (e.g. using a parabolic approach as approximation of a circular arc), we can calculate the course offset for

$$d_{yc} = d_{yv} - d_{yvCourse}$$

1 Determining the lateral (sideways) object position d_{yc} relative to the course (course offset)

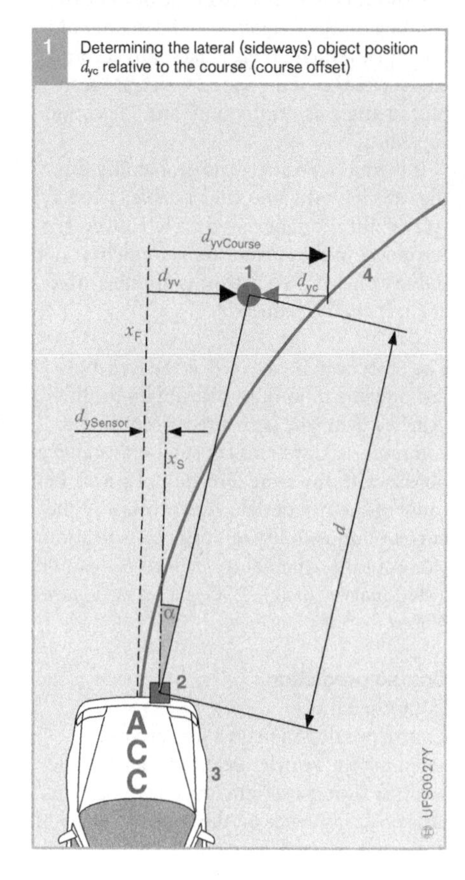

Fig. 1
1 Object
2 Sensor
3 ACC vehicle
4 Course

d_{yv} Lateral offset
d_{yc} Course offset
$d_{yvCourse} = k_y \cdot d^2/2$ Predicted course with d measurement distance to object
k_y Current curvature
$d_{ySensor}$ Sensor offset
x_F Center axis of the vehicle
x_S Sensor axis
α Angle of deviation of the object from the sensor axis

Thus the determination of d_{yc} depends on the type of description of the vehicle's own course, for which various methods exist. Some of them are described in greater detail below.

In a *second step*, a lane probability *"lp"* is calculated for each measuring cycle. This specifies the probability with which the radar object ahead is in the vehicle's own lane. Here, the vehicle's own lane is described using geometric principles that take into consideration both the "lane width" and variables such as "uncertainty of course definition".

The lane probability *"lp"* is the input variable for the integral variable "plausibility" of an object *"plaus"*. This variable is a ratio that determines the relevance of the object depending on the frequency and certainty of the measurement. It also takes into account characteristics of the sensors such as "accuracy of angle determination" and "detection capability."

If there is a positive lane probability for the vehicle's own lane, the variable *"plaus"* (plausibility) can be based on it. However, if the object in the current measurement is not in the vehicle's own lane or is not measured at all, *"plaus"* is reduced.

The object can be selected as the target object only if a minimum plausibility for the vehicle's own lane is ensured. Accordingly, common ACC systems consider *only moving* objects with the same direction of travel. Because of the risk of detection errors and the current impossibility of object classification (for example, whether it is a beverage can or a stationary vehicle), ACC *ignores stationary objects.*

Course prediction

Control quality

Course prediction plays a decisive role in assigning the vehicles detected ahead to the vehicle's own course and thus has a particularly great influence on the control quality of ACC.

In the example shown in Fig. 2, a vehicle driving in the left lane in a stationary curve with the curvilinear *course A* has an ACC control system that homes in on *object 1,* a vehicle driving ahead of it. The ACC vehicle follows this vehicle as desired by the driver.

The straight *course B*, on the other hand, mistakenly considers a slower *object 2* in the right-hand lane, for example just prior to a curve. As a result, the driver of the ACC vehicle experiences an inconvenient and implausible deceleration of his or her own vehicle.

Therefore, to lessen the risk of incorrect object selection illustrated in this example, reliable curve prediction is of great benefit.

The basic variable for determining the course is the "trajectory curvature". It describes the change of direction of the ACC vehicle as a function of the path the vehicle has already driven. To determine the future

2 Course prediction and object selection

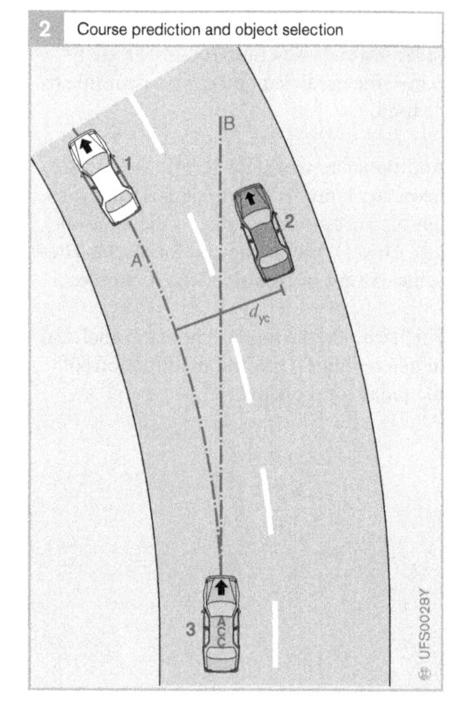

Fig. 2
1 Object 1
2 Object 2
3 ACC vehicle

A Course A
B Course B
d_{yc} Course offset

course, this information can be supplemented with the current and past positions of moving or stationary objects.

Future ACC systems will use navigation systems and video systems with image processing capabilities to determine the curvature.

Determining the curvature

The curvature k describes the change of direction of a vehicle as a function of the path already traveled. It is calculated as follows:

$$R = 1/k$$

The curvature of the vehicle trajectory can be determined by various sensors on the vehicle, but the prerequisite for all calculations is that they are used outside certain limit ranges of vehicle dynamics. Therefore, they do not apply to situations in which the vehicle skids or a greater wheel slip occurs.

To determine the course, currently available ACC systems use an offset-corrected yaw rate. This is obtained either directly from the ESP system from the signals of the steering wheel angle sensor, lateral acceleration sensor, wheel speed sensor and yaw rate sensor or determined by the ACC system itself using an offset correction.

The yaw rate $d\psi/dt$ as the rotation of the vehicle around its vertical axis, describes the current curvature k_y as the driving trajectory with the driving speed v_x:

$$k_y = (d\psi/dt) / v_x$$

Generally, the trajectory curvature is averaged, for example using simple low-pass filtering.

ESP sensor data for calculating the curvature

Common ESP systems have, in addition to the yaw rate sensor, three other sensors that allow the following curvature calculation:

To calculate the curvature k_s from the steering wheel angle δ, two other parameters are required, the steering gear reduction ratio i_{sg} and the wheelbase d_{ax}. These parameters allow a very good approximation of k_s under the typical conditions for ACC:

$$k_s = \delta / (i_{sg} \cdot d_{ax})$$

Also, to calculate the curvature k_a from the lateral acceleration a_y, the driving speed v_x is used:

$$k_a = a_y / v_x^2$$

To calculate the curvature k_v from the wheel speeds, the relative difference of the wheel speeds $\Delta v / v_x$ and the lane width d_{ay} are needed. To keep influences from the drive to a minimum, the difference $\Delta v = (v_l - v_r)$ and the driving speed are also measured on the non-driven axis.

$$k_v = \Delta v / (v_x \cdot d_{ay})$$

Although all of the specified methods can be used to determine the curvature, not all are equally well suited to different operating conditions. They particularly differ in crosswind, banked roads, and tolerances of the wheel radius and with regard to measurement sensitivity in various speed ranges. As Table 1 shows (grid), the curvature k_y from the yaw rate is the best suited for the methods considered as a whole.

However, there is a further increase in signal quality if one or all signals are used for mutual comparison. Specifically, this is possible if the ACC is equipped with a vehicle dynamics control system (such as ESP). Then, all of the sensors listed above are components of the system.

Curve prediction

For stretches of road with pronounced changes in curvature (such as winding stretches of expressway), a potentially incorrect object selection results from the ESP-assisted determination of the curvature, which describes the vehicle's current trajectory. Using the following approaches, predictive determination of the curvature in a given distance is, in principle, possible:

Prediction using radar data

There are two different methods in which radar data could be used:

1. *Analyzing the transverse movement of vehicles driving ahead and predicting a curve based on that information.*

Here, a collective transverse movement of multiple vehicles in front indicates that a curve begins ahead. Corresponding misinterpretations caused by vehicles changing lanes must be prevented.

2. *Analyzing stationary objects on the roadside in order to describe the future course of the road.*

Here, approximation methods can be used, but in this case, objects located further from the edge of the road must be reliably ignored.

Navigation systems

Predictive curvature information in defined intervals (of either time or distance) can, in principle, be determined in advance if this information is obtained based on digital maps with data points along the road (data points in the digitized road map) in increments of no more than 100 m. Here the curvature is calculated, for example, 50 m before the beginning of a curve using interpolation methods with reference to the existing data points.

Problems in this measuring principle are caused by such things as inaccuracies in the digital maps themselves or maps that do not correspond to the current course of the road. In the future, additional information (such as the number of lanes or type of road) will allow other applications.

Video image processing

A powerful but expensive method is lane identification using a video camera and image processing. This technique was used in the first ACC system on the Japanese market. Since then, however, no other ACC systems have provided additional sensors with video.

3 Experimental vehicle featuring "road sign recognition using video sensors"

1 Comparison of various methods for determining the curvature

Method	Curvature			
	from steering wheel angle k_s	from yaw rate k_y	from lateral acceleration k_a	from wheel speeds k_v
Robustness against crosswind	– –	+	+	+
Robustness against banked roads	– –	+	– –	+
Robustness against wheel radius tolerances	o	+	+	–
Measuring sensitivity at low speeds	+ +	o	– –	–
Measuring sensitivity at high speeds	–	o	+ +	–
Suitability of each curvature	+ + very well suited, + well suited, o moderately well suited, – not well suited, – – not at all suited			

Table 1

4 Test setup in Bosch research lab: Using radar sensors to detect moving objects

ACC control

Controller structure

Fig. 1 is a diagram of the basic structure of the ACC control system. Levels 1 through 3 have already been described in detail in the chapter on "Detection on object selection". This chapter, "ACC control", gives particular emphasis to Levels 4 through 6 (Fig. 1):

Level 1

In this top level of signal processing (function level), available physical variables are first measured (e.g. frequencies, echo propagation times, amplitudes etc.).

ACC measures some of this sensor information (such as radar data) itself and uses some sensor values from external sensors (such as values of the wheel speed sensors of the electronic stability program ESP for vehicle dynamics control).

Level 2

In this level (plausibility level), the physical variables measured previously are processed further, already related to the application in the specific vehicle. This results in an initial list of radar objects of interest for the ACC control system with the attributes "distance", "relative speed" and "lateral position".

To determine the curvature of the course, this level also evaluates the vehicle dynamics sensors with regard to the "trajectory curvature".

Level 3

In this next processing step (control level), the objective is to select, from the objects of interest, the object that is meaningful for the control system. This is normally the "target vehicle". In almost every case, this target vehicle is the one directly ahead of the ACC vehicle in the lane. Exceptions to this rule

1 Basic structure of the ACC controller

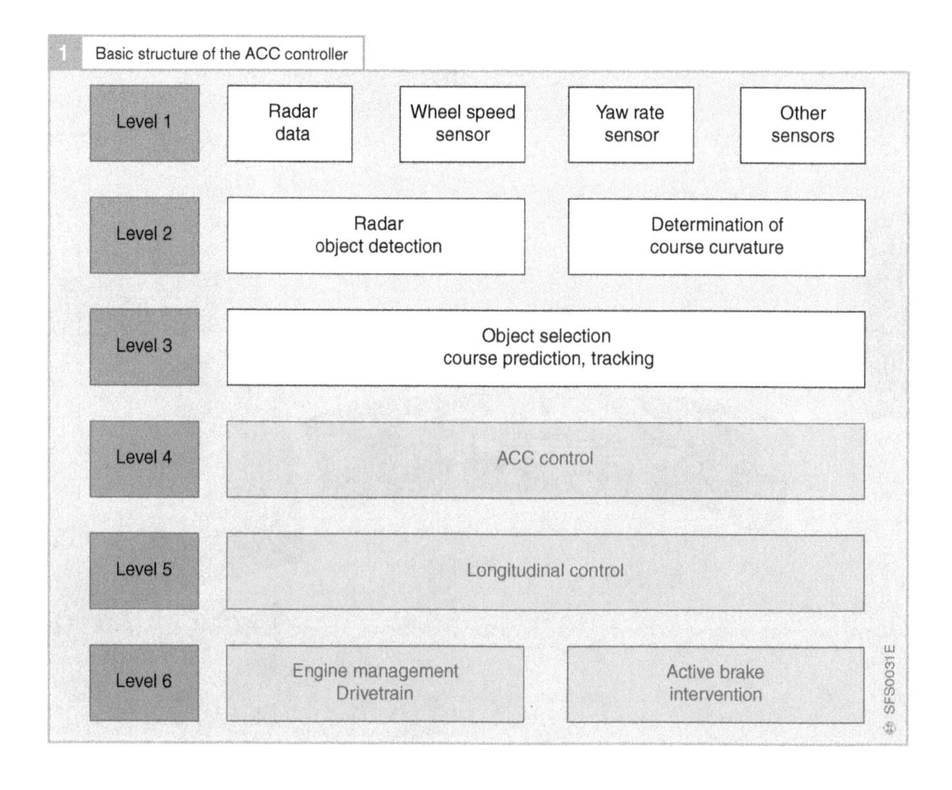

occur primarily when either the vehicles ahead or the ACC vehicle changes lane. For such cases, it is advantageous to have a list with more than one possible target object.

The decision is then shifted to the next level. For correct selection of the target vehicle(s), powerful "course prediction" and good "tracking" is indispensable (refer also to the "Detection and object selection" chapter).

Level 4

After the target object is selected, the actual ACC control takes place in this level. The result is a "vehicle setpoint acceleration".

If, according to the classification in Level 3, more than *one* target vehicle remains in the list, the control algorithms for multiple potential target objects can be calculated and subsequently evaluated. This processing step can also include the vehicle speed control and curve control.

Level 5

In level 5, the "vehicle setpoint acceleration" output variable from the fourth processing step is implemented by the longitudinal control. To do so, the specific branch is first selected that can set the desired acceleration. For positive and slight negative accelerations, this is the drive train.

If the deceleration attained with the engine drag torque is not sufficient, a switcher to the deceleration control branch takes place, which then makes use of active braking force generation.

In both branches, there is compensation for disturbance values caused by changes of total running resistance (particularly due to changing road gradients).

Level 6

This last level of processing steps is concerned with the generation and modulation of the wheel forces. In the drive branch, this primarily affects the engine control, where a modulation originating in the transmission (drive train) is also possible.

The deceleration is primarily caused by pneumatic or hydraulic actuator systems. They build up braking force actively, with no need for the driver to do anything (active brake intervention).

Controller functions

The ACC controller (Fig. 2, next page) includes the following controller functions described in detail below:

- Cruise control
- Tracking control
- Curve control
- Acceleration control.

The actuator signals are output to the actuator systems via a torque or acceleration interface.

Cruise control

The driver sets the desired vehicle speed using the control elements. Then, in the first step, the control system calculates a setpoint value for adjusting the current vehicle speed until it is equal to the desired cruising speed. It is important to note here that the speed displayed by the gauge moves ahead of the actual speed.

A byproduct of the display and operating concept of current ACC systems is the following two situations in which there can be (without the driver's intention) a large difference between the current vehicle speed and the desired set speed:

- Resuming cruise control by pressing "Resume" (resume function) activates the last set desired cruising speed as the current desired cruising speed. In some circumstances, the driver can reach a much higher speed than the set speed by simply applying the gas pedal before pressing the "Resume" button to resume ACC.

- While ACC is operating, the driver can disable the ACC control by depressing the accelerator. Thus he or she can attain a much higher speed than the set desired cruising speed.

In both situations, the driver may not be aware of the great difference between the actual and set speeds. The ACC vehicle speed control system assists the driver in these situations by means of a moderate control response.

Tracking control

The second step selects the vehicle driving ahead that is to be used for the measurement. To do so, the object data are compared to the geometry of the vehicle's own predicted (predefined) course. If more than one vehicle is in the range of this predicted course, usually the vehicle directly ahead is used for the control system.

Ideally, the vehicle selected is the one that provides the lowest setpoint acceleration at the controller output. However, this requires a feedback loop linking the controller value to the target vehicle selection. Once the target vehicle is selected, a setpoint acceleration is calculated based on the distance and relative speed. The setpoint distance results from the desired or setpoint time gap τ_{Set} set by the driver:

$$d_{Set} = \tau_{Set} \cdot v_F$$

The setpoint time gap is usually in the range from 1 to 2 seconds; the values tend to be larger at slower speeds.

2 ACC control loop with controller functions

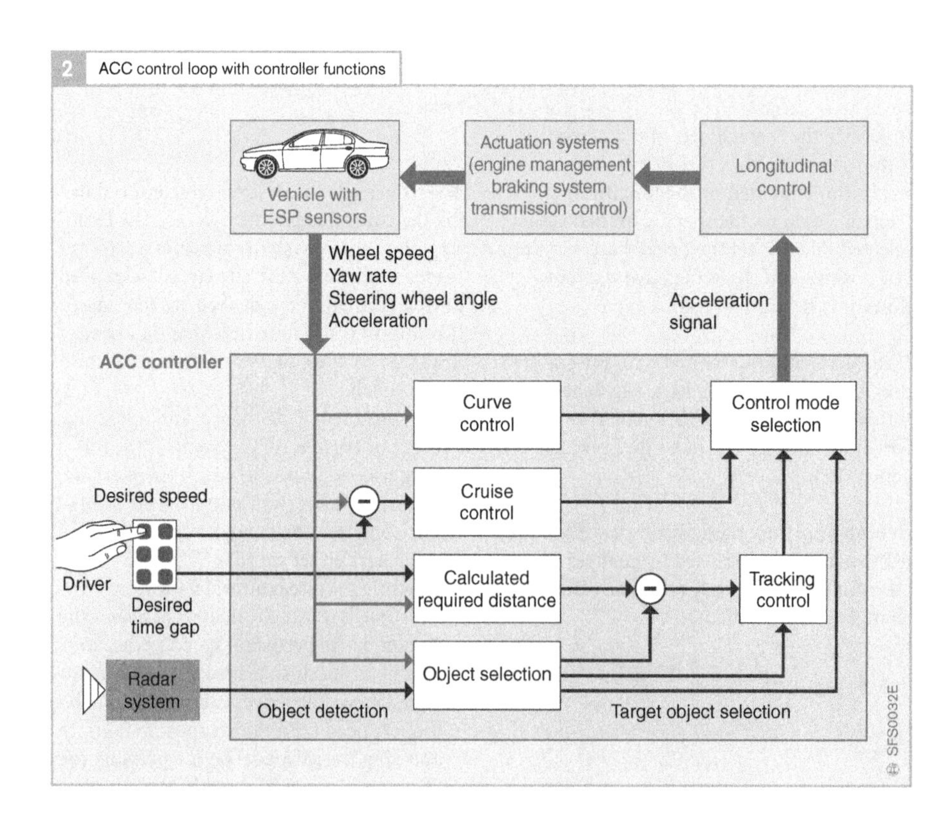

This range can be logically divided into three steps so that the driver can be offered three time gap settings:

- "Close"
- "Medium"
- "Far"

An example of these is illustrated in Fig. 3.

The selection of the control parameters is a compromise between optimizations in two opposing directions:

The first optimization is the fastest possible correction of the deviations from the setpoint defined by the relative speed $v_{rel} = 0$ and the setpoint distance.

The other optimization is for comfort purposes, whereby the system is to react as slowly as possible to small deviations of distance and changes in the speed of the vehicle ahead.

A nonlinear controller approach solves this optimization problem by being more sensitive to changes of relative speed than changes of distance. The rule of thumb is that a relative speed of 1 m/s triggers approximately the same setpoint acceleration as a deviation of 5-10 m from the setpoint distance.

Curve control

Although the ACC system was primarily designed to be used on expressways (with relatively large radii of curvature), it can also be used on other winding roads. In this respect, a few important features need to be considered:

- As a comfort system, ACC must not surprise the driver with uncomfortable linear accelerations when driving through curves.
- Due to the limited angle range of the radar sensor (Fig. 4), ACC adapts the possible acceleration of the system to the limited detection range in tight curves.
- The limited detection range of the sensor in tight curves also leads to situations in which the tracking control can no longer detect a selected target vehicle. In this situation, the curve control of ACC prevents the car from immediately accelerating again.

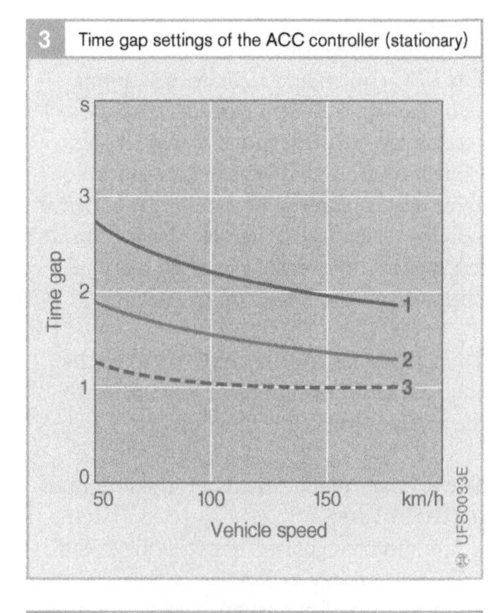

Fig. 3
Time gap settings:
1 "Far"
2 "Medium"
3 "Close"

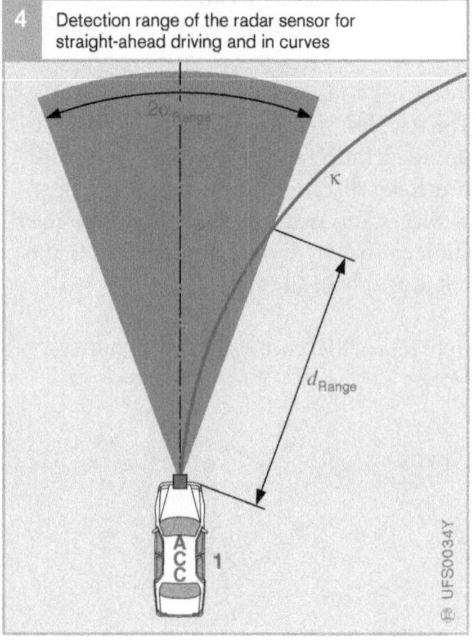

Fig. 4
1 ACC vehicle
k Curvature
$2\alpha_{Range}$ Radar detection range
$d_{Range} \cong 2\alpha_{Range}/k$

Selecting the control mode

The setpoint values for the tracking control, desired cruising speed and curve control are all calculated in parallel. The subsequent minimum selection process processes the setpoint values and ensures that an ACC vehicle never follows another vehicle at a faster speed than the desired cruising speed set by the driver.

Longitudinal control

The ACC controller calculates a setpoint value based on accelerations. A separate acceleration control converts this setpoint value into the actual vehicle acceleration. To do so, the suitable branch of the actuator system – specifically, the drive train or braking system – is selected depending on the setpoint value and the current status.

Then, these actuator systems calculate the actuator setpoint values corresponding to the acceleration setpoint value.

The primary requirements for the longitudinal control are:

- Smooth changeovers between drive and brake (and vice versa)
- Correction of disturbance values, particularly for an uphill or downhill gradient

Interfaces to actuator systems

The simplest solution would be to pass on the acceleration setpoint value of the ACC controller directly to the actuator systems. However, this requires a subordinate acceleration control in the actuator systems that is not available in all cases.

In practice, there are primarily two interfaces:

Torque interface

Most engine management systems operate based on engine torque. They handle ACC in the same way as conventional cruise control. Therefore, only minor adaptations of the engine management system are required in order to implement an external setpoint torque from ACC.

However, the setpoint acceleration must be converted into a setpoint torque within ACC. To do so, it is necessary to calculate the forces in the drive train and estimate the uphill or downhill gradient of the road or execute a subordinate torque control within ACC.

The torque interface is useful only if an actuator can also reliably implement the required torque.

Acceleration interface

Most common braking systems (smart booster, hydraulic braking intervention in ESP, SBC) support the acceleration interface.

Functional limits

Speed range

An ACC system is primarily intended for expressway and highway driving. The sensors currently used for ACC only cover the area of the vehicle's own lane beginning at an approximate distance of 40 m, with the result that vehicles driving ahead cannot always be detected in city traffic and on roads with tight curves.

For this reason, the lower speed limit of ACC systems (depending on the system design) is in the range from 30 to 50 km/h (refer also to Fig. 4 on the previous page).

Because of the comfort design of common ACC systems, the top speed limit is in the range from 160 to 200 km/h.

Longitudinal dynamics

Since an absolutely correct controller response cannot be guaranteed, the effects of the control (i.e. the vehicle acceleration and deceleration) have to be kept within certain limits. These limits can pertain to both the absolute acceleration and to its change-over time.

While the upper acceleration limits pertain to values that are also common with conventional cruise control (approximately 0.6 to 1.0 m/s^2), for ACC with active braking, there is a fixed deceleration limit value of typically 2.5 m/s^2. In many cases, this value is sufficient for the speed change. Although the deceleration in this case is clearly noticeable to the driver, it is still only one quarter of the maximum possible deceleration on a dry road.

However, the limited deceleration capacity, together with the equally limited range of the radar sensor, results in a maximum differential speed for which ACC is able to compensate without the intervention of the driver (Table 1). Though it would seem that a larger range is needed so that reaction can begin earlier, this is not feasible for the following reasons:

- The accuracy for correct lane assignment decreases significantly with increasing distance.
- The probability of a passing maneuver increases at a higher differential speed and can be confirmed only in the immediate vicinity of the target object.
  This results in a conflict of objectives: on the one hand, an early reaction is necessary at a high differential speed; on the other hand, particularly in this case, the likelihood of a passing maneuver is quite high, making early braking deceleration undesirable.
- In case of a lane change of either the vehicles driving ahead or the ACC vehicle, the beginning of the reaction is defined not only by a distance, but also by the beginning of the object assignment to the vehicle's own lane. Thus during a lane change, the driver must take into account the fact that ACC cannot compensate for the speed difference.
- In curves with a radius of less than 1000 m, roadside objects and vehicles in the adjacent lane can limit the detection range of the ACC sensor with the result that, although following is still possible in the curve, the forward detection range is not sufficient for an early reaction when approaching a newly appearing vehicle.

Although a higher degree of precision is possible in a few cases, this frequently is at the cost of transparency and thus can diminish the ability for assessment by the driver.

Stationary objects

ACC is fundamentally capable of distinguishing between stationary and moving objects. The radar system measures the relative speed $v_{rel,j}$ of an object, and the comparison with the ACC vehicle's own speed v_F yields the absolute speed v_j of the object:

$$v_j = v_F + v_{rel,j}$$

However, stationary objects are generally excluded by the ACC tracking control.

There are two primary reasons for this:

- ACC is a convenience system. Therefore, its deceleration capacity ACC is not designed to brake the vehicle in a timely manner to avoid colliding with stationary objects.

- It is currently technically impossible to make a sufficiently accurate decision as to whether or not an object is in the vehicle's own lane. Therefore, considering the large number of stationary objects on the roadside, it is highly probable that ACC would react erroneously to such an object.

For these reasons, the Bosch ACC system works according to the following strategy:

- The sensors consider and evaluate stationary objects at low speeds only.
- Only driving or stopped objects are considered for the tracking control. This makes an erroneous deceleration due to a stationary object on the roadside almost impossible.
- ACC prevents the vehicle from accelerating if it detects stationary objects in the vehicle's own lane.

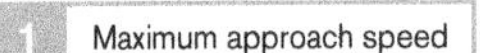

1 Maximum approach speed

At a distance d, a uniform deceleration a is used to reduce the relative speed as follows:

$$v_{rel} = -\sqrt{2d \cdot a}$$

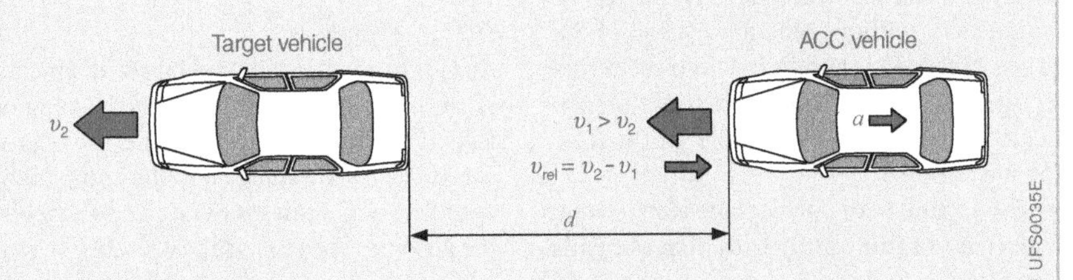

For various value pairs of d and a, the table shows the maximum approach speed that can be compensated for. Medium deceleration must be based on a smaller value than the maximum deceleration, as the deceleration is generally attained slowly.

Table 1

d m	a m/s²	$-v_{rel}$ m/s	$-v_{rel}$ km/h
50	−1	10	36
100	−1	14	51
150	−1	17	62
50	−2	14	51
100	−2	20	72
150	−2	24	88

Further developments

Sensors

Preparation for more widespread use of ACC systems is in the foreground of further development of ACC sensors.

Components that could be produced in large volumes, particularly for the high-frequency circuitry of the radar sensors, were until now primarily used in aviation, military and radio link system applications and associated with small production runs and excessive costs. Their further development is particularly important in order to make ACC affordable for many drivers. Second-generation ACC will achieve this goal with the following measures:

- Reducing the outer dimensions of the sensor & control unit: Its current size needs to be reduced by half, as in many cases, it is difficult to install the unit in the front area of vehicles. Advances in the area of electronic components allow the unit to be even more compact and highly integrated.
- Expanding the detection range:
  So that ACC can also be used on highways with tight curves, the angle range that can be evaluated by the sensors is being roughly doubled (16°). This is achieved by making the lens antenna smaller and increasing the number of radar lobes from three to four. Expanding the detection range in this way means that the radar energy is spread into a wider area. To prevent this from causing an unwanted reduction of the range, the circuitry for increasing the signal quality is being improved; this way, the range can actually be increased.

In addition to these further developments in the area of ACC sensors, Bosch is working on the development of additional new environment sensors for use in future driving assistance systems (Fig. 1, next page).

Function

Perfecting the present functional range

Perfecting the present functional range will allow future systems to be more reliable in their target selection and have even more adaptive control systems. The latter will be evident in more harmonious control behavior in lane change situations and in curves.

In the future, ACC will also work at driving speeds from 30 kilometers per hour down to a complete stop, and provide automatic assistance to the driver when restarting. Thus ACC relieves the driver of tasks even in very dense traffic, including traffic jams.

ACCplus for the entire speed range

Until now, ACC shut itself off in slow-moving traffic with speeds under 30 kilometers per hour. The newly developed ACCplus offers a significant improvement: the system controls the distance from the vehicle driving ahead in a way that varies with that vehicle's speed, all the way down to a complete stop. If the vehicle in front restarts, the driver is notified with a visual and – depending on the application – acoustic signal.

However, the decision as to whether or not to restart the vehicle after the stop remains with the driver. To follow the car ahead, all he or she needs to do is briefly push the ACC control element on the steering wheel or gently depress the accelerator pedal. In this way, ACCplus is particularly helpful to the driver in slow-moving traffic all the way down to a complete stop. Of course, the driver can intervene in the system at any time and manually accelerate or decelerate the car as necessary.

All ACC systems offer these opportunities for driver intervention, since ACC is intended to relieve the driver of tasks, not to strip the driver of decision-making power.

1 Environment sensors

Fig. 1
Assistance systems with multisensors gain widespread use in motor vehicles

- a Measurement trip with video sensors that recognize traffic signs
- b Driving with ACC on expressways or in "stop-and-go" traffic.

1. Long range: The 77-GHz radar maintains contact to the next vehicle ahead in the lane; it detects distance and relative speed as a basic function for driving with ACC (range 150 m, detection angle ±8°).
2. Close range: One or more close-range sensors (enhanced ultrasonic sensors, radar, lidar) measure the wide environment in front of the vehicle; this provides good detection of tight cut-in maneuvers.
3. Medium range: A camera measures the course of the lane ahead of the car for driving with ACC; it also recognizes traffic signs and measures object dimensions.

Fully automatic starting with ACC full speed range (FSR)

An even more future-oriented system than ACCplus is ACC full speed range (FSR) from Bosch. In addition to the signals from the long-range radar, the system processes information from a video camera and, in some cases, close-range sensors.

The system detects obstacles in front of the vehicle – particularly those at close range – even more quickly than ACCplus. Thus ACC FSR is even better suited to stop-and-go traffic, even on city thoroughfares.

When the vehicle in front restarts, the additional information obtained using the video camera allows fully automatic startup without driver confirmation. For legal reasons, however, this function will be restricted to a relatively short period, up to approximately ten seconds after the vehicle comes a complete stop. After that, the driver must restart manually without any assistance.

ACC will continue to gain market share, particularly due to the expanded functional range and added safety. The range of vehicles equipped with ACC also continues to expand: after starting in the premium class, ACC has already become available in upper midsize models and will soon be introduced in compact models.

Predictive safety functions

ACC2 is the core of future predictive safety systems (PSS). If ACC detects a critical traffic situation in the first level (PSS1), the brake linings are applied to the brake disks and the brake assist is set for a possible panic braking. Then, if the driver hits the brakes, fewer valuable fractions of a second are required for full deceleration effect.

Further expansion stages of the predictive safety system will include functions for warning the driver of impending collisions by means of a short, sharp brake impulse (PSS2) and automatic emergency brake interventions to reduce the severity of unavoidable accidents (PSS3).

Parking systems

On virtually all motor vehicles, the bodies have been designed and developed in such a way as to achieve the lowest possible drag coefficient values in order to reduce fuel consumption. Generally speaking, this trend has resulted in a gentle wedge shape which greatly restricts the driver's view when maneuvering. Obstacles can only be poorly discerned – if at all.

Parking aid with ultrasonic sensors

Application

Parking aids with ultrasonic sensors provide drivers with effective support when parking. They monitor an area of approx. 30 cm to 150 cm behind or in front of the vehicle (Fig. 1). Obstacles are detected and brought to the driver's attention by optical and/or acoustic means.

System

The system comprises the following components: ECU, warning element and ultrasonic sensors.

1 Scanning range of parking systems with all-around monitoring

Vehicles with rear-end protection normally have only 4 ultrasonic sensors in the rear fender. Additional front-end protection is provided by a further 4 to 6 ultrasonic sensors in the front bumper (Fig. 3).

The system is automatically activated when reverse gear is engaged or, for systems with additional front protection, when the speed falls below a threshold of approximately 15 km/h. During operation, the self-test function ensures continuous monitoring of all system components.

Ultrasonic sensor

Following a principle that is similar to echo depth sounding, the sensors transmit ultrasonic pulses at a frequency of approx. 40 kHz, and measure the time taken for the echo pulses to be reflected back from obstacles. The distance of the vehicle to the nearest obstacle is calculated from the propagation time of the first echo pulse to be received back according to the equation:

$$a = 0.5 \cdot t_e \cdot c$$

t_e propagation time of ultrasonic signal(s)
c velocity of sound in air (approx. 340 m/s).

Fig. 2 shows another example for the distance calculation.

The sensors themselves consist of a plastic housing with integrated plug-in connection, an aluminum diaphragm with a piezoceramic wafer attached to the inside, and a printed circuit board with transmit and evaluation electronics (Fig. 4). They are electrically connected to the ECU by three wires, two of which supply the power. The third, bidirectional signal line is responsible for activating the transmit function and returning the evaluated received signal to the ECU. When the sensor receives a digital transmit pulse from the ECU, the electronic circuit excites the aluminum diaphragm with square wave pulses at the resonant frequency so that it vibrates, and ultrasound is emitted. The diaphragm, which has meanwhile returned to rest, is made to vibrate again by the sound reflected back from the obstacle. These vibrations are converted by the piezoceramic wafer to an analog electrical signal which is then amplified and converted to a digital signal by the sensor electronics.

To detect the widest possible range, the detection characteristics must meet special requirements (Fig. 5 and 6). In the horizontal range, a wide detection angle is desirable. In the vertical range, however, it is necessary to have a smaller angle in order to avoid interference from ground reflections. A compromise is needed here so that obstacles can be reliably detected.

4 Block diagram of sensor

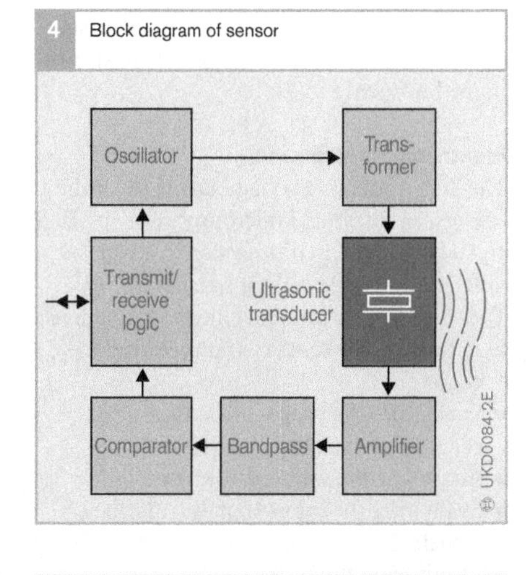

2 Calculating the distance from a single obstacle (example)

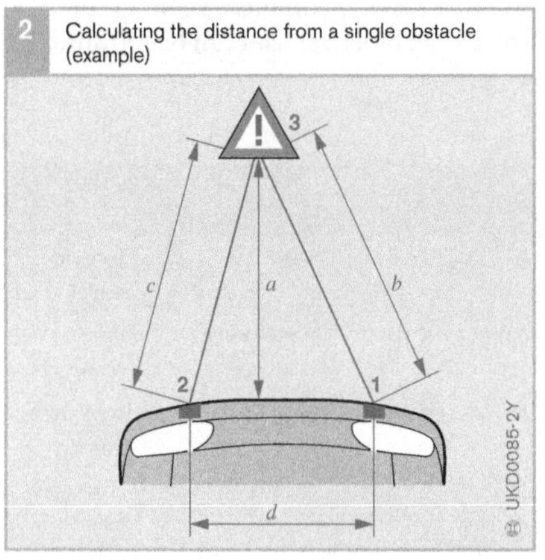

5 Antenna radiation diagram of an ultrasonic sensor

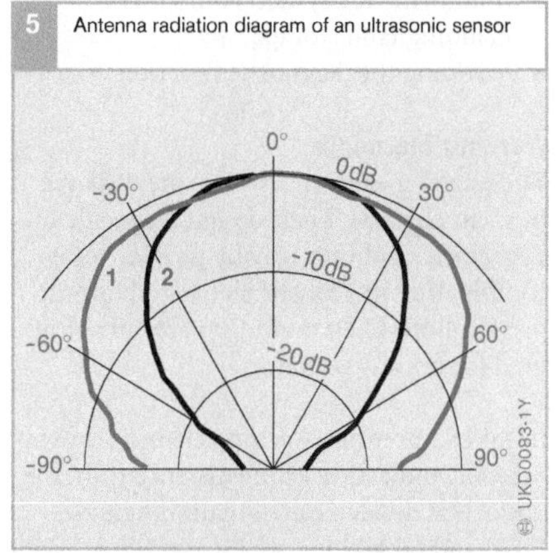

3 Installation principle of the ultrasonic sensor in the bumper

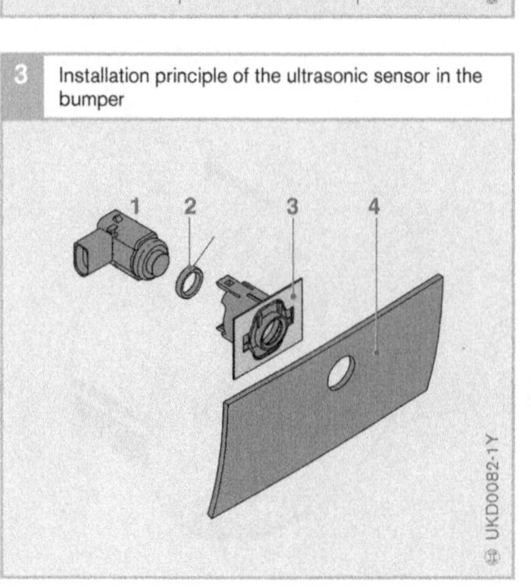

6 Simulation of an ultrasonic pulse ($t \approx 0.2$ s) for the parking aid

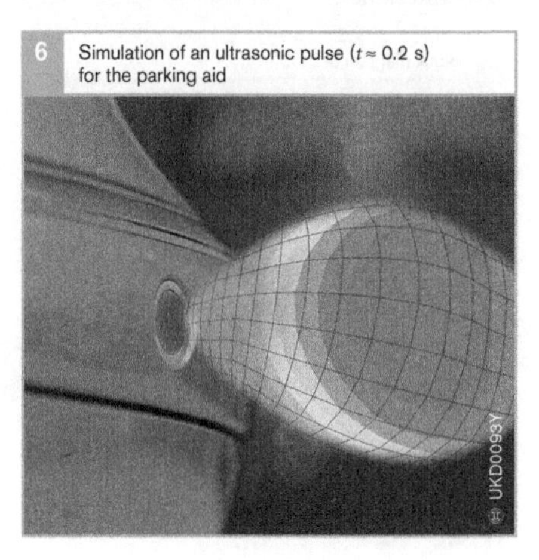

Fig. 5
1 Horizontal
2 Vertical

Fig. 2
a Distance between the bumper and the obstacle
b Distance sensor 1 to obstacle
c Distance sensor 2 to obstacle
d Distance sensor 1 to sensor 2
1 Transceiver sensor
2 Receiver sensor
3 Obstacle

$$a = \sqrt{c^2 - \frac{(d^2 + c^2 - b^2)^2}{4d^2}}$$

Fig. 3
1 Sensor
2 Decoupling ring
3 Installation housing
4 Bumper

Specifically adapted mounting brackets secure the sensors in their respective positions in the bumper (Fig. 3).

Electronic control unit

The ECU contains a voltage stabilizer for the sensors, an integrated microprocessor (µC) and all interface circuits needed to adapt the different input and output signals (Fig. 7). The software assumes the following functions:

- Activating the sensors and receiving the echo
- Evaluating the propagation time and calculating the obstacle distance
- Activating the warning elements
- Evaluating the input signals from the vehicle
- Monitoring the system components including fault storage
- Providing the diagnostic functions

Warning elements

The warning elements display the distance from an obstacle. Their design is specific to the vehicle, and they usually provide for a combination of acoustic signal and optical display. Both LEDs and LCDs are currently used for optical displays.

In the example of a warning element shown here, the indication of the distance from the obstacle is divided into 4 main ranges (see Fig. 8 and table).

Protection area

The protection area is determined by the range and number of sensors and by their emission characteristic.

Previous experience has shown that 4 sensors are sufficient for rear-end protection, and 4 to 6 for front-end protection. The sensors are integrated into the bumper, and thus the distance to the ground is fixed (Fig. 9 and 10).

The installation angle of, and the gaps between, the sensors are measured on a vehicle-specific basis. This data is taken into account in the ECU's calculation algorithms. At the time of going to press, application engineering had already been carried out on more than 200 different vehicle types. Thus, even older vehicles can be retrofitted.

Range	Distance s	Visual indicator LED	Acoustic indicator
I	< 1.5 m	green	beeping sound
II	< 1.0 m	green + yellow	beeping sound
III	< 0.5 m	green + yellow + red	continuous sound
IV	< 0.3 m	all LEDs flashing	continuous tone

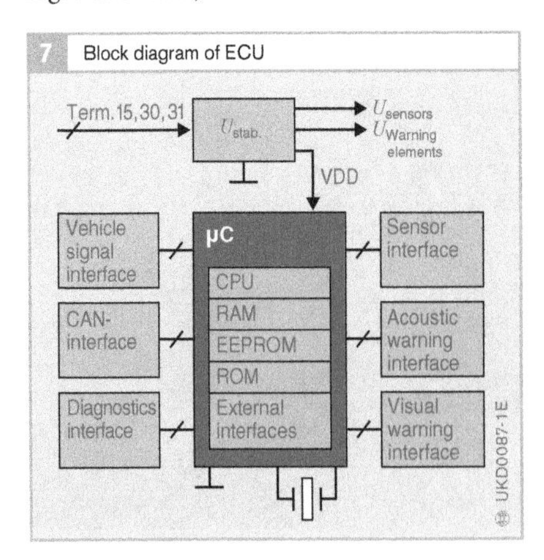

7 Block diagram of ECU

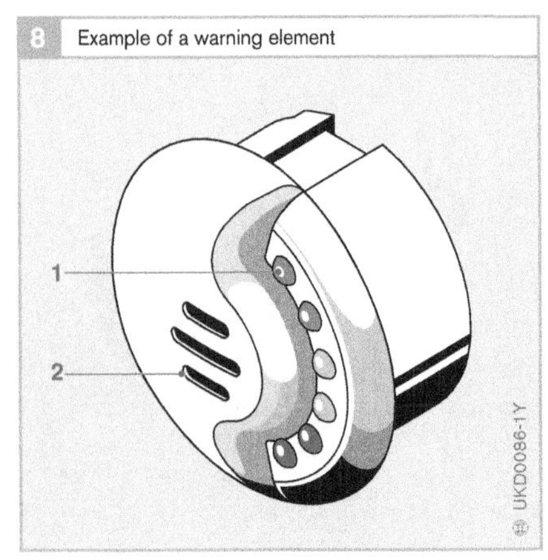

8 Example of a warning element

Fig. 8
1 LED warning light
2 Sound signal opening

9 Rear-end protection of a car (example)

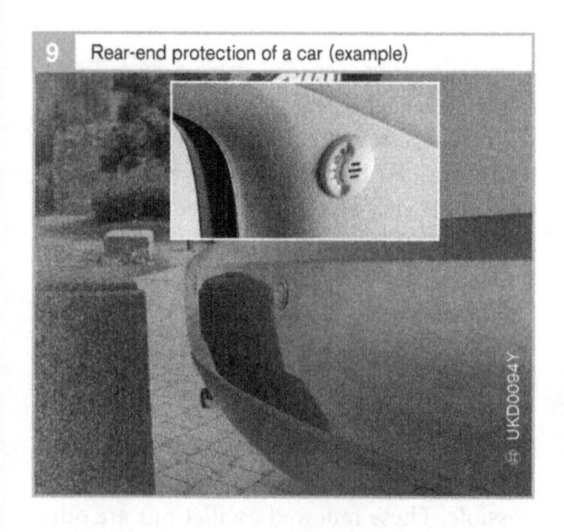

10 Rear-end protection of a bus (example)

Further development

Extended range

The present sensor range of about 150 cm is sometimes perceived to be too short by accustomed parkers. For this reason, a new sensor with a range up to 250 cm is currently under development. Due to greater levels greater packaging density of the electronics, it is also much smaller in size than the present generation of devices. This is a welcome improvement, particularly with regard to increased pedestrian safety requirements for fenders.

Parking space measurement

Another possible application for ultrasonic sensors is measuring the size of a parking space.

After the driver activates the system, a sensor mounted on the side of the vehicle measures the length of the parking space. After comparing the measured length with the signals from the wheel speed sensor as a plausibility check, the system indicates to the driver whether the parking space is long enough (Fig. 11).

A further refinement will make it possible for the system to suggest to the driver the optimum amount of steering lock required in order to complete the parking maneuver with the least effort. Systems with electric steering activation are also in development.

11 Parking space measurement

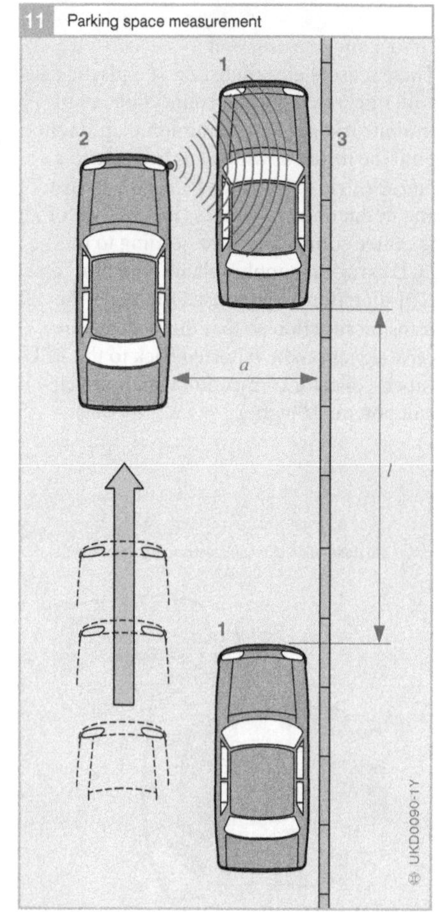

Fig. 11
1 Parked vehicles
2 Parking vehicle
3 Parking space boundary

a Measured distance
l Length of the parking space

The driver then needs only to be concerned with the linear control of the vehicle.

Ultrasonic sensors

Application

Ultrasonic sensors are integrated in the vehicle's bumpers for determining the distance to obstacles, and for monitoring the area to the front and rear of the vehicle when entering or leaving a parking lot or when manoeuvring. Using "triangulation", the very wide sensing angle which results when a number of sensors are used (4 to the rear, 4...6 to the front), can be applied to calculate the distance and angle to an obstacle. Such a system has a detection range of about 0.25 m...1.5 m.

Design and construction

These sensors are comprised of a plastic case with integrated plug-in connection, an ultrasonic transducer (aluminum diaphragm onto the inside of which has been glued a piezoelectric disc), and a PCB with transmit and evaluation electronics (Fig. 1). Two of the three connecting wires leading to the ECU carry the supply voltage. The third line is bi-directional and is used to trigger the transmit function so that the evaluated receive signal can be reported back to the ECU (open-collector connection with open-circuit potential "high").

Operating concept

These ultrasonic sensors operate according to the pulse/echo principle in combination with triangulation. Upon receiving a digital transmit pulse from the ECU, within typically about 300 μs, the electronic circuitry excites the aluminum diaphragm with square-wave pulses at resonant frequency and causes this to transmit ultrasound. The reflection from the obstacle hits the diaphragm and causes it to go into oscillation again (it had in the meantime stopped oscillating). During the time taken for it to stop oscillating (approx. 900 μs) no reception is possible. These renewed oscillations are outputted by the piezoceramic as analog electrical signals and amplified and converted to a digital signal by the sensor electronics (Fig. 2). The sensor has priority over the ECU, and when it detects an echo signal it switches the signal connection to "low" (<0.5 V). If there is an echo signal on the line, the transmit signal cannot be processed. The ECU excites the sensor to transmit when there is less than 1.5 V on the line.

In order to be able to monitor as extensive a range as possible, a wide sensing angle is used in the horizontal plane. In the vertical plane, on the other hand, only a narrow angle is required in order to avoid disturbance due to road-surface reflections.

1 4th generation ultrasonic sensor (cross-section)

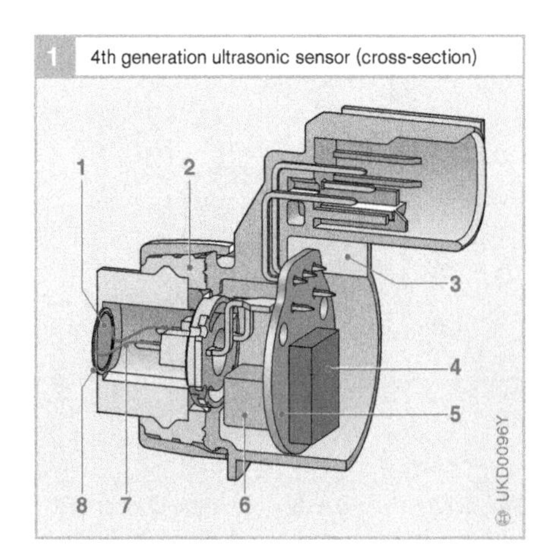

Fig. 1
1 Piezoceramic
2 Decoupling ring
3 Housing with plug
4 ASIC module
5 Printed circuit board (PCB)
6 Transformer
7 Wire bond
8 Membrane

2 Ultrasonic sensor: Block diagram

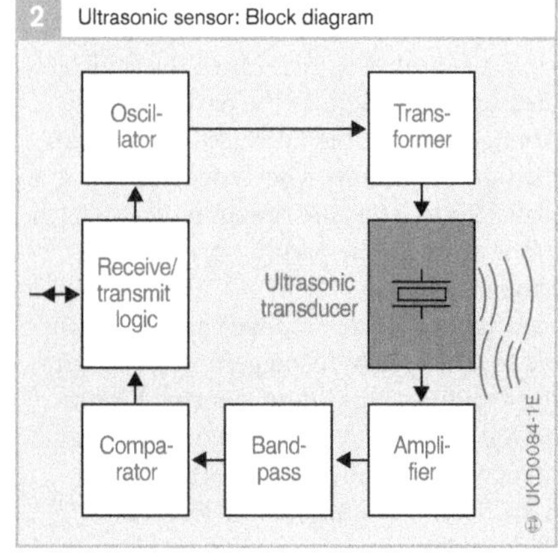

History of the seat belt

When in 1902, Ohio native Walter C. Baker rolled over in his electric car, he was the first racecar driver to have his life saved by a "seat belt".

The seat belt – in Baker's day, a simple leather belt used primarily by racecar drivers to literally strap themselves to the vehicle – has had a consistent evolution over the last hundred years. The goal has always been to make the belt safer and more comfortable and convenient to use.

Although the first shoulder belts used in cars held the occupants in their seats, they did not protect their head and upper body from impact against the steering wheel. To prevent this "jackknife" effect, suspender-like belts crossed over the chest came into use at an early stage. However, these were not only awkward to use, they sometimes allowed the occupants to slip through underneath the belt in the event of an accident.

Help in both respects was provided by the three-point belt, patented in 1958, which today remains the basis for all modern belt systems.

As early as the 1960s, experiments proved the usefulness of the seat belt. In addition, the invention of the automatic seat belt retractor allowed occupants to fasten their seat belts in mere seconds. However, even after seat belt use was mandated by law in Germany in 1976, only half of front seat passengers buckled up. Only when a fine was introduced in 1984 to remind drivers and passengers of the benefits of seat belts did their use rate rise to over 90 %.

Milestones in the development of seat belts

1903
The French inventor Lebeau is awarded the first patent for a seat belt for automobile and airplane occupants.

1949
American automaker Nash equips its Ambassador model with front seat shoulder belts as standard equipment.

1953
The seat belt arrives in Europe: the Spanish Pegaso Z-102 sport coupe is equipped with a two-point belt.

1956
The belt is introduced in Germany: the Porsche 356A features a shoulder belt as a special option.

1958
Sweden's Nils Bohlin, safety engineer for Volvo, patents the three-point belt.

1959
Volvo launches its P 121 "Amazon", which features static three-point belts as standard equipment.

1969
The belt becomes more convenient: Volvo presents the three-point seat belt with automatic seat belt retractor for the front seats.

1979
The Mercedes-Benz S-Class features variable-height belts.

1980
Mercedes-Benz offers the steering wheel airbag for the driver and the seat belt with seat belt pretensioner for the passenger.

1995
The first belt force limiter models from Mercedes-Benz decrease the risk of injury from the belt.

Instrumentation

Drivers have to process a constantly increasing stream of information originating from their own and other vehicles, from the road, and from telecommunications equipment. All this information must be conveyed to drivers in the information and communications areas of the vehicle on suitable display and indicating equipment that comply with ergonomic requirements. In the future, in-car cellular phones, navigation systems, and distance control systems will join automotive sound systems and vehicle monitoring systems as standard equipment in motor vehicles.

Information and communication areas

In any vehicle, there are four information and communication areas which must satisfy different requirements in terms of their display features:

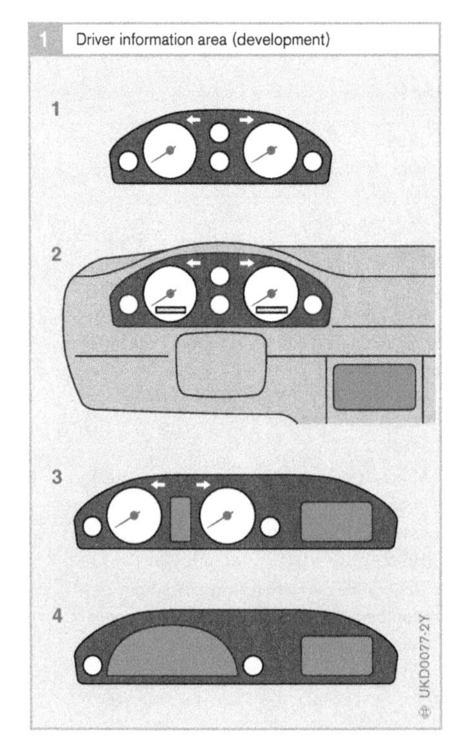

Fig. 1
1 Needle instrument
2 Needle instrument with TN-LCD and separate AMLCD in the center console
3 Needle instrument with (D)STN and integrated AMLCD
4 Programmable instrument with two AMLCD components

- Instrument cluster
- Windshield
- Center console
- Vehicle rear compartment

The display features in these areas are determined by the available range of information and the necessary, useful, or desirable information for the occupants.

1. Dynamic information and monitoring information (e.g. fuel level), to which the driver should respond, is displayed on the *instrument cluster* as close as possible to the driver's primary field of vision.

2. A head-up display (HUD), which projects the information onto the *windshield*, is ideally suited to engage the driver's attention and displaying information such as warnings from a radar distance control system (ACC) or route directions. The display is supplemented acoustically by voice output.

3. Status information or dialog prompts are mainly displayed in the vicinity of the control unit in the *center console.*

4. Information of an entertainment nature belongs in the *vehicle rear compartment,* far away from the primary field of vision. This is also the ideal location for a mobile office. The backrest of the front passenger seat is a suitable installation location for the monitor and operator unit of a laptop computer.

Driver information systems

The driver information area in the vehicle cockpit and the display technologies used have gone through the following stages of development (Fig. 1):

Individual and combined instruments

Conventional individual instruments for the optical output of information were initially superseded by more cost-effective instrument clusters (combination of several information units in a single housing) with good illumination and antireflection qualities. So that a continually increasing amount of information could be accommodated in the available space, today's instrument cluster evolved over time, featuring several needle instruments and numerous indicator lamps (Fig. 1, Item 1).

Digital displays

Digital instruments

The digital instruments fitted up to the 1990s displayed information using vacuum fluorescence display (VFD) technology and, later, liquid crystal (LCD) technology, but they have now largely disappeared. Instead, conventional analog needle instruments are used in combination with displays. At the same time, there is an increase in the size, resolution and color representation of the displays.

Central display and operator unit in the center console

With the advent of automotive information, navigation, and telematics systems, screens and keyboards on the center console are now becoming widespread. Such systems combine all the additional information from functional units and information components (e.g. cellular phone, car radio/CD, controls for heating/air conditioning (HVAC) and – important for Japan – the "TV" function) into a central display and operator unit. The components are interconnected in a network and are capable of interactive communication.

Positioning this terminal, which is of universal use to driver and passenger, in the center console is effective and necessary from both ergonomic and technical standpoints. The optical information appears in a graphics display. The demands placed by TV reproduction and the navigation system on the video/map display determine its resolution and color reproduction (Fig. 1, Item 2).

Graphics modules

Fitting vehicles with a driver's airbag and power-assisted steering as standard has resulted in a reduction in the view through the top half of the steering wheel. At the same time the amount of information that has to be displayed in the installation space available has increased. This creates the need for additional display modules with graphics capabilities and display areas that can show any information flexibly and in prioritized form.

This tendency results in instrumentation featuring a classical needle instrument but supplemented by a graphics display. The central screen is also at the level of the instrument cluster (Fig. 1, Item 3). The important issue for all visual displays is that they can be easily read within the driver's primary field of vision or its immediate vicinity without the driver having to look away from the road for long periods, as is the case, for instance, if the displays are positioned in the lower area of the center console.

The graphics modules in the instrument cluster permit mainly the display of driver and vehicle-related functions such as service intervals, check functions covering the vehicle's operating state, as well as vehicle diagnostics as needed for the workshop. They can also show route direction information from the navigation system (no digitized map excerpts, only route direction symbols such as arrows as turnoff instructions or intersection symbols). The originally monochrome units are now being superseded on higher-specification vehicles by color displays (usually TFT screens), which can be read more quickly and easily because of their color resolution.

For the central display monitor with an integrated information system, the tendency is now to switch from a 4:3 aspect ratio to a wider format with a 16:9 aspect ratio (film format), which allows additional route direction symbols to be displayed as well as the map.

Individual module with computer monitor

Beginning around 2006, TFT displays will be used to represent analog instruments for the first time (Fig. 1, Item 4). For cost reasons, however, this technology will only gradually replace conventional displays.

Instrument clusters

Design

Microcontroller technology and the ongoing networking of motor vehicles have meanwhile transformed instrument clusters from precision mechanical instruments to electronically dominated devices. A typical instrument cluster (LED-illuminated, with TN-type segment LCDs using conductive rubber, see Fig. 1) is a very flat component (electronics, flat stepping motors), and virtually all the components (mainly SMT) are directly contacted on a printed circuit board.

Method of operation

While the basic functions are the same in most instrument clusters (Fig. 2), the partitioning of the function modules in (partly application-specific) microcontrollers, ASICs, and standard peripherals sometimes differs significantly (product range, display scope, display types).

Electronic instrument clusters indicate measured variables with high accuracy thanks to stepping motor technology, and take over "intelligent" functions such as speed-dependent oil pressure warning, prioritized fault display in matrix displays, or service interval indicator. Even online diagnostic functions are standard and take up a significant part of the program memory.

1 Instrument cluster (design)

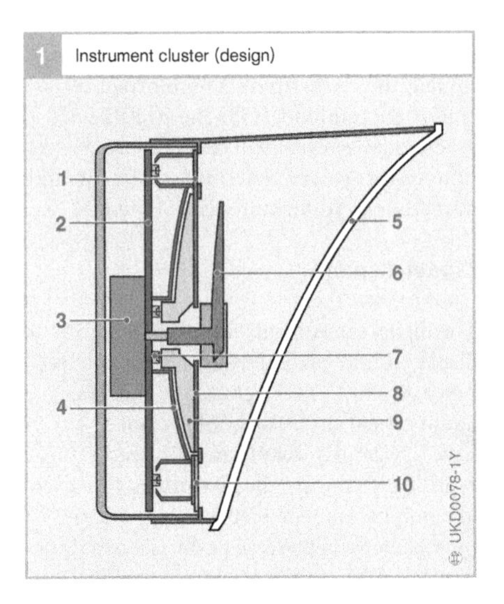

Fig. 1
1 Warning lamp
2 Circuit board
3 Stepper motor
4 Reflector
5 View cover
6 Needle
7 LED
8 Dial face
9 Optical waveguide
10 LCD

Since instrument clusters are part of the basic equipment of any vehicle, and all bus systems come together here in any case, they also incorporate gateway functions to a certain degree; in other words, they act as bridges between different bus systems in the vehicle (e.g. engine CAN, body CAN, and diagnostics bus).

Measuring instruments

The vast majority of instruments operate with a mechanical needle and a dial face. Initially, the compact, electronically triggered moving magnet quotient measuring instru-

2 Instrument cluster (schematic)

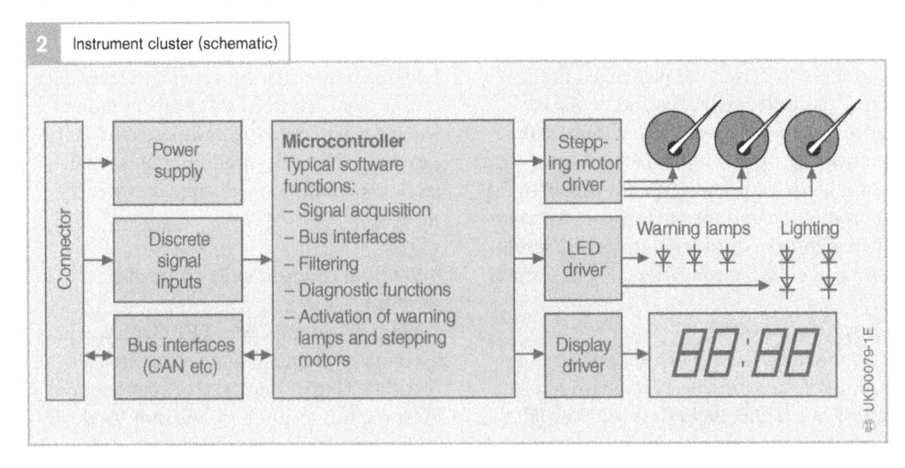

ment replaced the bulky eddy current speedometer. Nowadays, more durable geared stepping motors, which are very slim-fitting, have become the preferred choice. Due to a compact magnetic circuit and (mostly) 2-stage gearing with a power output of only about 100 mW, these motors allow swift and highly accurate needle positioning.

Lighting

Instrument clusters were originally lit by *frontlighting technology* in the form of *incandescent lamps. Backlighting technology* has meanwhile gained acceptance on account of its attractive appearance. Bulbs have been replaced by long-lasting *light emitting diodes* (LEDs). LEDs are also suitable as warning lamps and for backlighting scales, displays and, via plastic optical waveguides, needles (refer to Table 1, "Overview of lighting sources").

The efficient yellow, orange, and red InAllnGaP technology LEDs are now in widespread use. The more recent InGaN technology has produced significant efficiency improvements for the colors green, blue, and white. Here, white is obtained by combining a blue LED chip with an orange-emitting luminescent material (yttrium-aluminum granulate).

However, special technologies are also being used for specialized configurations:

- *CCFL (cold cathode fluorescent lamps):* Mainly for "black screen" instruments, which appear black when they are deactivated. When combined with a tinted view cover (with e.g. 25% transmission), these very bright lamps (high luminance, high voltage) produce a brilliant appearance with outstanding contrast. Since color LCDs have very low transmission (typically about 6%), it is imperative that CCFLs are used to backlight them in order to obtain good contrast even in daylight.
- *EL (electroluminescent) film:* This flat film, which lights up when an alternating voltage is applied and achieves an extremely even light pattern, has only been available in a form suitable for automobiles since about 2000. It offers extensive freedom of design for color combinations and/or for superimposing dial faces on display surfaces.

Table 1

1 Overview of lighting sources

Lighting source	Possible colors	Typical data [1]	Technical suitability for	Conventional inst. clust.	Black screen instruments	Service life [2] in h	Activation
Bulb	White	2 lm/W	Dial face	+	–	$B_3 \approx 4500$	No special
	(any color poss.	65 mA	Needle	o	–		activation
	with filter)	14 V	Display	o	–		required
SMD-LED	Red, orange, yellow	8 lm/W	Dial face	+	o	$B_3 \gg 10{,}000$	Series resistors
luminescence	(AllnGaP)	25 mA	Needle	+	+		or regulation
diode		2 V	Display	+	–		required
	Blue, green (InGaN),	3 to 12 lm/W	Dial face	+	o	$B_3 > 10{,}000$	
	White (with converter)	15 mA	Needle	+	+		
		3.6 V	Display	o	–		
EL film	Blue, violet, yellow,	2 lm/W	Dial face	+	–	approx. 10,000	High voltage
Electro-	Green, orange, white	100 V~	Needle	–	–		required
luminescence		400 Hz	Display	o	–		
CCFL	White (any color	25 lm/W	Dial face	+	+	$B_3 > 10{,}000$	High-voltage
cold cathode	poss. depending on	2 kV~	Needle	–	o		required
lamp	fluorescent material)	50 to 100 kHz	Display	+	+		

1) Efficiency in lm/W (lumen per watt), current in mA, voltage in V or kV, activation frequency in kHz.
2) B_3 time at which 3% of the components can have failed. Suitability: + Preferred, o Qualified, – Not suitable

Display types

TN-LCD

TN-LCD ("twisted nematic liquid crystal display") is the most commonly used form of display. The term stems from the twisted arrangement of the elongated liquid crystal molecules between the locating glass plates with transparent electrodes. A layer of this type forms a "light valve", which blocks or passes polarized light depending on whether voltage is applied to it or not. It can be used in the temperature range of –40°C to +85°C. The switching times are relatively long at low temperatures on account of the high viscosity of the liquid crystal material.

TN-LCDs can be operated in positive contrast (dark characters on a light background) or negative contrast (light characters on a dark background). Positive contrast cells are ideal for frontlighting or backlighting modes, but negative contrast cells can only be read with sufficient contrast if they receive strong backlighting. TN technology is suitable not only for smaller display modules but also for larger display areas in modular, or even full-size, LCD instrument clusters.

Graphics displays for instrument clusters

Dot matrix displays with graphics capabilities are needed to display infinitely variable information. They are activated by line scanning and therefore require multiplex characteristics. Under the conditions prevailing in a motor vehicle, conventional TN-LCDs can today produce multiplex rates of up to 1:4 with good contrast and up to 1:8 with moderate contrast. Other LCD display technologies are needed to achieve higher multiplex rates. STN and DSTN technologies are in current use for modules with moderate resolution. DSTN technology can be implemented to provide monochrome or multicolor displays.

1 Thin film transistor LCD (TFT-LCD)

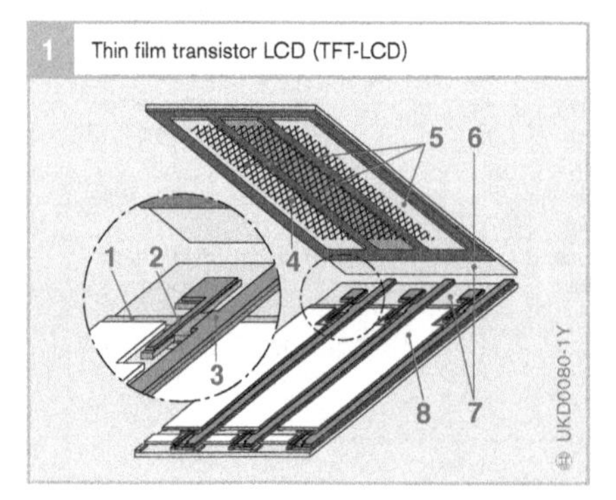

Fig. 1
1 Row conductor
2 Thin film transistor
3 Column conductor
4 Front plane electrode
5 Color layers
6 Black matrix
7 Glass substrate
8 Pixel electrode

STN-LCD and DSTN-LCD

The molecular structure of a super twisted nematic (STN) display is more heavily twisted within the cell than in a conventional TN display. *STN-LCDs* allow only monochrome displays, usually in blue-yellow contrast. Neutral color can be obtained by applying "retarder film", but this is not effective throughout the entire temperature range encountered in the vehicle. The *DSTN-LCD* (double-layer STN) features considerably improved characteristics that permit neutral black and white reproduction over wide temperature ranges with negative and positive contrast. Color is created by backlighting with colored LEDs. Multicolor reproduction is created by incorporating red, green, and blue thin film color filters on one of the two glass substrates. Under automotive conditions, shades of gray are only possible to a very limited extent. The result of this is that the range of colors is limited to black, white, the primary colors red, green, and blue, and their secondary colors yellow, cyan, and magenta.

AMLCD

The task of the visually sophisticated and rapidly changing display of complex information in the area of the instrument cluster and the center console can only be performed effectively by an active matrix liquid crystal display (AMLCD) which has high-resolution liquid crystal monitors with video capabilities. The best developed and mostly widely used are the thin film transistor LCDs (TFT LCDs) addressed by thin film transistors. Display monitors with diagonals of 4″ to 7″ in the center console area and an extended temperature range (–25°C to +85°C) are available for motor vehicles. Formats of 10″ to 14″ with an even wider temperature range (–40 °C to +95 °C) are planned for programmable instrument clusters.

TFT LCDs consist of the "active" glass substrate and the opposing plate with the color filter structures. The active substrate accommodates the pixel electrodes made from indium tin oxide, the metallic row and column conductors, and semiconductor structures. At each intersecting point of the row and column conductors, there is a field effect transistor, which is etched in several masking steps from a previously applied sequence of layers. A capacitor is also fabricated at each pixel (Fig. 1). The opposite glass plate accommodates the color filters and a "black matrix" structure, which improves display contrast. These structures are applied to the glass in a sequence of photolithographic processes. A continuous counter electrode is applied on top of them for all the pixels. The color filters are applied either in the form of continuous strips (good reproduction of graphics information) or as mosaic filters (especially suitable for video pictures).

Head-up display (HUD)

Conventional instrument clusters have a viewing distance of 0.8 to 1.2 m. In order to read information in the area of the instrument cluster, the driver must adjust his vision from infinity (observing the road ahead) to the short viewing distance for the instrument. This process of adjustment usually takes 0.3 to 0.5 s. Older drivers find this process strenuous or even impossible, depending on their physical capabilities. HUD, a technology involving projection, can eliminate this problem. Its optical system generates a virtual image at such a viewing distance that the human eye can remain adjusted to infinity. This distance begins at approx. 2 m, and the driver can read the information with very little distraction, without having to divert his eyes from the road to the instrument cluster.

Design

A typical HUD (Fig. 2) features an activated display for generating the image, a lighting facility, an optical imaging unit, and a "combiner", on which the image is reflected to the driver's eyes. The untreated windshield can also take the place of the combiner. Green vacuum fluorescence displays (VFDs) are most commonly used for HUDs with modest levels of information content, whereas more sophisticated displays generally use TFTs based on polysilicone technology. There are also projection systems under development that allow a wider field of vision, and therefore permit a step toward contact analogous display, i.e. a warning about an obstacle which is located below the driver's actual field of vision, for instance.

Display of HUD information

The virtual image should not cover the road ahead so that the driver is not distracted from the traffic or road conditions. It is therefore displayed in a region with a low road or traffic information content. In order to prevent the driver from being overwhelmed with stimuli in his primary field of vision, the HUD should not be overloaded with information, and is therefore not a substitute for the conventional instrument cluster. It is, however, particularly well suited for displaying safety-related information such as warnings, safety distance, and route directions.

2 Head-up display (HUD) (schematic)

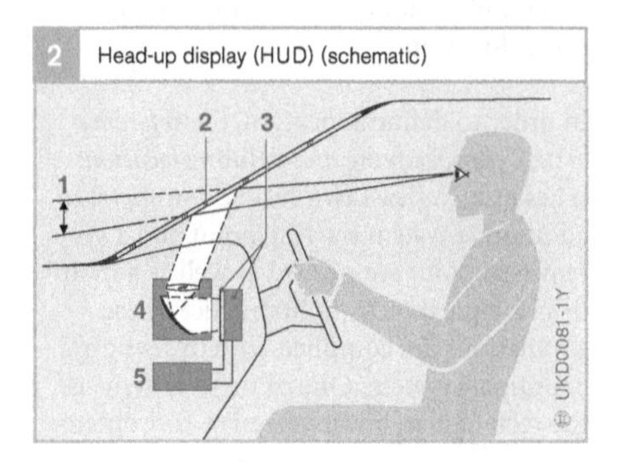

Fig. 2
1 Virtual image
2 Reflection in windshield
3 LCD and illumination (or CRT, VFD)
4 Optical system
5 Electronics

Orientation methods

In order to be able to understand various terms applied in road traffic, such as position-finding by means of satellite, and vehicle navigation, a number of the important basic terms associated with orientation and navigation methods are dealt with below.

Orientation

The word orientation is derived from the Latin word "oriens" (rising [sun]). It stands for navigation using the four points of the compass.

Position-finding

Assignment

Position-finding is used to determine one's own position or that of a searched-for objective by means of measurement, direction finding, or radiolocation techniques.

Reference systems

If considerable distances and areas are involved, *directions and locations* can only be determined using a reference system that provides suitable, repoducible reference directions (for instance, the starry sky (firmament) and the direction of the sun or of a mountain peak which is visible from a long distance).

The *reference system for direction* must be defined referred to a single, main direction (North, for instance, which can easily be defined using the Pole (North) Star). A name or a number suffices for the input of any other direction.

In order to define a location, the *reference system for specifying and defining locations* relies upon at least two related numbers. In addition, a system starting point and a system zero point are needed as well as a main or principle direction. In practice, these stipulations are complied with by using a coordinate system. One of these systems is the rectangular, linear cartesian coordinate system, which is familiar from maps and town/city street plans.

The earth's spherical shape means that when a very large area is concerned, locations can no longer be defined using cartesian coordinates. A *geographic coordinate system* is needed with a network of circular lines:

The coordinate lines from North to South all start and end at the poles and are termed "Meridians". At the equator, depending upon angular distance, these are divided into East and West degrees of longitude (the prime meridian passes through Greenwich near London). The second group of coordinates run from East to West and cut all the Meridians at right angles. The Equator marks the starting point for counting in the North and South degrees of latitude.

Slight modifications are needed before these coordinate lines apply on the actual terrestrial globe. They are then no longer circles but rather complicated curves by means of which every point on the earth's surface can be defined using two digits.

Navigation

Applications

Navigation is the continual definition of location and direction as needed to arrive at a desired destination. Position-finding is such an essential component in the process of navigation that very often no difference is made between them and one simply speaks of navigation, although navigation goes far further.

Among other things, navigation requires that the globe be represented on a smaller scale and that individual locations are clearly shown in their respective positions. Compared to a map though, a globe is far too unwieldy for entering routes, directions, and distances. The problem is that spherical figures cannot be transferred to a flat surface without distortion of the distances involved.

Coordinate systems

There are suitable coordinate systems available for representing areas of the earth's surface on maps. For instance, these include the Mercator Projection system and the related UTM (Universal Transversal Mercator Projection). These are particularly suitable for land navigation. In both systems, projection "beams" radiate from the globe's center point and project onto every point of the earth's spherical surface on a cylindrical generated surface. In one case it is applied to the equator, and in the other case turned through 90° and applied to a meridian (Figs. 1 and 2 respectively). In the subsequent calculations, the unavoidable distortions are compensated for by means of a distortion factor.

Developing the generated surface results in a flat map sheet. Further-going procedures are needed in order to define the location data (described using coordinate systems in appropriate maps) in the vehicle while it is on the move.

Composite navigation

Composite navigation must be applied when landmarks are missing and permanent position-finding by astronomical means (astronavigation) is impossible (Fig. 3). Composite navigation can be applied if it is possible to determine the direction taken by the vehicle by means of a compass for instance, and when the distance travelled can be measured. Composite navigation is the basis for all independent or autonomous navigation methods: A given travelled distance which is so small that the vehicle has not changed its direction perceptibly, is regarded as an oriented road element (arrow or vector). If, by means of computation (or on a drawing), a large number of these road elements, beginning with the coordinates of the starting location and up to the coordinates of the present position, are continually added to each other, one is said to have "coupled" the location.

1 Mercator projection

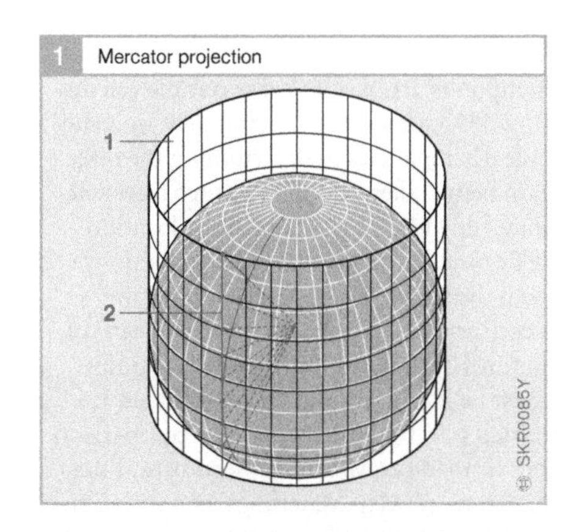

Fig. 1
1 Projection cylinder (based on the equator)
2 Projection lines/beams

2 UTM Universal Transversal Mercator Projection

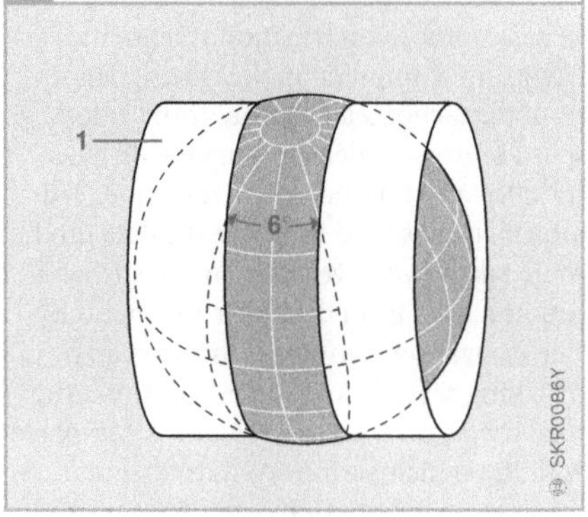

Fig. 2
1 Projection cylinder (based on a meridian)

3 Composite navigation (principle)

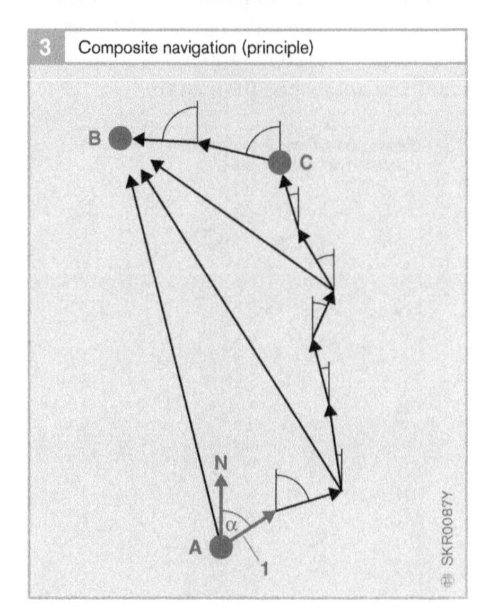

Fig. 3
A Known starting point
B Destination
C Present location (calculated position)
N North axis
1 Oriented route element (deviation from the North axis in angle α)

This is not a new method, although today computers are used to carry out the calculations electronically. Unfortunately, unavoidable discrepancies in determining the direction being taken and the distance travelled inevitably lead to an error in the location definition. Since the error increases along with the length of the journey, it becomes necessary to carry out corrections from time to time. That is, a specific reference point must be identified and its coordinates inputted into the system so that a comparison can be made between the location and the digital map (Map Matching), and the accumulated errors compensated for.

In an automotive navigation system, the navigation computer applies the signals from the tachometer sensor, or the wheel-speed sensors, in determining the distance travelled and the changes in direction. This information is used to derive the data on the route taken by the vehicle. Composite navigation results from the above interaction. The navigation computer performs map matching several times per second, and the road map stored on a CD-ROM is compared with the vehicle's route. As a result, accuracy increases to ± 5 m (15 ft) within towns and villages covered digitally in the CD-ROM road map, and ± 50 m (150 ft) on country roads and autobahns/motorways.

4 Satellite positioning system GPS (Global Positioning System)

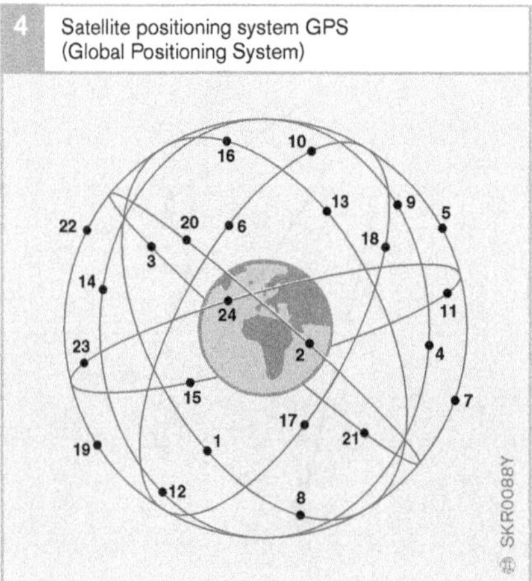

Fig. 4
1...24
24 satellites are used in defining the vehicle's position

Map matching is impossible in areas which are not covered digitally, and here navigation takes place solely by means of satellite data. The display indicates the direction to be taken, and the distance to destination is shown as a straight-line distance. The display also indicates "OFF-ROAD", in other words an indication that the vehicle is not driving on roads covered by the digital map.

Satellite navigation system GPS

At present, the GPS (Global Positioning System) navigation system is used by all automotive navigation systems for determining the vehicle's position. The GPS is based on a network of 24 intercommunicating American satellites which are used all over the world for this purpose (Fig. 4).

These satellites are uniformly distributed at an altitude of approx. 20,000 km. They orbit the earth every 12 hours in six different paths, and 50 times per second transmit special position, identification, and time signals. Since May 2000, civilian users have also been able to take advantage of accuracies of about ± 10 m.

Due to the differences in transit times, the signals from the different satellites reach the vehicle with a given time offset. Once the signals from at least 3 satellites are received, the navigation system's computer calculates its own geographical position (at least two-dimensionally).

If the signals from at least 4 satellites are received, 3-dimensional positioning is possible. Depending upon satellite position, a vehicle's navigation system is able to receive up to as many as 8 satellite signals simultaneously.

Reception of GPS signals can be interfered with, or even interrupted, by the following influences:

- Ionospheric and atmospheric interference,
- Mismatch of the combination antenna for GPS and telephone,
- Signal shadowing in valleys, due to

houses, trees, tunnels, high-rise buildings etc. (Fig. 5),
- Multipath reception due to the transit-time differences of reflected signals,
- Influencing of the satellite clocks.

Even though these interference factors can lead to inaccuracies in calculating position, the vehicle's own information sensors can nevertheless determine the vehicle's position.

Vehicle navigation

Since composite navigation and satellite navigation each have their specific advantages and disadvantages, a combination of both methods is used in the vehicle (Fig. 6).

In *composite navigation*, which is independent of infrastructure, when the stretches of road driven by the vehicle are added, the errors accumulate. The positioning uncertainty therefore increases as a function of time. These errors must be corrected by comparing the location and the digital map, and by checking the composite navigation by means of the calculated GPS position.

During longer journeys in areas not covered digitally, the vehicle-sensor errors accumulate so that it becomes impossible to perform map matching. The GPS receiver though provides the navigation computer with the position together with the degree of latitude and longitude

Since GPS can be received all over the world, this means that as soon as the vehicle enters a "digitalised" area again, the appropriate data from the particular section of road map in the navigation CD-ROM are entered into the navigation computer (this also applies when the system is taken into operation for the first time, and when the vehicle is transported by train or ferry). Map matching again becomes possible, and the starting point is soon found.

Even though *satellite navigation* is available worldwide, and is also very accurate, it can happen that reception interference causes gaps in the positioning. In such cases, the composite navigation is able to bridge the GPS reception interference by applying the information from the vehicle sensors to improve the positioning accuracy.

This means that on the one hand the composite navigation is used to overcome the GPS reception interference, and on the other the GPS position is applied to check the composite navigation.

5 Shadowing of a GPS signal (example)

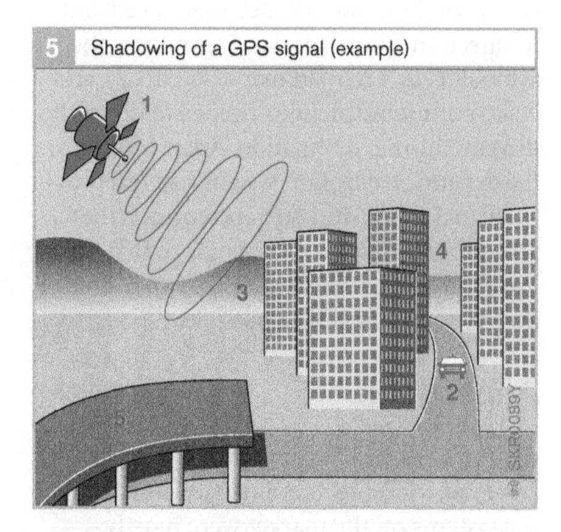

Fig. 5
1 GPS satellite
2 Vehicle
3 Valley
4 Rows of high-rise buildings
5 Tunnel and underground car parks

6 Combination of composite navigation and satellite positioning (GPS) for vehicle navigation

Composite navigation: Yaw-rate sensor, Wheel sensor or tachometer sensor, CD-ROM, Navigation ECU

+

Satellite position-finding: Satellite, GPS antenna and receiver, Navigation ECU

→ Vehicle navigation

SKR0090E

Navigation systems

Navigation in the vehicle is defined as the process of directing the driver to his/her intended destination by means of direction arrows and voice-output instructions. Guidance along the route uses a digital road map and navigation satellites. Dynamic navigation also makes use of digitally coded traffic reports.

Assignment

Navigation systems for automotive applications must continually provide the driver with acoustic and visual recommendations on the route to follow. In the process, the system relies upon data received from the GPS navigation satellites, the vehicle's road speed and its direction. From this data it generates the symbols (Fig. 1a), road maps (Fig. 1b, c), and voice-output instructions needed to direct the driver to his/her destination. Apart from this, the dynamic navigation system also reacts to the actual traffic situation and corrects the recommended route accordingly so that the driver is directed past the obstacle to his/her destination as quickly as possible.

Application

In recent years, navigation systems have achieved widespread popularity. Initially, the systems on the market were mostly retrofit systems, but have since become available as standard equipment or as an option for integration into new vehicles. This means that the sensors from a number of different systems can be used jointly and networked with other components. In the driver's primary field of view, instrument-cluster displays present him/her with the important information needed to navigate the vehicle to its destination.

Many vehicle manufacturers have incorporated the navigation system into a comprehensive driver information system featuring audio and telephone functions. This trend will continue into the future.

Method of operation

The basic functions "Position finding", "Selection of destination", "Route calculation", and "Navigation to destination" are common to all systems. Devices at the upper end of the scale also provide colored map presentation. All functions require a digital road map, which is generally stored on a CD-ROM or DVD.

1 Example of visual-signal navigation

Fig. 1
a Large pictograms support the voice-guided navigation
b Colored road maps provide information on parking lots, gas stations, and points of interest (POI)
c Traffic information (e.g. traffic jams) is shown in the map as symbols and is automatically taken into account when the route is calculated

Position finding

Compound navigation is used for position finding. Here, so-called road elements are added cyclically with regard to sum and angle (compounded). This of course leads to an accumulation of errors which, however, are compensated for by continually comparing the vehicle's position with the course of the road on the digital map (this is known as Map Matching, Fig. 2).

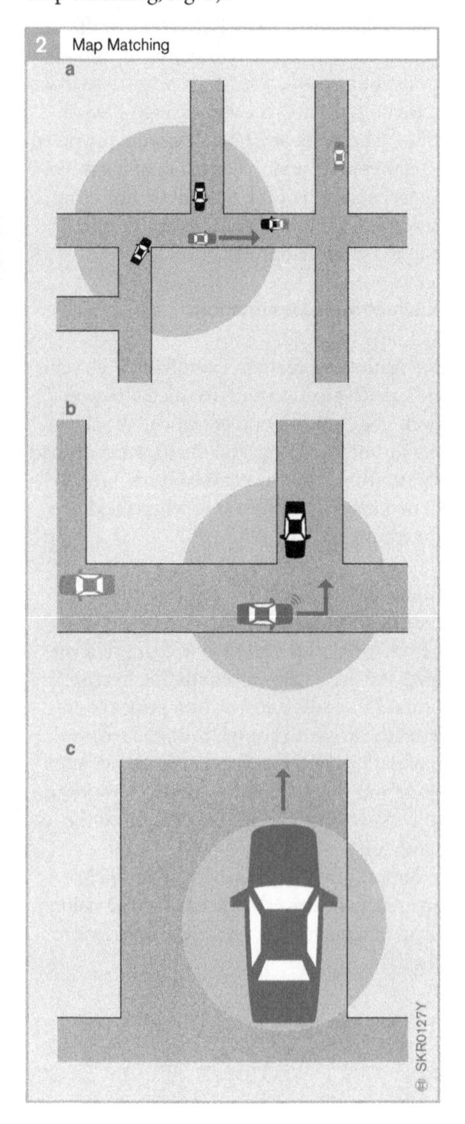

Fig. 2
a Initial, rough localization using the GPS system
b The road is identified after only a few meters. The yaw sensor registers when the vehicle turns off
c The vehicle's position is precisely defined and is kept up to date by continually comparing the sensor data with the digital road map

The GPS (Global Positioning System) satellite positioning system has increased in importance now that artificial errors are no longer introduced into the signal for military reasons. GPS allows the systems to function without any trouble, even after temporarily driving outside of roads included in the digital road map or transporting the vehicle by ship or rail. The GPS antenna together with the receiver is an essential navigation-system component.

Sensors

With the first-generation navigation systems, position finding often used two inductive *wheel-speed sensors* to determine the distance traveled and the changes in direction, and an *earth's-field sensor* for determining the absolute driving direction. For the most part, GPS serves to compensate for serious sensor errors, and for the re-entry into the stored road network following prolonged periods on roads outside the digital road map.

For current systems, it suffices to use the simple distance-traveled signal from the electronic tachometer. This signal is already commonly used in many vehicles for automatic speed-dependent volume control on the car radio. This *tachometer sensor* outputs a train of pulses the frequency of which is proportional to the vehicle's speed. These pulses are then evaluated by the navigation computer.

The vehicle's change of direction is registered by a *yaw sensor*. When the sensor position changes, this change is converted into a voltage from which the computer determines whether the vehicle is accelerating, braking, or changing direction. This sensor is very small and features a high level of insensitivity to the interference fields commonly found in the vehicle.

The earth's-field sensor is no longer required, as the vehicle's absolute direction of movement is determined from the GPS signals by means of the Doppler effect.

Selecting the destinations

Map directories

The CD's digital map contains directories which are used for inputting the destination in the form of an address. This necessitates lists of all available locations. In turn, all these locations need lists of the stored streets. Further precision in the definition of the destination is attained by the inclusion of street crossings/intersections and house numbers.

Normally, the driver is unaware of the addresses for such destinations as airports, railway stations, gas stations, and car parks etc. Therefore, to make it easier to find them, thematic directories are provided that list these destinations, which are frequently referred to as POIs (Points of Interest). These directories allow POIs such as a gas station in the area of the vehicle to be located (Figure 1b on preceding pages).

It is also possible to select a destination by directly marking it on the map display or by calling it up from a destination memory in which it had already been stored.

Guidebooks

The logical consequence is the provision of guidebooks on CD, which has resulted from the cooperation between publishing houses and the producers of the digital road maps. This allows searches for POIs such as hotels near the destination. Information is also available about the size, prices, and furnishings/equipment level of the POIs (Fig. 3).

Route calculation

Standard calculation

The calculation of routes can be adapted to the driver's wishes. This includes the settings for optimizing the route according to driving time or distance as well as the ability to circumvent expressways, ferries or toll roads. Driving recommendations along the route are expected within about 30 seconds after entering the destination.

The recalculation of the route when the driver leaves the recommended route is even more critical with regard to time. By the time the driver reaches the next intersection, he or she must have received updated recommendations. A "traffic jam" button must make it possible to exclude a certain road ahead and calculate an alterative route.

Dynamic routes

The evaluation of RDS-TMC coded traffic reports provides the basis for automatically circumventing traffic jams and congestions. Such coded reports are received through RDS or GSM. The necessary TMC codes are restricted to expressways and major national highways.

An extension of the possibilities for using new methods to make the system more dynamic is in development.

Navigation to destination

Defining the route

Navigation is a matter of comparing the vehicle's actual position with its calculated position. The stretches of road which the vehicle has just driven along, and the stretches of road still in front it on the planned route, are used in making the decision as to when the driver must turn off.

Route and direction recommendations

During the journey itself, and in good time before turn-off points or lane changes, a voice gives the driver the corresponding instructions. This audible instruction principle enables the driver to comply with the recommendations without distracting his/her attention from the surrounding traffic. A route and direction arrow appears on a display at the same time.

Simple graphics, which as far as possible are inside the driver's primary field of vision (instrument cluster), provide for even more clarity.

The conciseness of these acoustic and visual recommendations is of prime importance for the navigation quality. Due to the danger of distracting the driver, there is no question of using the symbols on the street-card display as the primary instruction medium. This is a "think-ahead" system and, as a function of vehicle speed, provides enough time for driver response.

Dynamic routes

The evaluation of coded traffic reports received through the Traffic Message Channel of the Radio Data System (RDS-TMC) parallel to the radio broadcast provides for the automatic circumvention of traffic jams and congestion. RDS-TMC is already available to the GSM services.

Route calculation centers around the expected average driving time for each section of road. Using the traffic reports, the navigation computer determines which sections of road are affected by obstructions and takes these into account when recalculating the route. Here, the system inserts a lower average speed depending upon the severity of the obstruction.

In case of traffic obstruction, if there is a time-saving, alternative route available with little congestion, the dynamic navigation automatically selects it and directs the driver onto it by means of spoken commands.

Map presentation

Depending upon the particular system, the road map can be displayed on the color monitor with a scale of between approx. 1:8000 and 1:16 million. This is a great help in obtaining an overview of the routes in the nearer vicinity or further away. Orientation is made easier by background information such as lakes, built-up areas, railway tracks and wooded areas.

Road-map memory

The CD is in widespread use as the road-map memory. The DVD, with more than seven times the storage capacity, can hold road maps of much larger areas and thus is increasingly displacing the CD.

The structure of the stored data is manufacturer-specific knowledge and has considerable influence on the system's performance. This is the reason why CDs and DVDs for the systems of different manufacturers are usually not compatible with each other.

In the future, vehicle-compatible hard discs will be used for road-map memory. In addition to their higher capacity, these make it possible to continually update the map.

3 Guidebooks on disk (CD or DVD) provide both navigation and travel information

Navigation software

Digital cartography

Digitalization

Digitalization is based on highly accurate maps, and satellite and aerial photographs which have been officially approved and released. If, in a given area, only inadequate documentation is available, measurement and surveying is performed on-site, and highly trained experts ("road inspectors") carry out the digitalization by hand using special digitalization computers. Subsequently, the names and classification of the objects concerned (roads, streets, boundaries, lakes, rivers, and canals etc.) is integrated in the database.

Road inspection

As well as checking the data from the initial digitalization, the road inspectors register all supplementary "traffic-relevant attributes" (e.g. one-way roads, bridges, subways, tunnels, and right-of-way rulings at difficult road crossings) (Fig. 8).

The results of this road inspection are integrated in the database, and used in the production of digitalized road maps in CD format.

Additional POI

POI (Points Of Interest) are hotels, restaurants, sightseeing objectives, public authorities and institutions whose names are well known but not their addresses. These POI include objects in the sector of public authorities and institutions, as well as those of interest from the touristic/cultural aspect. These form the basis for a wide variety of guidebooks and special guides.

Quality check

At every production step, a quality check is performed with special testing software, so that errors in data collection can be immediately remedied. Moreover, sample on-site checks are continually made to check geographical accuracy, correctness and actuality.

Data formats

The data format is a regulation covering the storage of data in line with a set of organizational criteria governing their processing. In the case of data which has been digitally registered, one differentiates between the registration format (standardised exchange format such as GDF), and the application format (e.g. TravelPilot format).

8 Road inspectors carry out detailed on-site checks of the traffic routing

GDF
GDF (Geographic Data Files) is a standardized international exchange format for presenting geographic features in the form of vectors (refer below to "Vector map"). Among other things, the GDF format ensures that the vehicle navigation unit outputs the familiar driving instructions notwithstanding the peculiarities of the particular country's road network. The "Tele-Atlas Co.", played a decisive role in the drawing up of the GDF format.

Vector map
The vector map generates geographic elements by means of a succession of straight lines (vectors). The beginning and end of each vector is clearly defined by geographic coordinates and by specific attributes (e.g. name, classification etc.). Vector maps are imperative for mathematical calculation of the route.

GIS
GIS (Geografic Information System) is a software application which applies geographic information for analytic and planning purposes. For instance, the calculation of radio cells for mobile-radio networks would be impossible without GIS.

Geo coding
By allocating them a coordinate pair (latitude and longitude), Geo coding can incorporate additional objects or POI in the digital maps.

TravelPilot format
This is an application format for the navigation CD of the Bosch/Blaupunkt navigation systems, as covered by the TravelPilot logo. The data of the GDF exchange format must be converted to the TravelPilot format.

Basis CD
For a number of European countries, Blaupunkt and Tele Atlas have cooperated in producing basic CDs which serve as the basis for navigation in the particular region. They incorporate the complete road network in digital form (minor roads and overland or cross-country roads), and all traffic-related information such as one-way roads, "no turns", limited-access roads etc.).

For instance, for Germany, for all the major economic areas and for all cities with more than 50,000 inhabitants, all road/street names are incorporated on the basis CD. And for Berlin. Hanover, Munich, and Stuttgart, even the house numbers are included. In addition, many other destinations such as railway stations, airports, car-hire companies, hospitals, and holiday/vacation areas are also available.

Guidebooks
Together with Varta, Michelin, Merian scout, ANWB, and De Agostini, Tele Atlas has also issued "Travel Guides" for selected countries and cities/towns. Using such a Travel Guide, it is possible to compile a selection of destinations according to one's personal tastes.

As a rule, the information is listed under specific categories (e.g. overnight stays, food and drink, the arts, architecture, touristic information, entertainment etc.). Information is also provided on hotels and restaurants (together with special quality awards such as "stars" and "chef's hats" etc.), and points of interest (POI). Of course, the system is also able to navigate the driver to these destinations.

Special guides
Such guides restrict themselves to points of special touristic interest. For instance, together with Merian scout, Tele Atlas has drawn up a special "Golf" guide.

Piezoelectric tuning-fork yaw-rate sensor

Application

In order that it can use the digital road map stored on the CD-ROM to calculate the distance driven, the computer in the vehicle's navigation system needs information on the vehicle's movements (composite navigation).

When cornering (for instance at road junctions), the navigation system's yaw-rate sensor registers the vehicles rotation about its vertical axis. With the voltage signal it generates in the process, and taking into account the signals from the tachometer or the radar sensor, the navigation computer calculates the curve radius and from this derives the change in vehicle direction.

For navigation systems, the piezoelectric tuning-fork yaw-rate sensor is increasingly being displaced by the micromechanical yaw-rate sensor.

Design and construction

The yaw-rate sensor is a steel element shaped like a tuning fork, and incorporates four piezo elements (two above, two below, Fig. 1) and the sensor electronics. It measures very accurately and is insensitive to magnetic interference.

Method of operation

When voltage is applied, the bottom piezo elements start to oscillate and exite the upper section of the "tuning fork", together with its upper piezo elements, which then starts counter-phase oscillation.

Straight-ahead driving

With the vehicle being driven in a straight line there are no Coriolis forces applied at the tuning fork, and since the upper piezo elements always oscillate in counter-phase and are only sensitive vertical to the direction of oscillation (Fig. 1a) they do not generate a voltage.

Cornering

When cornering on the other hand, the Coriolis acceleration which occurs in connection with the oscillation (but vertical to it) is applied for measurement purposes. The rotational movement now causes the upper portion of the tuning fork to leave the oscillatory plane (Fig. 1b) so that an AC voltage is generated in the upper piezo elements which is transferred to the navigation computer by an electronic circuit in the sensor housing. The voltage-signal amplitude is a function of both the yaw rate and the oscillatory speed. Its sign depends on the direction (left or right) taken by the curve.

1 "Tuning-fork" piezo yaw-rate sensor

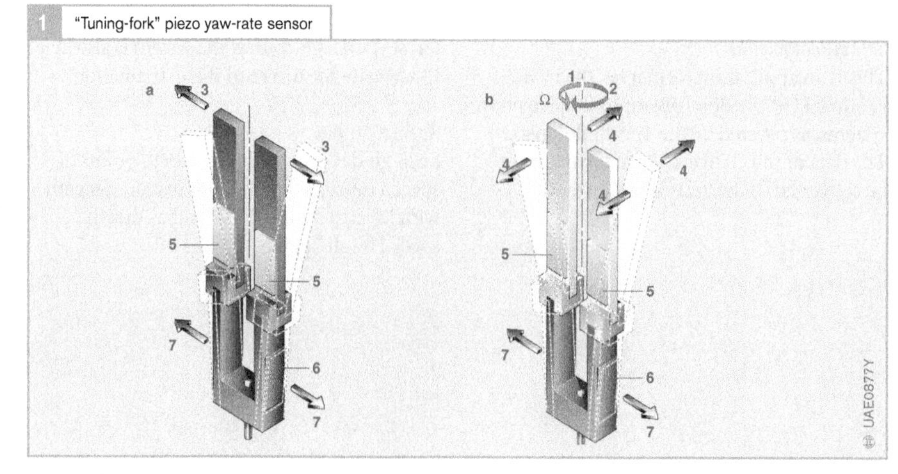

Fig. 1
- a Excursion during straight-ahead driving
- b Excursion when cornering
- 1 Tuning-fork direction of oscillation resulting from cornering
- 2 Direction of rotation of the vehicle
- 3 Direction of oscillation resulting from straight-ahead driving
- 4 Coriolis force
- 5 Upper piezo elements (sensing)
- 6 Bottom piezo elements (drive)
- 7 Excitation oscillation direction
- | Yaw

▶ Micromechanics

Micromechanics is defined as the application of semiconductor techniques in the production of mechanical components from semiconductor materials (usually silicon). Not only silicon's semiconductor properties are used but also its mechanical characteristics. This enables sensor functions to be implemented in the smallest-possible space. The following techniques are used:

Bulk micromechanics

The silicon wafer material is processed at the required depth using anisotropic (alkaline) etching and, where needed, an electrochemical etching stop. From the rear, the material is removed from inside the silicon layer (Fig. 1, Pos. 2) at those points underneath an opening in the mask. Using this method, very small diaphragms can be produced (with typical thicknesses of between 5 and 50 µm, as well as openings (b), beams and webs (c) as are needed for instance for acceleration sensors.

Surface micromechanics

The substrate material here is a silicon wafer on whose surface very small mechanical structures are formed (Fig. 2). First of all, a "sacrificial layer" is applied and structured using semiconductor processes such as etching (a). An approx. 10 µm polysilicon layer is then deposited on top of this and structured vertically using a mask and etching. In the final processing step, the "sacrificial" oxide layer underneath the polysilicon layer is removed by means of gaseous hydrogen fluoride. In this manner, the movable electrodes for acceleration sensors (Fig. 3) are exposed.

Wafer bonding

Anodic bonding and sealglass bonding are used to permanently join together (bonding) two wafers by the application of tension and heat or pressure and heat. This is needed for the hermetic sealing of reference vacuums for instance, and when protective caps must be applied to safeguard sensitive structures.

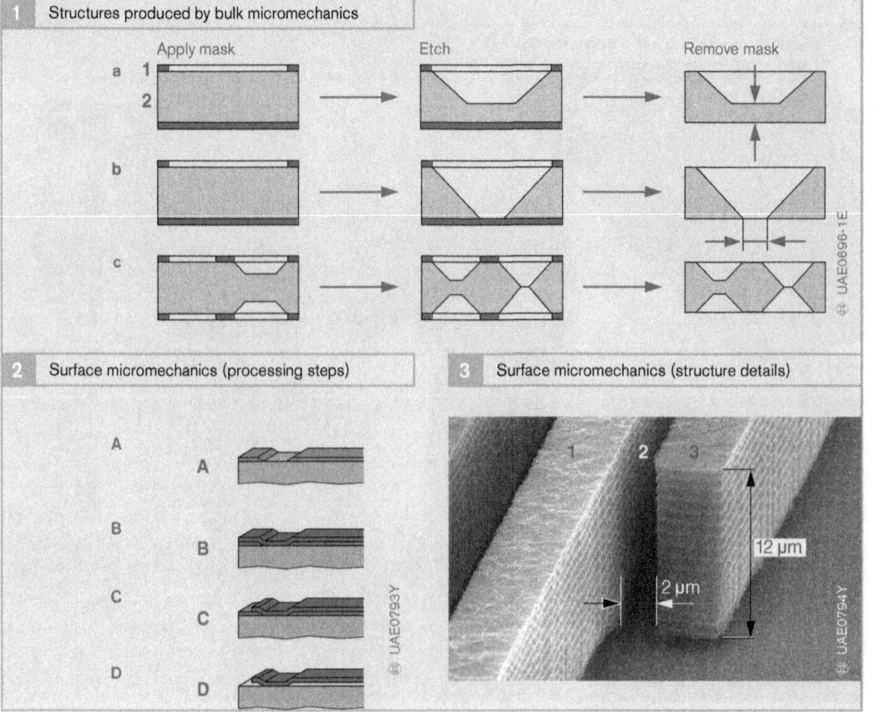

1 Structures produced by bulk micromechanics

2 Surface micromechanics (processing steps)

3 Surface micromechanics (structure details)

Fig. 1
- a Diaphragms
- b Openings
- c Beams and webs
- 1 Etching mask
- 2 Silicon

Fig. 2
- A Cutting and structuring the sacrificial layer
- B Cutting the polysilicon
- C Structuring the polysilicon
- D Removing the sacrificial layer

Fig. 3
- 1 Fixed electrode
- 2 Gap
- 3 Spring electrodes

Workshop technology

More than 30,000 garages/workshops around the world are equipped with workshop technology, i.e. test technology and workshop software from Bosch. Workshop technology is becoming increasingly important as it provides guidance and assistance in all matters relating to diagnosis and troubleshooting.

Workshop business

Trends

Many factors influence workshop business. Current trends are, for example:

- The proportion of diesel passenger cars is rising
- Longer service intervals and longer service lives of automotive parts mean that vehicles are being checked into workshops less frequently
- Workshop capacity utilization in the overall market will continue to decline in the next few years
- The amount of electronic components in vehicles is increasing – vehicles are becoming "mobile computers"
- Internetworking of electronic systems is increasing, diagnostic and repair work covers systems which are installed and networked in the entire vehicle
- Only the use of the latest test technology, computers and diagnostic software will safeguard business in the future

Consequences

Requirements

Workshops must adapt to the trends in order to be able to offer their services successfully on the market in the future. The consequences can be derived directly from the trends:

- Professional fault diagnosis is the key to professional repairs
- Technical information is becoming the crucial requirement for vehicle repairs
- Rapid availability of comprehensive technical information safeguards profitability

1 Diagnosis on a vehicle with a diagnostic tester

- The need for workshop personnel to be properly qualified is increasing dramatically
- Investment by workshops in diagnosis, technical information and training is essential

Measurement and test technology

The crucial step for workshops to take is to invest in the right test technology, diagnostic software, technical information, and technical training in order to receive the best possible support and assistance for all the jobs and tasks in the workshop process.

Workshop processes

The essential tasks which come up in the workshop can be portrayed in processes. Two distinct subprocesses are used for handling all tasks in the service and repair fields. The first subprocess covers the predominantly operations and organization-based activity of *job order acceptance*, while the second subprocess covers the predominantly technically based work steps of *service* and *repair implementation.*

Job order acceptance

When a vehicle arrives in the workshop, the job order acceptance system's database furnishes immediate access to all available information on the vehicle. The moment the vehicle enters the shop, the system provides access to its entire history. This includes all service work and repairs carried out on the vehicle up to that point. Furthermore, this sequence involves the completion of all tasks relating to the customer's request, its basic feasibility, scheduling of completion dates, provision of resources, parts and working materials and equipment, and an initial examination of the task and extent of work involved. Depending on the process objective, all subfunctions of the ESI[tronic] product are used within the framework of the *service acceptance* process.

Service and repair implementation

Here, the jobs defined within the framework of the job order acceptance are carried out. If it is not possible to complete the task in a single process cycle, appropriate repeat loops must be provided until the targeted process result is achieved. Depending on the process objective, all subfunctions of the ESI[tronic] product are used within the framework of the service and repair implementation process.

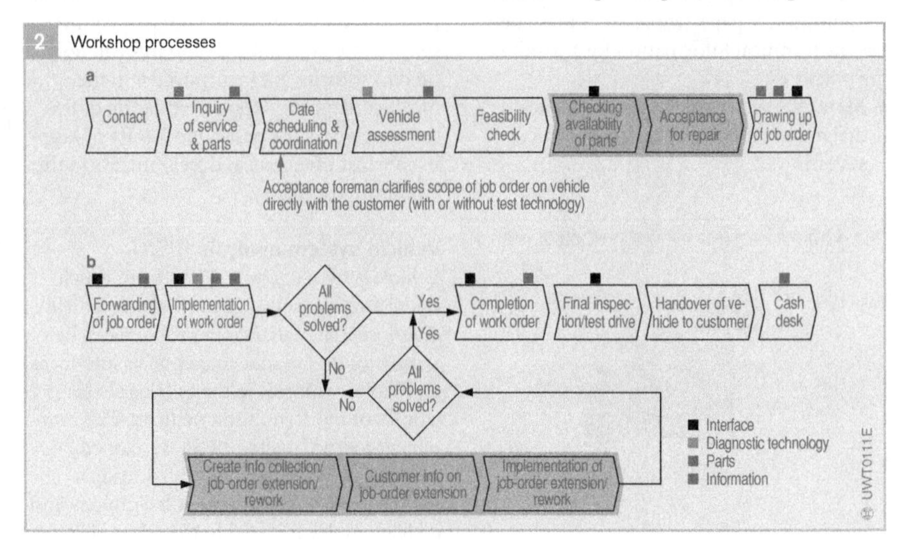

Figure 2
a Job order acceptance
b Service and repair implementation

Electronic service information (ESI[tronic])

System functions for supporting the workshop process

ESI[tronic] is a modular software product for the automotive engineering trade. The individual modules contain the following information:

- Technical information on spare parts and automotive equipment
- Exploded views and parts lists for spare parts and assemblies
- Technical data and setting values
- Flat rate units and times for work on the vehicle
- Vehicle diagnosis and vehicle system diagnosis
- Troubleshooting instructions for different vehicle systems
- Repair instructions for vehicle components, e.g. diesel power units
- Electronic circuit diagrams
- Maintenance schedules and diagrams
- Test and setting values for assemblies
- Data for costing maintenance, repair and service work

Application

The chief users of ESI[tronic] are motor garages/workshops, assembly repairers and the automotive parts wholesale trade. They use the technical information for the following purposes:

- Motor garages/workshops: mainly for diagnosis, service and repair of vehicle systems
- Assembly repairers: mainly for testing, adjusting and repair of assemblies
- Automotive parts wholesale trade: mainly for parts information

3 ESI[tronic] workshop software for all vehicle makes

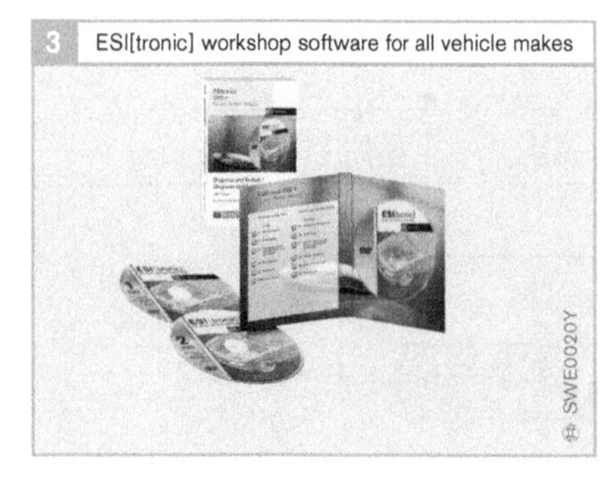

Garages/workshops and assembly repairers use this parts information in addition to diagnosis, repair and service information. Product interfaces enable ESI[tronic] to network with other (particularly commercial) software in the workshop environment and the automotive parts wholesale trade in order, for instance, to exchange data with the accounting merchandise information system.

Benefit to the user of ESI[tronic]

The benefit of using ESI[tronic] lies in the fact that the system furnishes a large amount of information which is needed to conduct and safeguard the business of motor garages/workshops. This is made possible by the broadly conceived and modular ESI[tronic] product program. The information is offered on one interface with a standardized system for all vehicle makes.

Comprehensive vehicle coverage is important for workshop business in that the necessary information is always to hand. This is guaranteed by ESI[tronic] because country-specific vehicle databases and information on new vehicles are incorporated in the product planning. Regular updating of the software offers the best opportunity of keeping abreast of technical developments in the automotive industry.

Vehicle system analysis (FSA)

Vehicle system analysis (FSA) from Bosch offers a simple solution to complex vehicle diagnosis. The causes of a problem can be swiftly located thanks to diagnosis interfaces and fault memories in the on-board electronics of modern motor vehicles. The *component testing* facility of FSA developed by Bosch is very useful in swiftly locating a fault: The FSA measurement technology and display can be adjusted to the relevant component. This enables this component to be tested while it is still installed.

Measuring equipment

Workshop personnel can choose from various options for diagnosis and trouble-shooting: the high-performance, portable KTS 650 system tester or the workshop-compatible KTS 520 and KTS 550 KTS modules in conjunction with a standard PC or laptop. The modules have an integrated multimeter, and KTS 550 and KTS 650 also have a 2-channel oscilloscope. For work applications on the vehicle, ESI[tronic] is installed in the KTS 650 or on a PC.

Example of the sequence in the workshop

The ESI[tronic] software package supports workshop personnel throughout the entire vehicle repair process A diagnosis interface allows ESI[tronic] to communicate with the electronic systems within the vehicle, such as the ESP electronic control unit. Working at the PC, the technician starts by selecting the SIS (service information system) utility to initiate diagnosis of on-board control units and access the ECU's fault memory.

The diagnostic tester provides the data needed for direct comparisons of specified results and current readings, without the need for supplementary entries. ESI[tronic] uses the results of the diagnosis as the basis for generating specific repair instructions. The system also provides displays with other information, such as component locations, exploded views of assemblies, diagrams showing the layouts of electrical, pneumatic and hydraulic systems etc. Working at the PC, users can then proceed directly from the exploded views to the parts lists with part numbers to order the required replacement components. All service procedures and replacement components are recorded to support the billing process. After the final road test, the bill is produced simply by pressing a few keys. The system also provides a clear and concise printout with the results of the vehicle diagnosis. This offers the customer a full report detailing all of the service operations and materials that went into the vehicle's repair.

4 ESI[tronic]: illustration of the installation position of the ESP hydraulic modulator

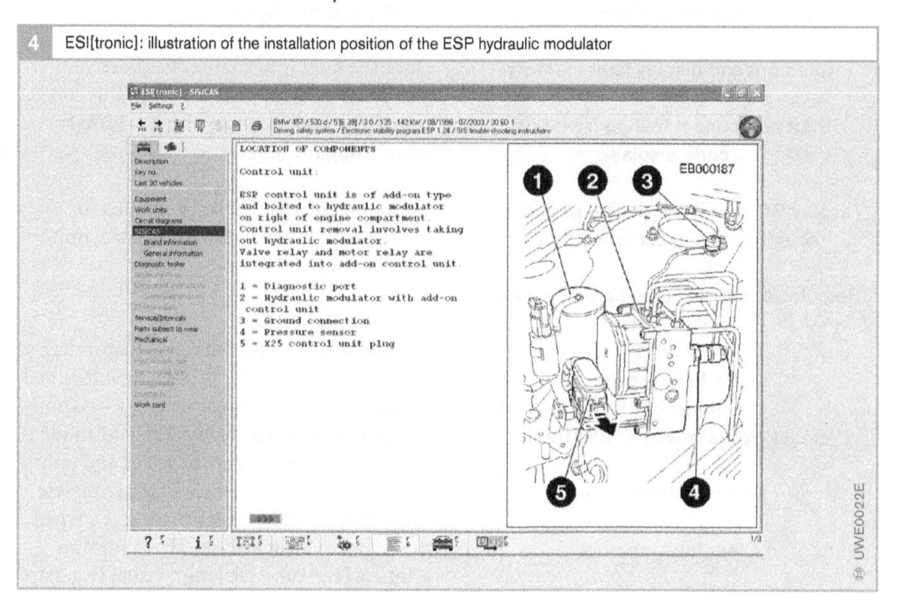

Diagnostics in the workshop

The function of these diagnostics is to identify the smallest, defective, replaceable unit quickly and reliably. The guided troubleshooting procedure includes onboard information and offboard test procedures and testers. Support is provided by electronic service information (ESI[tronic]). Instructions for further troubleshooting are provided for a wide variety of possible problems (for example, ESP intervenes prematurely due to variant encoding) and faults (such as no signal from speed sensor).

Guided troubleshooting

The main element is the guided troubleshooting procedure. The workshop employee is guided by a symptom-dependent, event-controlled procedure, which initiates with the symptom (vehicle symptom or fault memory entry). Onboard (fault memory entry) and offboard facilities (actuator diagnostics and offboard testers) are used.

The guided troubleshooting, readout of the fault memory, workshop diagnostic functions and electrical communication with offboard testers take place using PC-based diagnostic testers. This may be a specific workshop tester from the vehicle manufacturer or a universal tester (e.g. KTS 650 by Bosch).

1 Flowchart of a guided troubleshooting procedure with CAS[plus]

Identification

Troubleshooting based on customer claim

Read out and display fault memory

Start component testing from fault code display

Display SD actual values and multimeter actual values in component test

Setpoint/actual value comparison allows fault definition

Perform repair, define parts, circuit diagrams etc. in ESI[tronic]

Renew defective part

Clear fault memory

Fig. 1
The CAS[plus] system (computer aided service) combines control unit diagnosis with SIS troubleshooting instructions for even more efficient troubleshooting. The decisive values for diagnostics and repair then appear immediately on screen.

Reading out fault memory entries

Fault information (fault memory entries) stored during vehicle operation are read out via a serial interface during vehicle service or repair in the customer service workshop.

Fault entries are read out using a diagnostic tester. The workshop employee receives information about:

- Malfunctions (e.g. engine temperature sensor)
- Fault codes (e.g. short circuit to ground, implausible signal, static fault)
- Ambient conditions (measured values on fault storage, e.g. engine speed, engine temperature etc.).

Once the fault information has been retrieved in the workshop and the fault corrected, the fault memory can be cleared again using the tester.

A suitable interface must be defined for communication between the control unit and the tester.

Actuator diagnostics

The control unit contains an actuator diagnostic routine in order to activate individual actuators at the customer service workshop and test their functionality. This test mode is started using the diagnostic tester and only functions when the vehicle is at a complete stop below a specific engine speed, or when the engine is switched off. This allows an acoustic (e.g. valve clicking), visual (e.g. flap movement), or other type of inspection, e.g. measurement of electric signals, to test actuator function.

Workshop diagnostic functions

Faults that the on-board diagnosis fails to detect can be localized using support functions. These diagnostic functions are implemented in the ECU and are controlled by the diagnostic tester.

Workshop diagnostic functions run automatically, either after they are started by the diagnostic tester, or they report back to the diagnostic tester at the end of the test, or the diagnostic tester assumes runtime control, measured data acquisition, and data evaluation. The control unit then implements individual commands only.

Example

The assignment test checks that the electronic stability program (ESP) activates the wheel brake cylinders of the correct wheels. For this test, the vehicle is driven into the brake tester. After the technician starts the function, the diagnostic tester indicates how to proceed. After the brake pedal is activated, individual channels of the ESP hydraulic modulator are brought, one after another, to the pressure drop position. This allows a determination to be made of whether the corresponding wheel can be rotated. The diagnostic tester indicates the wheel for which the system has reduced the brake pressure. In this way, it can be determined whether the circuitry of the hydraulic modulator and wheel brake cylinders is correct.

Offboard tester

The diagnostic capabilities are expanded by using additional sensors, test equipment, and external evaluators. In the event of a fault detected in the workshop, offboard testers are adapted to the vehicle.

2 Functions of the KTS 650

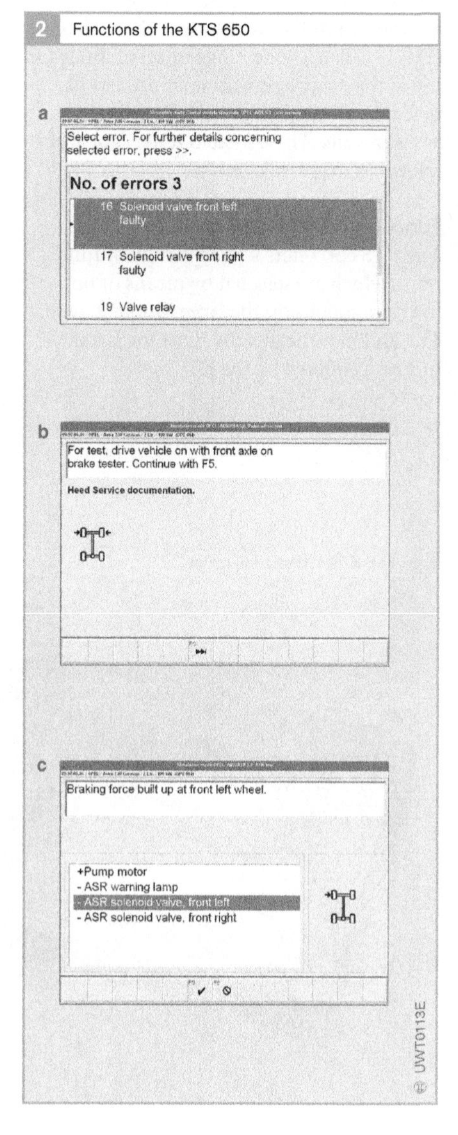

Fig. 2
- a Display of the fault memory contents
- b Procedure instructions for workshop diagnostic functions
- c Check of pressure maintenance function

Testing equipment

Effective testing of the system requires the use of special testing equipment. While earlier electronic systems could be tested with basic equipment such as a multimeter, ongoing advances have resulted in electronic systems that can only be diagnosed with complex testers.

The system testers of the KTS series are widely used in workshops. The KTS 650 (Fig. 1) offers a wide range of capabilities for use in the vehicle repairs, enhanced in particular by its graphical display of data such as test results. These system testers are also known as diagnostic testers.

Functions of the KTS 650

The KTS 650 offers a wide variety of functions, which are selected by means of buttons and menus on the large display screen. The list below details the most important functions offered by the KTS 650.

Identification

The system automatically detects the connected ECU and reads actual values, fault memories and ECU-specific data.

Reading/erasing the fault memory

The fault information detected during vehicle operation by on-board diagnosis and stored in the fault memory can be read with the KTS 650 and displayed on screen in plain text.

Reading actual values

Current values calculated by the ECU can be read out as physical values (e.g. wheel speeds in km/h).

Actuator diagnostics

The electrical actuators (e.g. valves, relays) can be specifically triggered for function testing purposes.

1 KTS Series testing equipment

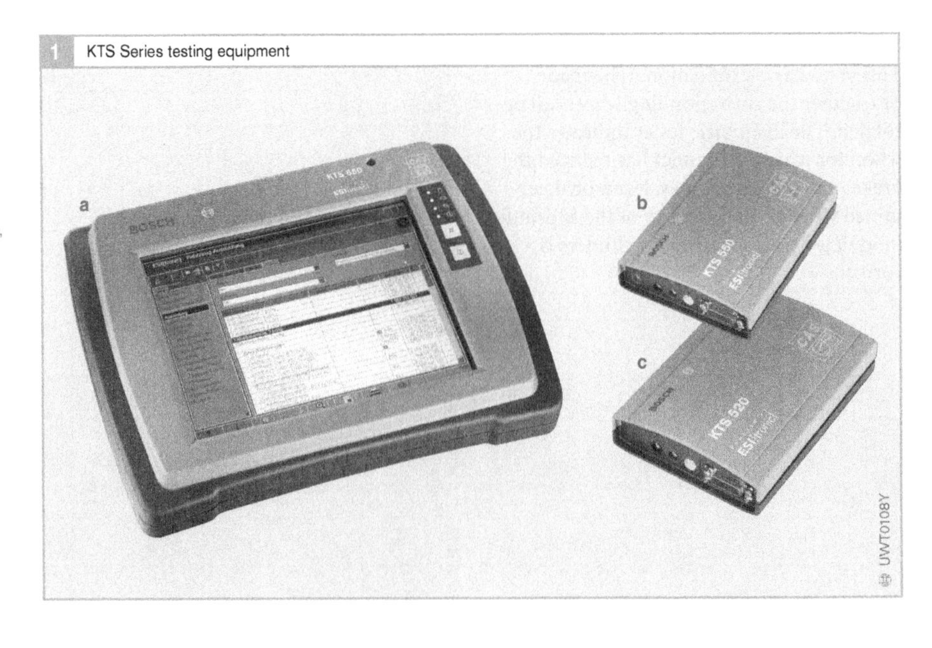

Fig. 1
- a Multimedia-capable, mobile KTS 650 diagnostic tester
- b Universal, convenient solution for vehicle workshops; KTS 550 in conjunction with PC or laptop
- c Universal solution for vehicle workshops; KTS 520 in conjunction with PC or laptop

Test functions

The diagnostic tester triggers programmed test procedures in the ECU. These allow testing of whether the channels of the ABS hydraulic modulator are correctly assigned to the wheel brake cylinders.

Multimeter function

Electrical current, voltage and resistance can be tested in the same way as with a conventional multimeter.

Time graph display

The continuously recorded measured values are displayed graphically as a signal curve, as with an oscilloscope (e.g. signal voltage of the wheel speed sensors).

Additional information

Specific additional information relevant to the faults/components displayed can also be shown in conjunction with the electronic service information (ESI[tronic]) (e.g. troubleshooting instructions, location of components in the engine compartment, test specifications, electrical circuit diagrams).

Printout

All data (e.g. list of actual values or document for the customer) can be printed out on standard PC printers.

Programming

The software of the ECU can be encoded using the KTS 650 (e.g. variant coding of the ESP ECU).

The extent to which the capabilities of the KTS 650 can be utilized in the workshop depends on the system to be tested. Not all ECUs support its full range of functions.

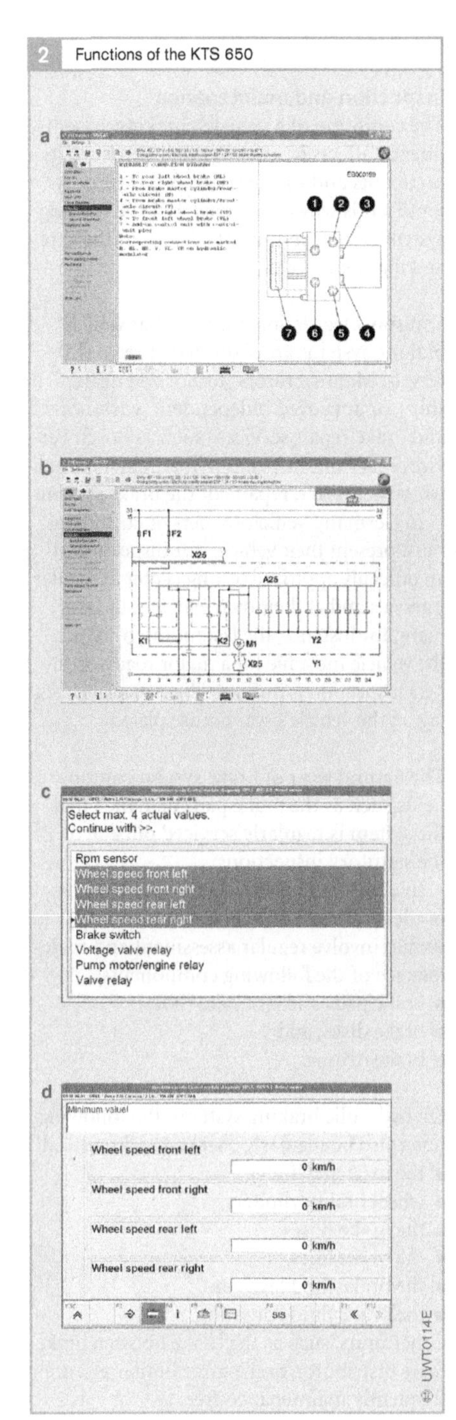

Fig. 2
- a Hydraulic connection diagram of the hydraulic modulator
- b Electrical connection diagram of the hydraulic modulator
- c Selection for measuring actual values
- d Measuring the wheel speeds

Brake testing

Inspection and maintenance

The condition of a vehicle's braking system directly affects its safety as well as that of its occupants and/or the goods it is transporting. That is why the servicing of the braking system is such an important part of the care and maintenance of a vehicle.

Transport legislation requires that vehicle braking systems are inspected at regular intervals. Manufacturers' authorized dealerships, or approved independent workshops and brake repair services (such as Bosch Service) carry out inspection, maintenance and, where necessary, repairs of the brake system.

In Germany, vehicle owners or custodians must present their vehicles for inspection at an officially approved testing center at regular intervals and at their own expense. In Germany, for instance, the last month by which the vehicle must file for a major roadworthiness inspection is indicated by a special check tag on the vehicle's rear license plate.

The natural wear of brake system components such as the brake pads demands that the system is regularly serviced outside of the statutory inspections.

In addition to checking the effectiveness of the brakes on a brake tester, servicing should involve regular assessment and maintenance of the following components:

- brake pads and/or brake shoes,
- brake disks, and
- brake drums.

On hydraulic braking systems, the following must also be regularly checked and serviced:

- the master cylinder,
- wheel brake cylinders,
- the brake hoses,
- the brake lines,
- the brake fluid level and
- the brake fluid condition.

Other units, such as the brake booster, brake force distributor, brake force limiter etc. are frequently maintenance-free.

For compressed-air braking systems, the following also need to be checked:

- air compressors,
- compressed-air cylinders,
- antifreeze unit,
- valves, cylinders,
- pressure regulators,
- braking force regulators,
- coupling heads and
- the air-tightness of the entire system.

Brake pads and shoes

The brake shoes and pads are the parts of the braking system that are subject to the greatest wear as the retardation of the vehicle is achieved by pressing the shoes/pads against the rotating drums/disks. Proper maintenance of these components is absolutely essential for the safety of a braking system.

Checking wear

Assuming they have been correctly fitted, the rate at which brake pads/shoes wear is dependent on the properties of the friction material (e.g. its frictional coefficient), the manner in which the vehicle is driven and the loads it carries.

On most vehicles, reliable checking of the brake pad wear on disk brakes requires the removal of the wheels. Attempting to assess the level of wear with the wheels in place risks inaccurate conclusions.

Checking the wear of brake shoes on drum brakes generally involves removing not only the wheels but also the brake drums.

On some more modern vehicles, inspection holes allow the brake shoe wear to be checked without the brake drums having to be removed, although they are inadequate for a thorough inspection of overall brake shoe condition.

Adjustment
There is normally a small gap (clearance) between the brake pad/shoe and the disk/drum that prevents continuous abrasion of the friction material against the disk or drum. As the friction material wears, that gap becomes larger and, in the case of drum brakes, necessitates regular readjustment of the shoes (assuming the brakes do not incorporate a self-adjusting mechanism).

Disk brakes with an integral parking brake mechanism automatically readjust themselves.

Straightforward disk brakes are likewise self-adjusting. This means that the brake pads automatically shift to take up the extra gap as they wear so that in effect the clearance between the pad and the disk never changes.

The need for readjustment of the brake shoes on drum brakes without a self-adjusting mechanism can be detected by the amount of free play when pressing the brake pedal.

If, for different brake systems (such as simplex or duplex brakes), the brake shoes are adjusted, the information from the brake manufacturer must be observed.

Nevertheless, the following basic principles will always apply:
Regardless of the type of drum brake, the brakes on both sides must always be adjusted at the same time. On vehicles with drum brakes all around, all four brakes must be adjusted at once.

The brakes must be cold before they are adjusted. The service brakes should be adjusted before the handbrake.

Replacing brake pads and shoes
Disk brake pads have to be replaced when the thickness of the friction material is worn down to 2 mm.

On systems with wear sensors on the brake pads, a warning lamp on the instrument panels indicates to the driver that the pads are in need of imminent replacement as soon as the remaining thickness is down to 3.5 mm.

On drum brakes the brake shoe friction lining thickness must not be less than 1.5 mm on cars and 4 mm on commercial vehicles. If the shoes are unevenly worn, or if the linings are cracked or chipped, they too must be replaced.

When replacing brake pads or shoes, it is important that the new pads/shoes conform to the specifications of the original equipment manufacturer.

Important!
The use of brake pads/shoes that do not match the specifications of the brake manufacturer may render the vehicle's insurance policy void.

Brake pads, disks, shoes and drums must always be replaced on both sides (i.e. both front or both rear wheels) at the same time, as otherwise the vehicle may "pull" to one side under braking.

Brake disks and drums
Brake disks and drums are made of steel or cast iron and therefore do not wear as quickly as the pads and shoes. Nevertheless, they still have to be maintained at regular intervals.

The contact surfaces of the brake disks and brake drums must be checked for:
- striations,
- cracks,
- corrosion,
- abrasion and
- differences of thickness.

For disk and drum brakes, these defects can be identified with the naked eye during a visual check.

Brake disks can also develop excessive runout or warping. The degree of runout at the outer edge of the disk must not exceed 0.2 mm and has to be checked using a dial gauge. Brake disks with more than the allowable runout must be replaced.

If scored or unevenly worn brake disks are reground, they must not be reduced to more than a minimum permissible thickness.

Brake drums may become misshapen (so that they are no longer perfectly circular) or develop hairline cracks. Loss of circularity is caused by overheating. It can be detected by pulsating feedback from the brake pedal or, of course, on a brake tester. Brake drums can be reground provided the degree of wear or damage is not excessive. When doing so, the maximum allowable internal drum diameter for the particular vehicle must not be exceeded. If the degree or nature of the damage is such that regrinding the drums is not possible, the only option is to replace them. Drums must always be reground or replaced on both sides (both front or both rear wheels) at the same time in order to ensure even braking.

Master cylinder

The wearing parts of the master cylinder are primarily the cup seals, which are made of a special rubber compound. They are responsible for creating the seal between the piston and the cylinder wall. Corrosion, which can develop as a result of water absorption by the brake fluid, causes pitting of the cylinder wall. That roughness then damages the piston seals by abrading them so that they start to leak.

Depending on the severity of the problem, this can result in partial or even total loss of brake pressure. The response of the brake pedal when depressed will indicate whether the primary or the isolating seal is leaking.

1 Testing brakes on a brake tester

Wheel brake cylinders

As with the master cylinder, the cup seals in the wheel brake cylinders are subject to wear. They can similarly develop leaks and cause corrosion on the cylinder walls. In addition, wheel brake cylinders can also develop leaks around the sealing caps. This can lead to brake fluid contaminating the brake pads/linings and reducing brake efficiency.

The following checks can be carried out to test the condition of the seals:

Low-pressure test

A pressure gauge is connected to the wheel brake cylinder and a pressure of 2 to 5 bar is applied and maintained using a special pedal positioner. There must be no drop in pressure for a period of 5 minutes.

High-pressure test

A pressure of 80-100 bar is applied. Over a period of 10 minutes, the pressure may not drop by more than 10% of its original level.

Pilot pressure test

The pedal positioner is removed and the pressure drops back to the pilot level (if applicable; only applies to cylinders with cup seals) of 0.4-1.7 bar. The pressure should not fall below 0.4 bar over a period of five minutes.

Brake hoses and lines

In theory, brake pipes and hoses are maintenance-free. Nevertheless, they are subject to environmental effects such as corrosion due to water and salt and impact damage from stones, grit and gravel.

Because of those factors, brake pipes and hoses should be regularly inspected. Brake lines should primarily be checked for corrosion, while the hoses should be inspected for abrasion and splits. The unions should be checked for leaks.

Brake fluid level and condition

The brake fluid level is checked on the brake fluid reservoir. The fluid level should be between the "MAX" and "MIN" marks. This check provides one means of detecting whether there are leaks in the braking system. If the fluid level is at or below the "MIN" mark, the system should be checked for leaks. On some vehicles, a warning lamp on the instrument panel indicates to the driver that the fluid level is approaching the minimum mark.

As brake fluid can absorb water by diffusion through the brake hoses, it should be completely replaced every one to two years.

This is absolutely essential for the safety of the braking system.

Maintenance checklist

The components of hydraulic brake systems are subjected to considerable stresses. Heat, cold and vibration can all lead to material fatigue in the course of time. Splash water, especially salt water, and dirt cause corrosion and diminish the ability of components and mechanisms to operate smoothly. Consequently, impairment of function can result.

For safety reasons, therefore, specific regular checks and maintenance work are absolutely essential.

The best time for carrying out such work is at the end of the winter season because the exposed components of the brake system are subjected to the most extreme weather conditions in the winter.

The checks and maintenance operations include

- visual inspections
- function checks
- leakage tests
- internal examination of brakes
- efficiency tests.

This maintenance checklist details the various components in alphabetical order and indicates the checks and tests required for each one. The abbreviations used are explained below.

Key to abbreviations:

A Remove
E Fit
F Lubricate
G Restore function
I Repair
N Replace/renew
NA Rework
P Check, assess
R Clean
S Adjust/align/correct

Maintenance tasks	
Brake fluid reservoir [1)]	
Cap	P/N
Reservoir	P/R/N
Attachment	P/I
Warning lamp switch (if present)	P/I/N
Brake fluid	
Level	P/S
Appearance, color	P/N
Moisture content	P/S
Handbrake lever (parking brake)	
Travel, no. of ratchet notches	P/S
Ratchet function	P/I
Freedom of action	P/G/F
Lever stop (if present)	P/S/I
Return spring (if present)	P/S/F
Braking force limiter	
External damage	P/N
Attachment	P/I/N
Pipe connections	P/I/N
Function	P/N
Limited pressure (observe testing conditions)	P/S
Braking force regulator	
External damage	P/N
Attachment	P/I/N
Pipe connections	P/I/N
Linkage, lever	P/I/F
Travel spring	P/N/F
Function	P/N
Limited pressure (observe testing conditions)	P/S
Brake servo unit	
External damage	P/N
Attachment	P/I
Hoses (splits etc.)	P/N
Function	P/N
Leakage	P/N
Brake pedal (service brakes)	
Pedal	P
Pedal rubber (wear, condition)	P/N
Pedal travel	P/S
Connecting rod play	P/S
Freedom of action of shaft	P/G/F
Pedal stop	P/S/I
Pedal return spring	P/S/F

[1)] **Caution:**
If the level of fluid in the reservoir is very low, simply adding more fluid must on no account be viewed as the solution. The cause of the fluid loss must be established and rectified. Dark or cloudy brake fluid must be replaced immediately.

Brake lines [2]	
External damage	R/P/N
Attachment	P/I/N
Corrosion	P/N
Brake hoses	
External damage	P/N
Attachment	P/N
Kinking, length	P/N
Routing (e. g. twisting)	P/I/N
Suitability for pressure medium	P/N
Age	P/N
Master cylinder	
External damage	P/N
Attachment	P/I/N
Pipe connections	P/I/N
Seal against brake servo unit	P/I/N
Low-pressure seal	P/I/N
High-pressure seal	P/I/N
Brake light switch	P/N
Brake lights	P/I
Brakes (general)	
Basic adjustment of drum brakes	P/S
Clearance adjustment on disk brakes	P/S
Non-return valve	
External damage	P/N
Attachment	P/I
Hoses (splits etc.)	P/N
Function	P/N
Leakage	P/N
Disk brakes (brake pads)	
Damage (cracks etc.)	P/N
Shining, hardening etc.	P/N
Friction pad thickness [3]	P/N
Pad guides	P/R/F
Suitability for vehicle	P/N
Disk brakes (brake caliper)	
External damage	R/P/N
Attachment	R/I/N
Brake pad channels	P/R
Guides	P/G/F
Piston freedom	P/I/N
Piston position	P/S
Dust seals	P/N
Small parts (expander springs, bolts etc.)	P/N
Bleed valve, dust cap	P/G/N

Disk brakes (brake disks)	
Damage (cracks etc.)	P/N
Thermal overload	P/N
Wear [4]	P/N
Wear pattern [4]	P/NA/N
Minimum thickness [4]	P/NA/N
Runout [4]	P/NA/N
Drum brakes (general)	
Backplate (damage)	R/P
Wheel brake cylinders	P/I/N
Dust seals	P/I/N
Parking brake mechanism and linkage	P/I/F
Adjusting mechanism	P/G/F
Handbrake cable	P/F
Brake shoes and linings	P/I/N
Shoe anchor bearings	R/F
Return springs	P/N
Drum brakes (handbrake cable, linkage)	
External damage (cable sheath)	P/N
Attachment	P/I/N
Correct routing and fitting	P/I/N
Guides, rollers etc.	P/I/F
Cable (fraying etc.)	P/N
Freedom of action	P/G/F
Adjusting mechanism	P/G/F
Basic adjustment	P/S
Drum brakes (brake drum)	
Damage (cracks etc.)	P/N
Thermal overload	P/N
Wear [5]	P/NA/N
Concentricity [5]	P/NA/N
Warping [5]	P/NA/N
Service brake efficiency test	
Braking force, front wheels	P/I
Braking force difference (front)	P/I
Braking force, rear wheels	P/I
Braking force difference (rear)	P/I
Actuating force	P/I
Parking brake efficiency test	
Braking force	P/I
Braking force difference	P/I

[2] **Caution:**
Do not use abrasive materials or tools on coated brake lines. Corroded or damaged lines must be replaced.

[4] **Caution:**
Refer to maximum wear limits.

[5] **Caution:**
Refer to maximum wear limits.

[3] **Caution:**
Minimum thickness for disk brake pads is 2 mm, excluding backplate.

Index

CPSIA information can be obtained at www.ICGtesting.com
Printed in the USA
LVOW02s1624270115

424571LV00004B/49/P